CASS SERIES: STUDIES IN AIR POWER

(Series Editor: Sebastian Cox)

RUSSIAN AVIATION AND AIR POWER IN THE TWENTIETH CENTURY

W0114819

CASS SERIES: STUDIES IN AIR POWER

(Series Editor: Sebastian Cox)

ISSN 1368-5597

RUSSIAN AVIATION
AND
AIR POWER
IN THE
TWENTIETH CENTURY

Edited by
Robin Higham,
John T. Greenwood and
Von Hardesty

FRANK CASS
LONDON • PORTLAND, OR

First published in 1998 in Great Britain by
FRANK CASS PUBLISHERS.
Newbury House, 900 Eastern Avenue
London, IG2 7HH

and in the United States of America by
FRANK CASS PUBLISHERS
c/o ISBS, 5804 N.E. Hassalo Street
Portland, Oregon, 97213-3644

Transferred to Digital Printing 2004

Website http://www.frankcass.com

Copyright of collection © 1998 Frank Cass & Co. Ltd.
Copyright of articles © contributors

British Library Cataloguing in Publication Data:

Russian aviation and air power in the twentieth century. –
(Cass series. Studies in air power)
   1. Aeronautics – Soviet Union   2. Aeronautics – Russia
(Federation)
3. Air power – Soviet Union   4. Air power –
Russia (Federation)
   I. Higham, Robin   II. Greenwood, John T.   III. Hardesty. Von,
1939–
629.1'3'0947'0904

ISBN 0-7146-4784-5 (cloth)
ISBN 0-7146-4380-7 (paper)
ISSN 1368-5597

Library of Congress Cataloging-in-Publication Data:

Russian aviation and air power in the twentieth century / edited by
Robin Higham, John T. Greenwood and Von Hardesty.
      p.   cm. – (Cass series – studies in air power; 7)
   Includes bibliographical references and index.
   ISBN 0-7146-4784-5 (cloth). – ISBN 0-7146-4380-7 (pbk)
      1. Aeronautics – Soviet Union – History.   2. Aeronautics, Military –
Soviet Union – History.   I. Higham, Robin D.S.   II. Greenwood,
John T.   III. Hardesty, Von, 1939–   .   IV. Series.
TL526.S65R87   1998
629.13'00947 – dc21   98-4629
                                                                                     CIP

Typeset by Regent Typesetting, London

Contents

Illustrations

Contributors

Thomas M. Alison

Tom Alison is Chief of the National Air and Space Museum's (NASM) Collections Division and also serves as the Aeronautics Department's curator for military aircraft. He joined the NASM staff upon retirement from the United States Air Force after a 27-year career. During his Air Force career Tom logged over 6,000 hours in high performance jet aircraft, including almost 1,000 in the Mach 3+ SR-71 Blackbird.

As Chief of the Collections Division Tom is responsible for the restoration, preservation, and conservation of all the artifacts in the collection as well as all collections processing. A specialist on the history of Soviet military aircraft, he is currently preparing a history of Soviet ground-attack aviation, with special attention to the Ilyushin Il-2 *Shturmovik* design, technical development, and wartime operations. He is responsible for the museum's addition of the Ilyushin Il-2 to the national collection.

John T. Greenwood

John T. Greenwood is currently Chief, Field Programs and Historical Services Division, US Army Center of Military History, Washington, DC. After obtaining his PhD at Kansas State University, he was an historian in the USAF and then Chief, Office of History US Army Corps of Engineers. He has published widely, in both official historical works and elsewhere, on a variety of topics, including *Soviet Aviation and Air Power* (1977). He has visited Russian archives and works constantly with Russian materials.

Von Hardesty

Von Hardesty is a curator in the Aeronautics Department of the Smithsonian's National Air and Space Museum. He is the author of *Red Phoenix: The Rise of Soviet Air Power, 1941-1945*, along with numerous books and articles on Russian and Soviet aviation.

Hardesty serves as editor of the Smithsonian History of Aviation book series, a scholarly series devoted to aerospace history. As editor and translator, in 1996 he published L. L. Kerber's classic *Stalin's Aviation Gulag: A Memoir of Andrei Tupolev and the Purge Era.*

Robin Higham

Robin Higham, Professor of Military History at Kansas State University since 1963, is the author of *Air Power: A Concise History* (1988), and was the co-editor of *Soviet Aviation and Air Power* (1977) as well as editor of *Aerospace Historian* (1970–1988).

Christopher C. Lovett

Christopher Lovett holds a PhD from Kansas State University and is an assistant professor of Modern European History at Emporia State. He has written articles on Soviet naval aviation, Vietnam, and the Second World War. Currently, Dr Lovett is completing a history of the Russian and Soviet naval air arm.

Dennis J. Marshall-Hasdell

Dennis Marshall-Hasdell was born in England. He joined the Royal Air Force in 1972 and following training as a navigator he was selected to fly the Phantom FGR2 in the air defense role. He served for three years on 29(F) Squadron at RAF Coningsby, Lincolnshire, and a further three years on 92(F) Squadron at RAF Wildenrath, Germany. He had flown over 1,000 hours on the Phantom when he volunteered to serve as a liaison officer with No. 3 Commando Brigade, Royal Marines. He passed the Commando Course and was awarded his Green Beret in mid-1980. He served nearly four years with Commando Forces on many operations in different theaters around the world, including active service during the Falklands War in 1982. In 1985 he was medically discharged from the RAF, holding the rank of squadron leader, as a result of injuries received while on active service.

In 1986 he entered the University of Wales, Aberystwyth, to read International Politics and Strategic Studies; he graduated in 1989 with an honours degree. He continued with postgraduate studies and, in 1994, was admitted as a Doctor of Philosophy to the University of Wales, where he was appointed a lecturer. He has been a Research Fellow at the Conflict Studies Research Centre, Camberley, since 1990. In December 1995 he was appointed Senior Editor (Strategic Studies) at Samaras Publishing Ltd.

Mark A. O'Neill

Mark O'Neill received his MA and PhD in Modern Russian/Soviet History from Florida State University. His research interests include the participation of Soviet pilots in the Korean War and the Soviet Air Force during the Second World War. Dr O'Neill is presently a Visiting Assistant Professor of History at Auburn University.

Reina Pennington

Reina Pennington is a former Air Force intelligence officer. During her nine years in service, she worked as a Soviet analyst with fighter squadrons, the Aggressors, the Defense Intelligence Agency, and the Alaskan Air Command. Pennington has published numerous articles on Soviet and US tactical air forces in *Air Force*; *The Fighter Weapons Review*; *Airpower Journal*; *Air & Space/Smithsonian*, and *The Journal of Slavic Military Studies*. She is currently a PhD candidate in history at the University of South Carolina. Her book, *Military Women Worldwide: A Biographical Dictionary*, is to be published in 1998 by Greenwood Press.

Series Editor's Preface

The startling collapse of the Soviet Union and the subsequent transformation in the political face of the former communist bloc have
wrought changes in Russia which may yet prove as fundamental as
those brought about by Lenin's seizure of power in 1917. As we move
towards a new millennium, it is arguable that the rise of Soviet-style
communism and the development of air power have been two of the
most potent influences on the military and political history of the
twentieth century. The extent to which the external relations of Soviet
communism differed from those of the old-style Russian nationalism
under the czars has been a source of fascination and debate for
Western historians for many years, and the uncertain path that lies
ahead seems likely to provide further opportunity to explore that
theme.

The contributors to this book have profited from the partial lifting
of the Iron Curtain and the subsequent increased access to original
documents on the Soviet regime. The ending of the Cold War also
allows us to view many of the events described here without the dis-
torting prism of Cold War suspicion. However, many questions
remain about the true nature of Russian power in the post-Cold War
era. The Russian Air Force played a prominent, but far from uni-
versally successful, role in the recent fighting in Chechnya, and the
potential for conflict along the southern periphery of the former
Soviet empire remains high, while many of the purely military
problems created for the Russian armed forces by changes in the geostrategic situation remain unresolved.

The original intention of this work was to produce a revised and
updated edition of *Soviet Aviation and Air Power* (1977), edited by
Robin Higham and Jacob Kipp. However, as both the opening and
the closing eras were Russian, and many of the contributors were new,
it was decided to start afresh. The result is an entirely new book,
which serves to expand on and to illuminate the scholarship of the
1977 volume. Chapters 6 and 7, on the aviation industry and the
design bureaux respectively, draw to some extent on the chapter from

Soviet Aviation entitled 'Patterns in the Soviet Aircraft Industry', but contain much original material and bring the story up to date.

In examining the story of Russian air power from its beginnings in czarist Russia through to the last adventures of old-style Soviet communism in Afghanistan, this book makes an important contribution to our knowledge and understanding of the subject. At a time when former Warsaw Pact countries are entering NATO, and the future direction of Russia is uncertain, these scholarly essays bring a timely historical perspective that should help to illuminate current thinking.

SEBASTIAN COX

Introduction

ROBIN HIGHAM, JOHN T. GREENWOOD
AND VON HARDESTY

The story of Russian aviation, and by extension, aerospace, is nearly co-extensive with the twentieth century. Only the United States possesses an aeronautical tradition with such continuity and influence. Language, geography, and politics have conspired to make Russian achievements in this important sphere of technology remote and, until quite recently, inaccessible. Few books in English have been devoted to Russian and Soviet aviation.

Such a lacuna in English-language historical literature, of course, has prompted myth and distortion – often at the expense of the Russians. One enduring perception of Russian aviation has been to see the typical Russian airplane as simple, if rugged, in design; a species of technology nearly devoid of technological sophistication. For the more hostile observer such 'primitivism' mirrors a larger scientific and industrial backwardness. Co-existing with this stereotype is another perception, one fueled by the Cold War and Soviet triumphs in rocketry and space: a view of Russian aeronautical prowess as a threat, and something to fear in the nuclear age.

Stereotypes do possess a measure of reality. Russian approaches to aviation give substance to both perceptions. Russian-designed aircraft, especially in the mass production of military airplanes, have been simple and rugged, austerely equipped with instrumentation, and seemingly inattentive to creature comforts or pilot safety. Less apparent in these flying machines has been the Russian priority on effectiveness. For Russian designers this priority has meant the studied effort to design aircraft that perform well. These designers have fashioned aircraft that routinely fly in some of the most inhospitable environments in the world; accordingly, Russians would pioneer cold weather flying in the twentieth century.

Only with the collapse of the Soviet Union did Westerners acquire a context to understand more fully the historical weaknesses and strengths of Russian aviation. For the first time, both Russians and

Western historians could assess accurately Russian aeronautical achievements. It appeared that everything that had been said about Russian aviation was more or less true; Russia's aeronautical establishment embodied both the primitive and the most modern. Innovative engineering mixed with a traditional dependence on certain forms of Western technology. If viewed over the entire sweep of the century, Russians designed many modern aircraft, but for decades remained behind the West in the design of aero engines.

A new anthology on Russian aviation is timely, especially as we approach the end of the century and reflect on the impact of the airplane, civil and military. This story cannot be told without reference to the Russian/Soviet contributions. As an assessment, the present anthology represents a revised and augmented version of Robin Higham and Jacob Kipp, *Soviet Aviation and Air Power: A Historical View* (Boulder, CO: Westview Press, 1977). This original anthology provided extraordinary insights into Soviet aviation to English readers in 1977, at a time when little was known about the internal workings of the Soviet aeronautical establishment. Two decades passed. Now, in the post-Soviet period, the comprehensive coverage of the 1977 anthology has been revisited and updated in light of new archival materials. *Russian Aviation and Air Power* brings to the reader a comprehensive view of Russian aviation, from its genesis in the late czarist period to the present era. The approach is essentially chronological with a major emphasis on the evolution of military aviation. The contributions are diverse, however, with appropriate attention to civilian and institutional themes. When viewed in its entirety, Russian aviation possesses a certain continuity that transcends politics.

AN HISTORICAL OVERVIEW

Looking at Russian aviation and air power in the twentieth century provides an excellent opportunity to compare and contrast developments within both the geography of that vast Slavic state and the development of aviation elsewhere within and without a monolithic system.

Most people have probably forgotten that there were times when the Russians led the world. In the years immediately preceding the First World War Igor Sikorsky developed the largest heavier-than-air passenger plane and then converted it into a bomber. The latter's principal problem was that there were few suitable targets within range and so it was employed on reconnaissance. Later in the 1930s, when Stalin's Five-Year Plans were in place, the Soviets led the world in many classes of aircraft, only to fall out of the pattern during the

Spanish Civil War when the latest German models appeared. But the Soviets recovered during the Great Patriotic War of 1941–45, not so much in quality as in quantity and in the effective use of tactical air power.

Then the rules of the game changed again. The Soviet Union found itself naked and fearful of the might of United States grand-strategic nuclear air power. But the Slavs responded to the challenge and by the 1960s they had equipment and electronics equal to that of the West. The problem thereafter until the collapse of the USSR in 1991 was lack of combat experience. Equipment was tested vicariously in Eastern bloc satellites and especially in the air forces of clients and allies such as those in the Arab world. The West and its allies did not fully appreciate the sophistication of Soviet aircraft, electronics, and missiles until the Gorbachov era at the tail-end of Soviet power when invitations were extended to visit Soviet bases and fly aircraft such as the MiG-29.

The latter showed that one constant in Russia has been the relationship of aviation to the people combined with the demand that all aircraft be able to operate off unprepared grass strips. Thus the MiG-29 had doors over its engine intakes to keep out foreign matter until after the nosewheel left the ground and the engine could breathe full ram air instead of through louvers on top of the intakes.

Only after the Great Patriotic War of 1941–45 did the USSR begin really to develop Aeroflot (the umbrella civil aviation side) into a major airline system. Even so, the early jet airliners were variants of bombers rather than purpose-designed machines. But this was also less obviously so in the West, where the De Havilland Comet and the Douglas DC-8 broke new ground.

Russian design differed from that in the West in carrying the large turbo-prop to much greater extremes as a long-range reconnaissance bomber.

On the other hand, Russian aircraft design and manufacturing remained in the pre-Second World War mold of design bureaux with their own ability to produce prototypes separate from the factories which produced the separate models. And this system was not forced to change until 1997 when the perilous state of the Russian economy, now market-driven, forced Moscow to take drastic action to reduce the mix of bureaux and factories to perhaps two competitive establishments with which to face Boeing-McDonnell Douglas and Airbus-DASA-British Aerospace.

IN THE BEGINNING

While the beginnings of aviation took place under the czarist regime, it was limited by finances, and natural and human resources in a country whose serious industrialization had only really accelerated after the Franco-Russian Alliance of 1894. During the 1914–17 war against Germany and Austria, the Russian military and naval air forces had been limited to what they could produce out of indigenous and imported designs as well as the small number of aircraft and engines that were imported through the White Sea ports of Murmansk and Archangel. Essentially, this meant that the Russian air forces were not as well equipped as those on the Western and Italian fronts where the stability of designs was constantly in flux as better aircraft, guns, and airmen were placed on an increasing number of airfields. The collapse of the czarist and then of the Kerensky regimes came at a time when in the West uniform fighting formations had been made possible by standardized machines, more reliable engines of higher horsepower, and better trained, disciplined and led front-line combat forces. Moreover, long-range bombing either by Zeppelins or by heavy bombers had become possible. But at the same time, the defenses had also reached wartime equilibrium after the long slope of rearmamental instability had been ascended.

While on the Western and Italian fronts, and to a much lesser extent in the Middle East, the fighting had been concentrated over only a part of the trench lines, on the Eastern Front where the Russians faced both the Germans and the Austrians, the space was so vast that aircraft were spread in penny packets as the eyes of the armies, and little air-to-air combat developed. Thus by the time that the Red Armies took over for the Civil and Polish wars that followed Lenin's seizure of power, aviation was still a hope rather than an ancillary. And it remained so until 1922 and the end of the various campaigns for Soviet survival.

THE INTERWAR YEARS

By contrast, in the West, the Treaty of Versailles blamed the Germans for starting the war in 1914 and their military activities were proscribed, forcing them to concentrate on the development of civil airlines and to refine their technology – to their ultimate advantage in 1935. Austria–Hungary disappeared as an empire and none of the successor states had the wherewithal to take even the parlous Austro-Hungarian industry's place. In Italy General Guilio Douhet and the new Fascist dictator Benito Mussolini made aviation independent and in theory relied upon a deterrent bombing force to be launched from behind the bastion of the Alps. In Britain, the exigencies of wartime production had forced the amalgamation of the Royal Naval Air Service and the Royal Flying Corps into the Royal Air Force under an independent Air Ministry. In 1918 the RAF was the second largest air force in the world. Postwar it was shrunk to a colonial policing force by 1922, when the idea of a Home Defence grand-strategic deterrent bomber force emerged as a cheap way to avoid another killing ground on the continent of Europe. The number of private aircraft and engine firms declined rapidly to those in 'the Ring' to whom the very limited orders were passed out. No infrastructure was created to make grand strategy workable, for it was assumed that there would be no major war for at least ten years.

France emerged from the war with the largest air force in the world and a determination to keep Germany from ever being the aggressor again. Unfortunately, the great depression and eternal squabbles between the Army and the Navy on the one hand, the airmen seeking an independent force not tied in penny packets to the army for tactical purposes on the other, and the politicians in the middle who wanted, as always, to save money and who ultimately nationalized the aircraft industry, all combined to make the great French Air Force impotent by 1940.

And the United States, in spite of having been the home of the Wright brothers and controlled flight, had overconfidently assumed in 1917 that it would mass-produce aircraft at a time of fluid designs and trans-Atlantic time lags. Instead the nascent US Air Service in France flew foreign aircraft, which Billy Mitchell concentrated for tactical use at St Mihiel and the Meuse–Argonne. After the Great War the same patterns as in France and Britain pertained. The Army and the Air Service were split over grand-strategic and tactical uses of air power, while Mitchell's sinking of anchored battleships aroused the US Navy, whose William A. Moffett created the carrier-borne naval aviation that would prove so effective, unlike the RAF-shackled Fleet Air Arm, in the Great Pacific War of 1941–45.

In the West, aircraft design and production was in private hands after the First World War, with the exception that the government retained control over experimental and test facilities such as the Royal Aircraft Establishment at Farnborough, UK, and the National Advisory Committee for Aeronautics facilities at Langley Field, Virginia, USA.

In sharp contrast, in the new Soviet Union, where aviation was seen as a symbol of modernization similar to the dynamo, a central design establishment was created in Moscow for aerodynamics (airframes) and one for engines. The TsAGI (*Tsentral'nyi Aero-*

gidrodinamicheskii Institut) and the TsIAM (*Tsentral'nyi nauchnoissle-dovatel'skii institut aviamotorstroeniya*) became both training grounds and technical think-tanks with enormous influence on the development of Soviet designs. Outside of these two central establishments were the limited number of factories with their own design teams such as Polikarpov and Tupolev. Tupolev doubled also under his initials as ANT.

Since Russia had almost no roads and a skeletal railway system, much damaged by the civil wars of 1918–22, aviation quickly came into its own, and Aeroflot, when it came into being in 1932 was not only a unified airline, but controlled all non-military air activities from crop-spraying to mail-carrying. Moreover, unemployment in the cities was anathema to Soviet doctrine, so that aircraft factories like any other approved by the State had to be kept in production, though in 1922–28 90 per cent of all Russian aircraft were imports or of foreign design. This meant that in contrast to the hand-to-mouth existence of aircraft constructors in the free world until rearmament began in the mid-1930s, Soviet aircraft design bureaux and factories were constantly employed. And they were ahead of the West because Stalin's Five-Year Plans from 1928, a sound management technique, were designed to rearm the USSR against the outside imperialist enemies which both doctrine and Stalin's paranoia took to be just waiting to crush the nascent state as they had attempted to do in 1918–22.

Great strides were made in Russian aircraft design and production from 1928 onwards. By then not only had the designers got on their feet technically, so to speak, but they had the means to turn their sketches into reality. Thus on the one hand there were the few great Antonov bomber-paratroop designs and on the other the very nimble biplane fighters such as the Polikarpov I-15 and the I-16. The latter were combat tested in the Spanish Civil War from 1936 to 1938, until the German Me-109s appeared, the next generation of high-speed monoplanes.

Ironically, both the Germans and the Russians had developed their equipment and doctrinal ideas at the illicit German base at Lipetsk, while Junkers manufactured at Fili, near Moscow. These arrangements were canceled for anti-semitic reasons in 1933 after Hitler came to power. In the meantime, the cross-fertilization, which included Soviet officers visiting Germany, led to the development of sound doctrines for the employment of tactical air forces, having little influence in a country which could trade space for time and which could generally not reach any worthwhile grand-strategic targets. The Germans, on the other hand, sought to avoid another stalemate of the 1914–18 type and so opted for blitzkrieg, in which tactical aviation would play a very important part.

THE ROAD TO WAR

Thus by the turning point of 1934, the beginning in most countries of rearmamental instability, the Soviet Union was already launched on the second Five-Year Plan and was approaching the third.

Most developments in Russian aviation were now tied into diplomatic, military, political, economic, scientific and technological, medical, social and ideological patterns. Moreover, these tended both to reflect Soviet attitudes and to mirror the Russian background. Very large armed forces were being prepared as a hedge against the activities of Nazi Germany, not to mention France and Britain – Stalin was suspicious of many entities. At the same time domestically there was the purge of the higher leadership of the armed forces including the air force, an action which was also seen in other countries' retiring of superannuated senior officers.

In addition to the war in Spain, from which the lessons were being absorbed throughout 1939–42, the armed forces of the USSR were engaged in the Nomohan and Khalkin Gol incidental wars (1937–38) with Japan in the Far East, in which the latter were defeated, and the Winter War with Finland (1939–40). In the Far East the Soviets dominated in spirit, technique, and equipment. But in the war against Finland, the Russians were held up by much smaller Finnish forces, in part because the purges had placed new commanders in the field who did not yet have the necessary doctrine, experience, or trained troops to win a quick victory in bitter weather. Success came both from sheer numbers and from the fact that the Finns were cut off from the West. Ironically, the Finns later allied with the Germans as they both had a common enemy to the east.

In the meantime the Soviets had watched the events at Munich in September 1938 with interest and alarm. France and Britain, which might have called Hitler's bluff if they had combined their strength with that of the Czechs, forced the latter to give way owing to Britain's weakness in the air and the aversion to another great war on the part of both themselves and their Gallic allies, not to mention the adhesion of Italy to the Axis.

Hitler's non-aggression pact of the summer of 1939 gave Stalin breathing room, and shortly part of Poland once more. The brief September 1939 campaign in Poland made it very clear that the Germans had perfected the blitzkrieg they had been contemplating in Russia in the 1920s. Stalin barely had time to seize his share of Poland as the *Luftwaffe* demonstrated its use as advanced scouts and mobile field artillery as well as showing how to destroy an enemy air force on the ground.

And if Stalin needed to have the handwriting on the wall pointed out more clearly, the blitzkrieg which overwhelmed the Low Countries and France between 10 May and 20 June 1940 illuminated it clearly.

Stalin was no fool, but his innately suspicious nature was almost his undoing. In spite of warnings from the British delivered by their socialist Labour ambassador in Moscow as to the German D-Day for Barbarossa, the assault on the Soviet Union, the Russian dictator refused to believe this intelligence. Thus just before dawn on 22 June 1941 Soviet forces were surprised by the German assault which very quickly wiped out some 1,500 aircraft on Russian airfields in Poland and the western USSR. These losses, considered staggering for many years, compare not unfavourably with the roughly 1,000 RAF aircraft lost in the month of the 1940 campaign in France and to the complete destruction at the same time of the French Air Force. Moreover, the French lost their entire aircraft industry in the German bloodsucking which followed.

What saved the USSR were a number of important developments. First, much of the equipment destroyed was older while the aircrews generally survived to fight another day. Second, the new commanders, blooded in Spain, Finland, and the Far East survived to form the backbone of the new high command and the leadership of the massive air armies which the USSR would soon field. Third, a massive effort was already underway when the war started to move all Soviet aircraft and other munitions factories from the Ukraine and the other areas in the west, to east of the Urals, where they would be safe from almost anything but very long-range German air attacks or from a panzer breakthrough. Britain had had a funk line from the Wash to Bristol, but placing factories west of this was vitiated when the Germans conquered Norway and Denmark in April 1940, as well as France and the Low Countries shortly thereafter. Moreover, the British had a sophisticated air-defence system which repelled the daylight assault in the summer of 1940 and was relieved from the nighttime blitz by Hitler's mobilization of all forces for the attack upon Russia. The United States was beyond all but U-boat range and that was limited to the Atlantic and the Caribbean coasts. Thus, as had been done in 1812 in response to Napoleon's invasion, the Russians once again heeded their history and traded space for time. By the winter of 1941–42 the weather was once more upon their side.

In contrast with the Western Desert, the North African and Italian campaigns, and the invasions of France in 1944, the war on the steppes was fought by massive armies across vast spaces. Whereas the Western Desert was over 1,200 miles in length, geography limited the armies to a narrow coastal strip. In contrast, on the Eastern Front it was some 1,350 miles broad from Leningrad to the Caucasus and the armies were engaged face-to-face all along that line.

At first the Red Air Force was at a disadvantage in terms of equipment and experience. But quite soon that changed as the Germans found themselves in the same position as the Russians in Spain – the opponent had newer and more up-to-date equipment and plenty of resources. Moreover, the Soviets were fighting over their own land and they knew well that Russia was still very much a workers' and peasants' society. The Red Air Force was, therefore, a tactical air force there to support the infantry, armor and artillery. There was never the dichotomy of doctrine or purpose that plagued the British and American air forces, divided as they were in Europe between round-the-clock bombing of Germany, tactical support of the ground forces, and the Battle of the Atlantic with all its ramifications. (The Soviet naval air force was limited to minor activities over the Black Sea, the Sea of Azov and the White Sea, which hurt it postwar.)

This singleness of purpose on the Eastern Front meant that the Russians were able to concentrate small aircrews in single and twin-engined tactical air armies flying simple, easily producible aircraft such as the prime ground-attack machine, the Shturmovik, which squadrons had made into a two-seater. In these short-range support missions, the Red airmen and women rarely flew above 3,000 feet.

And where in the West wings were the standard large formation of perhaps at the most 200 aircraft, over the steppes air armies of 4,000 frontline aircraft were committed against the increasingly decimated *Luftwaffe*, especially after its transport fleet was ruined in supporting von Paulus' army trapped in shattered Stalingrad (now Volgograd) in the winter of 1942–43. Yet the Red Air Force never went after enemy airfields and aircraft on the ground for the doctrine of air superiority was not an absolute, the space being so great that that ultimate objective could not have been held for more than a few minutes. Nor was air superiority a legacy of the Eastern Front in the previous war.

By the time that wartime equilibrium was reached for the Red Air Force late in 1942, it had become the dominant flying weapon over the Eastern Front at the same time as the RAF and the USAAF attacks in the West were just beginning to demolish German cities such as Hamburg, well within their range.

The massive Soviet air armies also contrasted sharply with the forces deployed in the concurrent Great Pacific War. There, however, the space was mostly water and so the key points were islands and the offensive forces of the American judo blitzkrieg were carriers and

amphibious forces striking at the pinpoint fortresses and airfields held
by the Japanese.

THE TECHNICAL REVOLUTION

In the meantime, the Soviet Union had been much less affected than
the West by the multiple revolutions in aviation of the years 1934–45
which had such a great impact on the invisible infrastructure.

The Soviets did adopt the technical revolution, for they read
Western literature, and that compelled them also to go over to all-
metal construction. But they were far behind in electronics, airfields,
jet aircraft, atomic bombs and computers. An early impact of this in
Russia was the licensed production of the Douglas DC-3 transport
which started just before the war, the arrival of Lend-Lease aircraft
from the United States and Britain; and the landing in the USSR in
1944 of several new American B-29 Superfortresses, which were
interned in the Far East and copied. On the other hand, at the end of
the war in 1945 when the *Luftwaffe* and the RAF had operational jet
fighters and the Italians and USAF (United States Air Force) were
testing their first ones, the Russians had none. However, they made
rapid progress through a combination of capturing German scientists
and designers of both aeroplanes and rockets, together with (as an ally)
obtaining new Rolls-Royce engines from Britain. Thus the Russians
were able to make up lost ground very rapidly, so that by the time of
the Korean War five years later they were able to test the new MiG-15
in combat and shortly thereafter to supply the Arabs with MiGs and
Sukhois for their battles against the Israeli Air Force.

More significant was the way in which air transport had emerged
from the 1939–45 War. It had come of age. The twin-engined DC-3s
and Lockheed 14s and 18s had been replaced with four-engined,
primarily US-built transports such as the Douglas DC-4 Skymaster
and the Lockheed 049 Constellation, both capable of spanning the
oceans and making use of the many airports which had been built as
wartime air bases. Both the recivilianized airlines and rival newcomers
entered a boom phase thanks on the one hand to wartime navigational
developments from sextant to Loran, the long-range radar-fixing
system, and on the other to the masses of well-trained and experienced
aircrew from which they could choose the best.

In contrast to the United States, the Soviet Union did not use air
transport lavishly during the war. It was limited to the DC-3 (Li-2)
and C-47s. It was not until 1944 that the Ilyushin design team began
work on the Il-12, which first flew in 1946 and soon went into mass
production for Aeroflot, which had to have aircraft that could land on

grass. The Il-12 was the first Soviet aircraft with a tricycle under-carriage and the first with full-feathering propellers. Rugged structure and simplicity of manufacture were qualities demanded in contrast to the West's need for a lighter more efficient aircraft. The Russians who had neglected four-engined heavy bomber design during the war had a piece of luck in 1944 when three USAF B-29s landed out of fuel near Vladivostok. Accustomed to studying foreign examples, the Soviets at once took the planes apart and analysed them acutely. From this came the Tupolev copy, the Tu-4 and its passenger counterpart, the Tu-70, which did not emerge as the twin of the Boeing B-29 derivative Stratocruiser.

Russian design bureaux and their attached factories for the next quarter-century followed the Western practice also of developing both bombers and transports from the same basic design work. Thus, just as the Boeing prototype Model 367-80 became the basis for both the 707 airliners and the KC-135 tankers, and the Vickers Valiant and the unfinished V-1000 transport followed the same pattern, so a string of Tupolev and Ilyushin designs had commonalities. This was especially evident in the 1950s when the transport versions still carried the glassed bomber nose compartment as in the Tu-16 and its civilian counterpart, the Tu-104, which first appeared in the West at London's Heathrow in 1956. Contemporaneously, when Bristol in the UK was developing the Britannia four-engined turbo-prop airliner, which was eventually produced in a maritime reconnaissance version for the Canadians, the Russians went ahead to the very long-range turbo-prop Bear (Tu-20) of 1955 with its civil counterpart the Tu-114, and to the all-jet Tu-16 derived from the Tu-104, which appeared in the maritime role. It would take the Royal Air Force into the 1960s to derive the Nimrod from the Comet airliner.

Compared with the 9,000 MiG-17 jet fighters produced, the number of Tu-104s was a modest 200. However, it was estimated that some 2,000 Tu-16s were built, but only 300 Tu-20s. These figures compare very favourably with 760 B-52s, 1,832 Boeing 727s and something over 5,000 F-4s manufactured in the USA.

Soviet civil aircraft development reached a plateau with the 17 Tu-144s of the 1960s (one of which crashed at the Paris Air Show in 1973), which bore a close resemblance to the 13 Anglo-French Concorde supersonic transports (SST), conceived in 1955 and operational in 1976. In the meantime plateau theory hit the United States. Variable geometry SST had been canceled by 1969 owing to political, economic, and social-ideological forces. Whereas the Concorde has been in visible North Atlantic service with Air France and British Airways since 1976, Aeroflot operated the Tu-144s only on freight service within the USSR and all had been withdrawn by the 1990s. Ironically, one was taken out of mothballs in 1996 to serve as a test bed for joint Russo-American tests for a next-generation SST.

Supersonic military aircraft, of course, continued to be evolved in competition with the West during the Cold War, but the development times lengthened as did time in production and service.

THE COLD WAR

Immediately after the war the Soviets made rapid progress in jet fighters and were able to deploy the MiG-15 and later the MiG-17 in combat over North Korea from 1950 to 1953, against USAF F-86 and then F-100 sweptwing fighters. Soviet fighters proved to be rugged, but they were outfought by better trained and more experienced American pilots. Thereafter, the Russians gained their combat experience vicariously against the Israeli Air Force thanks to the efforts of their Arab allies, but with serious losses which had to be replaced. Nevertheless, this foreign aid helped factories in the Soviet Union dispose of the quotas of aircraft which they produced. India became another client who tested MiGs and Sukhois against Pakistan's American and British-built machines, both sides finding that older aircraft were more reliable and useful because they could be kept serviceable. At the same time, China, always lacking an aircraft industry, was a market for both finished and license-built machines. Overall, the pattern was very similar to that in the free world where some aircraft were sold from home-manufactured stocks and others were licensed for off-shore manufacture or for shares in the work.

The influence of the Cold War on Soviet aviation has to be seen against the background of Russian suspicions of outsiders that went back at least to the struggles over Russification or Westernization started by Peter the Great at the end of the seventeenth century. This was compounded by the Communist ideology that expressed itself in the Comintern fear of imperialist powers after 1917, in itself simply a wider-world view of the Great Game that had been perceived all along the southern frontier from the Black Sea east. Then, too, there was the experience of the Second World War in which Leningrad, Moscow, Stalingrad, and Sevastopol had been besieged and the economy devastated with great loss of life, while leaving many surviving veterans of both sexes. Moreover, many of these in the victorious armies which had reached Berlin had seen the destruction wrought by the Allied Combined Bomber Offensive. All of these factors made the Kremlin very apprehensive, especially when the man who made the Iron curtain speech in Fulton, Missouri, was the same British leader, Winston Churchill, who had urged on the attack on the Reds in 1918 in alliance with the White armies.

Thus it should not have been at all surprising that the Russians put great effort into various forms of home defense from accelerated development of the German V-2 rocket into a ballistic missile system, beefed up air defenses from guns and radar to fighters, and a civil defense organization with more shelters than had been built in the Great Patriotic War. Western instigation of the North Atlantic Treaty Organization (NATO), public pronouncements and display of weapons and doctrines aimed solely at the USSR simply spurred efforts rather than leading to peaceful co-existence. The doctrine of containment of the USSR and its Warsaw Pact allies simply revived the Russian desire for warm-water adventures. At first these were limited to contact in the Atlantic and the Mediterranean, but eventually reached the Indian Ocean.

The later 1950s saw the Soviets making rapid strides in many fields. The Navy was accepted, after its dismal showing in the Great Patriotic War. Thus it was modernized in a long process that brought it VTOL (vertical takeoff and landing) aircraft-carriers in the 1970s and ocean-going full carriers just in time for the collapse of the nation. Sputnik, the first satellite, was put in orbit in 1957, and thereafter the Soviets made great progress with IRBMs (intermediate range ballistic missiles) and then ICBMs (intercontinental ballistic missiles) causing the USA and its allies to counter first with Thor IRBMs in Britain and Turkey and then with Polaris submarine-launched IRBMs, and finally with the major ICBMs culminating in the demonstration of American scientific and technical prowess with the landing of a man on the moon in 1969. While the Russians had been the first into space and would continue to place people in orbit, they never achieved a moon flight. In the end, of course, after the collapse of the Soviet Union and the destruction of nuclear missiles, the former enemies joined together to man Soyuz space stations.

In the meantime, the nuclear deterrent strategy had proved to be the unusable weapon because of mutually assured destruction (MAD) capabilities and a lessening interest in war as mutual fears were eroded by better communications and tourism, as well as by the achievement of a balance of power by the opening of relations between China and the USA in 1972 in the wake of long Sino-Russian hostilities that stretched back into the nineteenth century.

The weakening of Soviet suspicions became evident not only in the permission for Western airlines such as PanAm and BEA to fly to Moscow, but the 1960s opening up of the air routes across Russia to BOAC and Japan Air Lines, as the new longer-ranged four-engined jets allowed the airlines to make non-stop Moscow–Tokyo and Moscow–New York flights.

RUSSIAN AVIATION IN MAHANIAN TERMS

Another way of assessing Russian air power is to look at the country in Mahanic terms – geographic position, physical conformation, extent of territory, number of population, national character, and character of the government.

Positioned across the top of the Eurasian continent, Russia is a vast country that in the twentieth century has stretched from the middle of Poland eastward to Alaska, a distance of roughly 3,850 miles. Except for the great rivers in western Russia beyond the Urals, transportation for heavy commodities and raw materials was almost non-existent outside of the thin rail network. Yet the country was well supplied with raw materials, even having large oil-fields with which to fuel aviation, while there was hydroelectricity for power grids, factories, and for smelting aluminum. Much of the land, moreover, was very suitable for aviation; only in the few mountainous areas and in the forest was it necessary to carve out airfields. All aircraft were fitted either to land on water or to operate off grass, so the continual shortage of concreting equipment was not a disadvantage. Equally importantly, the vast distances within the country made aviation very time-effective, the miles from Moscow being such that very long-range flights could be undertaken without leaving the USSR, and climatic variations meant that aircraft, equipment, and crews could be tested in everything from hot and high, to sea level in sub-zero conditions. Only the United States enjoyed a similar advantage.

As time passed and pilots were no longer either first-class navigators or certified maintenance personnel, then aircraft had to be tested under all conditions and not be expected to be modified by pilots *en route*. For the British, then, this meant taking aircraft to Canada for winter trials in Manitoba and to Kenya for hot and high trials. Moreover, even in the limited competition between British and foreign-built aircraft, the race was primarily with the United States until the 1970s when the French challenges from the Dassault Mirage and the Airbus Industries airliners came to the fore. And even after 1945 when the British were able to operate the internal airline in Germany, and British European Airways was pushing out the range of its services as newer aircraft such as the Comet IV became available to it, the opportunities for airlines in Europe was limited. Though as the speed of airliners rose, their time-savings over the excellent railway network grew. However, in the 1980s that began to be reversed with the coming of the French TGV (high-speed train) and the following development of a high-speed rail network in Europe (and in Japan). In the Soviet Union and Russia, however, the advantage lay with the air services because trains in Russia have never been known for their high speed, nor, with so much capital being expended on defense, were funds available for the construction of high-speed rail lines, all hope for which flew out the window with the collapse of the USSR in 1991.

As to people, Russia had a growing population which made plenty of labour available, though the lateness of the industrialization, the costs of the Great War and of the subsequent Bolshevik activities, meant that it was a long time before recovery to 1913 levels was achieved. The Five-Year Plans did mobilize and train labour, and the Great Patriotic War demonstrated that peasants could be trained along with urban workers to produce large numbers of effective aircraft, and that men and women could be mobilized to man them. The legacies of war and the philosophy of full employment meant that there was a sufficient workforce which was probably more capable than anti-Soviet Western sources wished to admit, in part also because the known working conditions were not those in Western countries. Certainly, the average Russian or Soviet urban or peasant worker was patriotic, as demonstrated by their will and determination to fight for the Motherland throughout 1941–45, and their traditional fear of unknown outsiders.

The tourism made possible by aviation, in much of the rest of the world, which saw a million passengers a year crossing the North Atlantic in the early 1960s and climbing upward ever since, was not evident in the Soviet Union. Breaking down mental blocks by tourism has been a relatively recent development. On the other hand, for the internal development of aviation, the general mechanization of Russia, especially in the Second World War, created a sufficient supply of mechanics. How efficient they have been has been shrouded in some mystery since unlike in the West, accident reports have not been readily accessible. It can also be noted that the West for its own reasons has tended constantly to ignore or discredit Russian technical prowess as can be seen in the surprise with which close-up examination of the MiG-29 revealed the excellence of its systems as well as its superior flying ability.

Perhaps the critical question in such a Mahanian examination has to be the nature of the government.

At the time of the Wright brothers' flight, Russia was an autocracy upon which the Revolution of 1905 forced a measure of middle-class democracy through the newly elected Duma or parliament. However, the problems to be tackled were so great even without a war, that it was not surprising that the system collapsed in 1917 and had to be rebuilt by the urban-minded Soviets as a nineteenth-century industrial entity. In spite of purges after 1917 and in the late 1930s, as well as emigration, the Soviets made the economic engine run.

The Soviet government was a monolith imposed upon many

different nationalities speaking a host of different languages. Power and decision making resided in Moscow, with decisions influenced by the Communist Party members from various regions. Very like the Roman Catholic Church, it was essentially a top-down hierarchy. Thus, until the collapse after the unsuccessful Afghan War (1979–90), the military, especially the Army and its affiliate the Red Air Force, exercised a great deal of control over the budget and kept a very large force off the labor market, while also providing a solid consumer for aircraft production. After the collapse, and the split into the Commonwealth of Independent States (the CIS), as Oliver Wendell Holmes would say, the wonderful one-horse shay fell apart.

As a result, the armed forces have been starved of money, have had to discharge millions, and are still faced with the near impossibility of continuing any defense projects let alone procuring new types. The Su-37 variable-thrust fighter may have stolen the show at Farnborough in 1996, but the likelihood of orders is slight. In addition, an old trend has re-emerged. The civilian side of the Russian aircraft industry has turned to the West for engines and electronics in the hope not only of making their much cheaper aircraft saleable, but to make them also maintainable by the standards of world airlines.

At the same time, the privatization move that has been driving other governments to divest themselves of their state-run enterprises so as to make them self-supporting, has also hit the CIS. Aeroflot was broken up and privatized, and by sometime in 1997 it was expected that the main Moscow airport, Domodedovo, would also be run by a civilian corporation, perhaps even a Western one.

CONCLUSION

Although Russia and then the Soviet Union were by choice isolated societies with great suspicions of Westerners, the patterns of aviation development within the country have mirrored those outside. In part this was inevitable not only because in tackling common problems in the air a certain similarity of solutions was bound to occur, but also because the Russians both read foreign technical journals, saw foreign products at air shows and elsewhere, and received periodic injections of Western developments from the 1914 war through the Great Patriotic War and its aftermath, especially in jet technology. Moreover, periodically immigrants, willing or otherwise, have been attracted to Russia to supply expertise, as has been happening again after the collapse of the USSR.

Given the experience of France after 1944, it would be a mistake to believe that the Russians will not once again become important players challenging the Europeans, the Americans, the Japanese, the Chinese, and perhaps even the Indonesians in aviation.

THE CHAPTERS

The pieces that follow vary in length for a reason. The chapters on the early days by Von Hardesty, on the interwar years by Reina Pennington, on aviation and the transformation of combined-arms warfare 1941–45, by Tom Allison and Von Hardesty, on naval aviation by Christopher C. Lovett, on the defense of Russian aerospace by Dennis J. Marshall-Hasdell, and on air combat on the periphery by Mark A. O'Neill all conform to the original lengths suggested. But in the course of the development of the work, it became clear that four subjects required more space.

These lengthier chapters did not lend themselves to being divided easily into chronological sections. Thus John T. Greenwood's chapter on the Great Patriotic War (Chapter 3), and his piece on the aviation industry (Chapter 6), and on the designers (Chapter 7), together with David R. Jones' coverage of Aeroflot (Chapter 10) are of a greater length.

Apart from that irregularity, we hope the reader will find this work both informative and useful, offering as it does a look at one of the world's most potent air forces as well as the life history of an establishment.

1

Early Flight in Russia

VON HARDESTY

For Russia, with its 11 time zones, aviation has been a dynamic instrument to provide for national defense and to forge social cohesion. Under the Soviets, the airplane occupied an exalted place at the very center of public life, as a metaphor for progress and an enduring measure of national achievement. Whether flying over the North Pole in the 1930s or building Aeroflot, the largest airline in the world, Russians have endeavored to be at the cutting edge of aeronautics. Once human flight acquired a trajectory into outer space, the Soviet Union again allocated enormous human energy and economic resources in the ceaseless quest to maintain parity with the West.

One of the most striking features of Russian aviation in the twentieth century has been its separateness, its evolution for decades as a parallel universe in the realm of aeronautics. The fateful intersection of war and revolution between 1914 and 1917 assured that Russia, for most of the century, would exist as a garrison state. Aviation, as with other critical sectors of the economy, embraced a goal of self-sufficiency. Ideological commitments only reinforced this estrangement from the outside world. Over time, the Russian approach to aviation acquired a peculiar style of its own, a blend of Western technology and indigenous engineering.

At the very dawn of the air age, however, Russia enjoyed close ties with the wider European aeronautical world, then experiencing rapid and dramatic growth. This first incarnation of Russia's aviation community in the late czarist period, if small, possessed great vitality. Russian air enthusiasts were an ubiquitous presence in Europe. They flew in international air races, made substantial theoretical contributions to the new field of aerodynamics, pioneered new aircraft designs, and even leaped into the high-risk world of capital investment to build a Russian aviation industry. Aviation proponents enjoyed easy access to the highest echelons of the Russian government. The czarist regime, despite subsequent distortions by Soviet propagandists, actively supported the development of aeronautics in the period 1909–14. Aero

clubs spread rapidly across Russia, reflecting a blend of active government patronage and public volunteerism. Aviation plants quickly
emerged in major cities such as St Petersburg and Moscow to supply a
growing market for aircraft and spare parts. Many Russians traveled to
western Europe to train as pilots and to purchase aircraft. Military
links were established as early as 1910, when the Imperial Russian
Navy purchased an Antoinette airplane for testing at Sevastopol. The
following year the Russian Army purchased a French Bleriot monoplane to inaugurate army aviation training at Gatchina, near St
Petersburg.

This pattern of natural growth for Russian aviation was abruptly
shattered with the advent of the Great War in 1914. For a brief
interlude, Russia sustained the established ties with French military
aviation; as long as Russia remained active in the war effort there was a
steady, if uneven stream of logistical support for the tiny Imperial
Russian Air Force. But the war brought military defeat and social
upheaval. The year 1917 unleashed two revolutions – the first in
March toppled the Romanov dynasty, and the second in November
saw the Bolsheviks seize power. With the Bolshevik triumph in the
long struggle for control of Russia's destiny, Russian aviation found
itself in a profoundly altered context – one of technological isolation
and strict party control.

AVIATION VIEWED THROUGH THE SOVIET PRISM

Any historical reconstruction of early flight in Russia must begin with
a consideration of Soviet historiography. For over seven decades the
Soviet Union embraced an official history of aviation that became
normative for both the public and historians. Accordingly, Soviet
histories viewed the year 1917 as more than a political faultline; it
represented a technological watershed, the borderland between the
primitive and the modern. The czarist era, by definition, reflected a
backward stage of history, an era that stood in sharp contrast to the
future embodied in the Bolshevik revolution. All that preceded 1917 –
even the most impressive Russian national achievements in aviation –
fell victim to this script of history.

Accordingly, Igor Sikorsky, a Russian aviation pioneer of world
stature, slipped into a memory hole, his name being expunged from
many history texts for decades and his creative work as an aircraft
designer studiously ignored. Sikorsky had made the fateful decision
to reject Bolshevism and emigrate to the West where he promptly
inaugurated a second illustrious career. By contrast, Nicholas Ye.
Zhukovskiy, the famed Russian aerodynamicist, cast his lot with the

Bolsheviks in 1918, a decision that in the near term earned him the active patronage of Lenin and ultimately the title of 'Father of Soviet Aviation'.

Once in power, the Soviets engaged in a deliberate campaign to downplay the pre-1917 Russian aeronautical accomplishments. The process of subordinating the history of aviation technology to the ideological imperatives of the Soviet state meant more than seeking out those who were sympathetic to the revolutionary cause. The Soviets soon engaged in a systematic program of distortion, to fashion for political purposes a sequence of mythic technical achievements. Under Stalin the quest to identify the Soviet regime with modern technology, in particular aviation, meant legitimacy at home and respect abroad.

When the Soviets sought out the useful past, they took a special interest in Alexander Mozhaiskiy, an obscure Russian naval officer who had experimented with a steam-powered flying machine in the 1880s. Mozhaiskiy became a convenient rival in Soviet historical literature to the Wrights, a way to demonstrate that Russia had been the birthplace of one of the most important inventions in modern times. Few details of Mozhaiskiy's life survive to allow for a nuanced retelling of his life and career. The outline of his career as an engineer and aircraft designer, however, is known. Mozhaiskiy first built a series of flying models, to study how a flying machine might be designed. He then constructed a full-scale monoplane, which he equipped with a steam engine. Mozhaiskiy's attempts to fly the monoplane proved unworkable: he catapulted his machine down a ramp in the expectation that his steam powerplant would be sufficient to propel it skyward. The excessive weight of the engine precluded any sustained powered flight. While Mozhaiskiy's design did not lack sophistication, he fell short, as many others had, in the effort to fashion a flying machine that could be self-propelled and possess a workable system of flight controls.

Mozhaiskiy's experiments were largely unknown in pre-revolutionary Russia, even among air enthusiasts. However, during the Soviet era official history texts rescued Mozhaiskiy from the ash heap of history and heralded him as a precursor of the Wrights, a claim which amazed westerners and only reinforced the sense of separation between Russian aviation and the outside world. Sadly for Mozhaiskiy, his true accomplishments were obscured by the excesses of Soviet propaganda. It would take decades before his contribution to aviation would receive proper coverage. One can note that the residual strength of this Soviet-era myth endured even into the late Soviet period: the Central House of Aviation (*Tsentralnyy dom aviatsii*) in Moscow showcased Mozhaiskiy as the 'designer of the world's first airplane' as late as the 1980s, although the aviation community as a whole ignored this embarrassing episode in Soviet historiography under Stalin.

By 1989, however, the Russian aeronautical community displayed a new appreciation for the Russian aeronautical accomplishments in the late czarist period. That year a special exhibition was held in Moscow to celebrate the centennial of Igor Sikorsky's birth. This extraordinary exhibition, based in part on resources from the Smithsonian Institution's National Air and Space Museum, provided the first complete historical portrayal in Russia of Sikorsky's life and contributions. Russian historians openly embraced Sikorsky as a national hero, as part of a concerted effort to rehabilitate Sikorsky and, by extension, recast Russian aviation history anew.

THE GOLDEN AGE OF RUSSIAN AVIATION

The late czarist era constitutes a distinct phase in Russian aviation history, indeed a golden age, a reminder that the story of aviation in Russia is not co-extensive with the Soviet era. Russian participation in the saga of human flight goes back to the eighteenth century when Mikhail V. Lomonosov, the founder of the Imperial Russian Academy of Sciences, speculated on aeronautical theory and flying machines. Lomonosov became Russia's Leonardo da Vinci, a visionary for a field of technology yet to be born. A century later, another Russian scientist named Dmitri I. Mendeleyev became a well-known balloonist. Mendeleyev is best known today as the compiler of the periodic table of atomic weights, but his public persona as an aeronaut in the nineteenth century did much to promote airmindedness in Russia. One can trace the institutional basis for Russian aeronautics to Dmitri A. Miliutin, who, as War Minister under Alexander II (ruled 1856–81), took a keen interest in the military uses of balloons. Miliutin's foresight bore fruit with the establishment in 1885 of a special army 'flying school' for balloonists at Volkov Field near St Petersburg. By the time of the Russo-Japanese War (1904–05), the Russian Army fielded its own balloon battalion to conduct aerial reconnaissance. A decade later, Russia's lighter-than-air contingent included a total of 13 dirigibles.

The Wright brothers obtained a Russian patent for their flying machine in 1909. The Wright patent coincided with a burgeoning Russian interest in the new heavier-than-air technology. The first public demonstration of a Wright airplane took place in St Petersburg that same year, an event which included an inspection of the Wright flying machine by Sergei Witte, the famed builder of the Trans-Siberian railroad. Two of Russia's best known aviators – N. Ye. Popov and M. N. Yefimov – subsequently flew Wright aircraft in a series of air shows in 1910.

Russian interest in the Wrights, if intense, proved to be shortlived. Between 1910 and 1914 Russian aviation enthusiasts, civil and military, shifted their attention to France, then the epicenter of aviation in Europe. The names Farman, Bleriot, and Voisin quickly became the exemplars of modern aviation in the minds of most Russians. M. N. Yefimov became the first Russian to fly a Bleriot. In 1910, citizens of St Petersburg looked up in awe as G. Piotrovskiy in a Bleriot circled the Winter Palace and St Isaac's Cathedral, ending his trek over the imperial capital with a direct flight to the island naval base at Kronstadt. Ye. V. Rudnev, another prominent Russian aviator, made one breathtaking non-stop flight from St Petersburg to Gatchina – at the time a dramatic long-distance flight of 25 miles. The hub for flying was Komendantskiy field, located in the forest district north of St Petersburg.

At the core of early Russian aviation was the Imperial All-Russian Aero Club (IRAC). Founded in January 1908, the IRAC provided a powerful vehicle to organize and promote airmindedness in Russia. Being new and linked to western European aviation centers, the IRAC displayed a keen interest in heavier-than-air technology. As the IRAC set up a national network of flying clubs across the Russian Empire the airplane became the major focus. This new emphasis ran counter to the traditional Russian interest in balloons and more recently dirigibles. For a short period of time there was an intense rivalry between these two camps, based in part on the fact that the lighter-than-air proponents possessed strong institutional resources.

Russia's firm footing in aviation in the early years rested in part on a highly sophisticated scientific research program that pre-dated the Wright brothers flight at Kitty Hawk. Russia had established an experimental lab at Kuchino in 1901, a facility which would place Russia at the cutting edge of aerodynamics. Here Russia built a modern wind tunnel in 1904 (Russia's first wind tunnel had been constructed at Moscow University in 1902). Located near Moscow, Kuchino drew to its modern labs a cluster of talented scientists. As early as 1904, Nicholas Ye. Zhukovskiy – Russia's famed aerodynamicist – worked at Kuchino, taking full advantage of Kuchino's new and sophisticated wind tunnel, spacious workshops, and library. Zhukovskiy went on to contribute a sequence of theoretical works on aerodynamics in the years before the First World War. Kuchino drew other specialists such as D. P. Riabouchinskiy (the director of Kuchino facility), K. Ye. Tsiolkovskiy, the famed rocket and space flight visionary, and S. A. Chaplygin, an authority on high-speed aerodynamics.

Through Zhukovskiy, Kuchino maintained strong ties with the nearby Moscow Higher Technical School, which allowed Russia to attract a whole new generation to aeronautics, a group that would include Andrei N. Tupolev, the renowned aircraft designer of the Soviet Union. Later, the Bolsheviks would build on the Kuchino legacy by recruiting Zhukovskiy to organize TsAGI (*Tsentral'nyy aerogidrodinamicheskiy institut*), the Central Institute of Aero-Hydrodynamics. TsAGI became the Soviet Union's premier research facility for aviation.

The IRAC joined the International Federation of Aeronautics (FAI), a step that established official ties with the international aviation community. Even as these steps took place, many wealthy Russians traveled to France or Germany to attend air shows, to take flight instruction, and to purchase airplanes. European aircraft-manufacturing firms, mostly French, quickly forged commercial ties with the growing Russian aviation community, licensing and selling aircraft and engines in the growing Russian market. While substantial foreign investments did much to establish the Russian aviation sector, indigenous Russian entrepreneurial talent played a key role from the start.

Moreover, Russia developed parallel military links with the West, especially with France. The military alliance with France set the stage for numerous Russian military officers to visit French plants and purchase aircraft on behalf of the Russian government. While ties to the French were strong, Russians also established links with aviation firms in Britain and the United States; for example, much of Russian naval aircraft technology in the pre-1914 period came from the Curtiss firm in the United States. Consequently, Russia's aeronautical links to the West were broadly based and freely developed in the years before the First World War.

Vivid images of the pre-revolutionary Russian aeronautical community may be found in surviving copies of the many Russian aeronautical journals and magazines that appeared after 1909. Air events in western Europe were routinely chronicled in these journals, showcasing the latest aircraft designs. Advertisements in these journals allowed Russian air enthusiasts to learn about the latest technological breakthroughs. Most issues contained one or more practical articles with vital information on how best to construct a flying machine. Many aviation journals appeared in the provinces, as newly organized chapters of the IRAC took shape and launched their own air shows. Local clubs affiliated with the IRAC enjoyed real links with the larger world of aviation in Europe. IRAC publications also stressed technical understanding of the new flying machines with translations of key foreign articles on aircraft design and aero propulsion.

The mainstream Russian newspapers also took notice of the airplane and its potential impact on Russian life. As early as 16 January 1910, o.s. (old-style dates, from the Gregorian calendar, which was 10 days different from the Julian, not adapted in Russia until 1917), *Moskovskiye vedomosti* (Moscow Gazette) published an article on 'War and the Air Fleet', lamenting that this new weapon of war was little more than 'fun for a few' while in Europe air power had become a 'serious affair'. Later, on 17 June 1910, o.s., another article appeared on the nature of a future air war. Under the title, 'Bombing from the Air', this highly speculative article warned of 'murderous projectiles' directed by a 'steady hand', and suggested that the lethal potential of air power would inevitably have to be controlled by international law. M. Nikulin wrote an article in the same newspaper on 8 March 1911, o.s., 'The Military Significance of Aeronautics'. He concluded prophetically that the airplane would have two essential roles in the next war: reconnaissance and attack. More slowly, the Russian military began to define a rudimentary air theory to shape air operations in a future conflict. By the time of the First World War the Imperial Russian Air Force would possess an inventory of around 250 aircraft.

While the leadership of the IRAC initially was drawn from the aristocracy and the upper levels of the bureaucracy, the Russian aviation community quickly became a popular organization with members drawn from all classes. The IRAC abandoned any concerted effort to sustain a social pedigree among its members. The recruitment of engineers and mechanics – so essential to aviation activities – meant the active enrollment of individuals from the technical and commercial classes of Russia. Expediency dictated that merit should prevail over class origins as a criterion for membership. Still, at the heart of the Russian aviation community were the sportsmen-pilots. These individuals, as a rule, came from aristocratic families and possessed ample wealth and leisure time to pursue a career in flying as Russia's first generation of pilots. The IRAC, as part of the FAI, licensed all pilots in Russia, a policy that assured that all Russian pilots attained a minimal standard of proficiency.

Russian aircraft designers became known for their innovative work in large aircraft, seaplanes and early experimental work in vertical flight. Among these designers, Igor I. Sikorsky quickly established himself in the late czarist period as an outstanding designer. As early as 1908 Sikorsky had visited both Germany and France where he saw firsthand the latest developments in aeronautics. Returning home to Kiev, he decided to enter the pioneering field of aircraft design, working with a small group of fellow enthusiasts at the local polytechnical institute. His first effort was to design a helicopter, a quest which failed after intense work and experimentation. Three decades later, in his new home in the United States, Sikorsky would return to his youthful fascination with vertical flight and design the VS-300, the first successful helicopter.

Moving to the design of airplanes, Sikorsky demonstrated considerable skill. His 'S' series signalled his arrival as the pre-eminent Russian designer of the pre-World War I period. These single-engine aircraft designs culminated in the S-6A which won the Moscow military competition of 1912. Sikorsky's S-16, a fighter which became operational in the First World War, represented the logical development of the 'S' series. Sikorsky's work for the Russian military won him considerable support and patronage in the highest levels of the IRAC.

By 1913, Sikorsky was ready to launch his most ambitious design, the construction of the four-engine *Russkiy vityaz* (Russian Knight). Building a multi-engine biplane aircraft was a bold experiment for Sikorsky. At the time there was little theoretical work available on the aerodynamics of large multi-engine machines. Moreover, there was a widespread belief that such aircraft would be inherently unstable and unsafe. Aero engines were neither robust nor dependable, and the breakdown of one engine (always a possibility) could force an aircraft into a spin with tragic results. Single-engine aircraft, by contrast, were viewed as more maneuverable and safe. Against these perceived obstacles Sikorsky pitted his own notion of 'intuitive engineering' and considerable practical experience as a designer and pilot. Sikorsky's *Russkiy vityaz* did fly successfully, becoming the aerial wonder of early aviation, the pride of Russian engineering in aeronautics. This same aircraft served as the prototype for the *Il'ya Muromets* (named after a mythical folk hero), the world's first long-range bomber.

While Sikorsky garnered considerable attention in the West, he was not the only figure who made an impact on early Russian aviation. Rivals to Sikorsky were Joseph Gakkel, a talented aircraft designer who produced a sequence of inspired designs, and D. P. Grigorovich who built a series of seaplanes for the Russian Navy. Flying schools at Gatchina (near St Petersburg) and Sevastopol (the Black Sea naval base) allowed Russia to train pilots year round.

Aviation industrialists quickly made their appearance in Russia after 1909. As a growing class of entrepreneurs, they vied with aristocratic leadership of the IRAC for influence over the aviation community. Russia's first factory, the Shchetinin works, opened in St Petersburg in 1910. The Shchetinin plant, by any standard, was a rather small affair, being in reality a workshop to assemble foreign-built aircraft. At the Shchetinin plant workers assembled Bleriot monoplanes for a growing group of Russian air enthusiasts. In time, the Shchetinin facility acquired a highly skilled staff, which built the *Rossiya*, the first production aircraft designed in Russia. The Shchetinin plant was not alone; any number of shortlived 'factories' arose to build flying machines.

Sportsmen-pilots were often auto racing enthusiasts who became attracted to the thrills associated with aviation. Racing cars and flying airplanes co-existed as parts of the same daredevil ethos. The Imperial Russian Automobile Society (*Imperatorskoye Rossiiskoye Avto-mobil'noye Obshchestvo*) dated back to 1903, and sponsored its own exhibitions and races. The related notion of an 'air sport' activity first took place in 1911 with the establishment of the All-Russian Aero-nautical Union (*Vserossiiskiy vozdukhoplavatel'nyy soyuz*). The union functioned under the aegis of the Ministry of the Interior and, as a result, possessed a more direct connection with the government. Members of the union, however, worked freely with the IRAC throughout these years, suggesting that Russia's aviation community was diverse and loosely organized.

M. V. Shidlovskiy, the director of the Russo-Baltic Wagon Works (R-BVZ) was widely known as one of Russia's most energetic entre-preneurs, the director of a large plant in Riga for the manufacture of railroad stock. In time Shidlovskiy became enamored of automobiles, becoming one of Russia's pioneering auto barons with his own auto-mobile under the mark 'Russo-Balt'. The transition to aviation followed. Shidlovskiy's growing interest in aviation would be trans-lated through his patronage of Igor Sikorsky, a partnership that would profoundly influence the direction of early flight in Russia. Other firms began with automobile manufacturing only to turn to airplanes, once public interest in flying machines became manifest. The Lebedev and Duks plants, for example, offered both autos and flying machines for sale. The Duks plant in Moscow had been organized by Yu. A. Meller, and, astoundingly, it manufactured or assembled automobiles, motor-cycles, dirigibles, and airplanes.

Shidlovskiy had been impressed with the possibility that Russia could design its own aircraft. He viewed Igor Sikorsky's S-6A, the winner of the military design competition in 1912, as confirmation of his outlook. Ultimately, Shidlovskiy provided the necessary backing for Sikorsky to build the *Russkiy vityaz*. Others shared Shidlovskiy's optimism and took similar risks. A. Lebedev (St Petersburg) and A. A. Anatra (Odessa) established aviation firms, and traveled freely between France and Russia to promote the sales of French-designed airplanes. Anatra would later manufacture an aircraft design of his own. Moscow possessed the largest aircraft plant in Russia, the Duks factory, which became the major aviation assembly plant for the Russian military during the First World War.

Aerial spectaculars in prewar Russia fueled public interest in the airplane. The IRAC became the chief sponsor of air races and air exhibitions. Behind the boosterism, of course, were other, more somber considerations: the IRAC sought to place Russia at the forefront of the new air age. Air races and air shows helped to generate broad public interest and governmental patronage of all aeronautical institutions. Already Russians were learning the techniques of cold weather flying, an arena where Russians would excel later with historic flights over the North Pole. The success of the IRAC in czarist Russia rested on its ability to mobilize popular enthusiasm for aviation, a program that eventually saw the airplane replace the lighter-than-air technology as the chief object of government patronage. Government largesse combined with volunteerism in the public sector to give IRAC activities great vitality the years before 1914. The IRAC in its organization and activities reflected the blurred lines between government and private sector involvement in aviation. The czarist treasury provided subsidies for aerodrome construction, support for flying schools, awards for aircraft design competition, and the general promotion of aeronautical technology as a vital part of state policy. As in the days of Peter the Great, the state embraced a progressive posture toward technology in general and aviation in particular.

IRAC-sponsored aero clubs quickly spread throughout the Russian Empire. As a member of the International Federation of Aeronautics, the IRAC licensed all pilots in Russia. By the end of 1912, the IRAC with its headquarters in St Petersburg had established affiliates in the major cities of Moscow and Kiev, along with branches in such remote centers as Saratov, Irkutsk, and Vladivostok. One aero club, the 'Turkestan Society of Aeronautics' at Tashkent shared a dusty, windswept marching field with General A. V. Samsonov's cavalry and operated under the official patronage of the Emir of Bokhara. The IRAC operated from the spacious Komendantskiy field north of St Petersburg, boasting a modern complex of hangars, workshops, and facilities.

Russia became the locale for record-breaking flights, as part of an international quest to fly 'faster, farther, and higher'. There were individual exploits such as Rudnev's solo flight from St Petersburg to Gatchina (at the time a remarkable trek of 25 miles!) or Igor Sikorsky taking 16 people aloft over St Petersburg in his *Il'ya Muromets*. Also, Russians pioneered flights into the Arctic region, a harbinger of the grandiose Polar exploration feats with aircraft under the Soviets. But air shows became the more typical means to establish new records. In 1910 the IRAC sponsored the 'St Petersburg Aviation Week'. The following year, again at St Petersburg, the IRAC sponsored the first 'All-Russian Aeronautical Exhibition'. Foreign pilots and exhibitors attended, along with a strong contingent from Russia. The Russian press followed the exhibition with great enthusiasm, giving the event broad coverage. The marriage of government patronage and public volunteerism – the hallmarks of early Russian aviation – was evident at the 1911 aeronautical exhibition. Moscow rivaled St Petersburg as a center for aviation. *Moskovskiye vedomosti*, for example, gave detailed coverage to Moscow's 'Aviation Week', held from 29 May to 5 June 1911, with daily installments on new records and occasional air disasters.

That same year the IRAC sponsored its most ambitious project, the 'St Petersburg to Moscow Air Race' – an international competition involving 11 pilots. The 400-mile route from the Russian capital to Moscow's famed Khodinka Field pitted three Russian aviators against an experienced group of European fliers. Interestingly, all three Russian participants (A. A. Vasil'yev, G. V. Yankovskiy, and M. G. Lerkhe) had trained abroad, reflecting Russia's strong links to French and German centers of aviation. Vasil'yev was the most flamboyant Russian pilot, having flown in the Crimea and in the Caucasus, and held the Russian altitude record at the time (1,650 meters). Vasil'yev had announced earlier that his ultimate goal was to fly a Bleriot mono-plane from Tiflis in the Caucasus to Vladivostok in the Far East, a flight of over 6,000 miles! His ambitious flight to Vladivostok, how-ever, had been canceled for 'an indefinite period' due to general weather conditions and other considerations. For the St Petersburg to Moscow race, however, Vasil'yev was prepared to compete, finding the prize money (75,000 rubles or around $35,000 in American money) irresistible. No doubt Vasil'yev, now in a more sober mood, found the 400-mile course to Moscow much more manageable than the projected flight to distant Vladivostok.

The actual race captured the attention of the Russian public, and received news coverage in western Europe. The pilots followed a nearly direct route to Moscow, flying at low altitude nearly parallel with the railway tracks that connected the two cities. The race proved to be a grueling one, punctuated with numerous breakdowns and mishaps. There were crashes, forced landings with pilots being greeted by incredulous peasants who had never seen a flying machine before, and moments when airplanes lost all contact with their ground crews who followed in automobiles filled with spare parts. Only one pilot made it to Khodinka Field – the intrepid Vasil'yev flying a Bleriot. Enthusiastic crowds greeted Vasil'yev at Moscow, including city officials and the Governor General. The IRAC and the Russian press took undisguised pride in the fact that this epic race had been won by a Russian against a field of competitors from western Europe.

Russian pilots also flew in various international air meets in western Europe. One of the most successful Russian pilots, V. M. Abramovich, set a new altitude record of 2,100 meters in May 1912 at Berlin. Abramovich won the Gold Medal from the German Aero Club that same year. Flying a modified Wright aircraft, Abramovich later made the first flight from Berlin to St Petersburg. Abramovich studied aerodynamics in Germany, where he established a reputation as one of the most competent flight instructors. He trained a large number of pilots, including one of Russia's first female aviators. His untimely death in an air accident in 1913 was mourned in both Russia and Germany.

The most dramatic aerial spectacular in the early years of Russian aviation, one that would rival in significance Bleriot's cross-channel trek, was the flight of Igor I. Sikorksy in his giant *Il'ya Muromets* from St Petersburg to Kiev in late June 1914. Sikorsky with a crew of three flew to Kiev in the spacious *Il'ya Muromets*, an aircraft unique for its time with heated cabins equipped with tables and chairs, a forward observation balcony, and a wind-driven generator which supplied the behemoth aircraft with electric lights. For Sikorsky, the *Il'ya Muromets* was an incarnation of the flying ship portrayed by the visionary Jules Verne, a boyhood hero of Sikorsky. In practical terms the *Il'ya Muromets* possessed the configuration of a modern airliner with its capacity to carry over a dozen passengers over great distances. For the 800-mile trek to Kiev, Sikorsky took many spare parts and an ample supply of food, because the itinerary called for only one quick stop at Orsha for refueling.

Instrumentation on the *Il'ya Muromets* was primitive – a compass for navigation, four tachometers to monitor the engines, a bank indicator, and a horizontal tube on the nose to align the aircraft with the horizon and measure the angle of climbs and descents. The aircraft cruised at 65 mph and flew most of the trip at an altitude of around 5,000 feet. There was only one mishap eight hours into the flight – one engine caught fire, forcing one crew member to walk out on a plank on the lower wing to put out the fire with his greatcoat; the emergency forced a landing which allowed Sikorsky to repair a broken fuel line.

Sikorsky descended through a cloud layer above Kiev to appear over the golden spires of the Kiev Pechersk monastery along the Dnieper. As the large and lumbering *Il'ya Muromets* flew over the city, large crowds followed it to the local aerodrome. The landing of the *Il'ya Muromets* at Kiev became a legendary moment for the ancient city, also the birthplace of Sikorsky. After several days of celebration, Sikorsky made the return trip to St Petersburg without difficulty, setting a new record for long distance flight.

Nicholas II renamed the aircraft the *Il'ya Muromets Kievskiy* in honor of this extraordinary flight. Sikorsky received many awards and the Russian press gave the flight enormous attention. At this moment in time Sikorsky held a total of nine world records in various categories of distance, duration, and altitude. Russian aviation led the world in the summer of 1914 in the design of large multi-engined aircraft – in no small part due to the intuitive genius of Igor Sikorsky. Arguably, the epic flight of the Il'ya Muromets ranks as one of the most significant flights in the history of early aviation. Sikorsky demonstrated in real terms the lethal potential of air power, the ability of large multi-engine aircraft to reach distant targets and urban centers. Later, in fact, the Russian military would realize this possibility when they organized the *Il'ya Muromets*-type aircraft into the world's first strategic bombing squadron. Ironically, the advent of the First World War – within less than a month of the Sikorsky flight – meant that the West only received fragmentary news of the *Il'ya Muromets*. When the aircraft later captured the attention of the West, it would be as a bomber, not as the first modern air transport.

THE CRUCIBLE OF WAR AND REVOLUTION

The genesis of the Imperial Russian Air Force may be traced to one event – the flight of Louis Bleriot across the English Channel in 1909. The Grand Duke Alexander Mikhailovich, then in Paris on holiday, witnessed the epic Bleriot flight. His reaction was one of awe and enthusiasm. Prior to the Bleriot flight the Grand Duke had viewed airplanes as a mechanical marvel, to be understood as one of the novel spinoffs of the new technological age; now he comprehended the military potential of the airplane. Bleriot's fragile flying machine had broken England's isolation, demonstrating the airplane's capacity to transcend natural barriers such as an ocean. Now as an 'air theorist', the Grand Duke saw clearly that the airplane could be a weapon, the means to transcend fortified barriers and geographical distance, to make any nation vulnerable to air attack in a future war.

Returning to Russia, the Grand Duke Alexander Mikhailovich organized the 'Committee for Strengthening the Air Fleet'. His new committee functioned as a quasi-governmental entity to promote military aviation and reinforced the work of the IRAC. Russia's embryonic air force took shape under the tutelage of the Grand Duke, who oversaw aircraft purchases and training. His role would be a powerful one in shaping the direction of Russian aviation in the late czarist period.

Government procurement of aircraft, in part financed by Grand Duke Alexander Mikhailovich's volunteer committee, quickly expanded the inventory of Russian military airplanes. At the outset of the First World War in 1914, Russia possessed around 250 aircraft in its fledgling air force, a figure more or less on par with the major powers in the West. No less important was Russia's evolving naval air arm, which pioneered in the war years a special naval air strike force for the Black Sea; by means of transports Russian seaplanes were launched at sea for attacks against the enemy. Army aviation consisted of pilots and ground crews organized into small units attached to the front. As in the West, the administration and development of the air force began in the engineering branch of the army and then elevated to a distinct armed force over time.

The articulation of a coherent air theory proceeded slowly, always behind the purchase of aircraft and the organization of air units. Within the military establishment there was a powerful inertia against the acceptance of a new technology. The airplane in the pre-1914 era, of course, added fuel to those skeptics who doubted the potential military utility of air power: flying machines constructed of wood and fabric flew slowly with numerous engine and structural breakdowns. Still, many saw the airplane as the weapon of the future.

During the first years of the Great War air operations on the Eastern Front were sporadic at best, and at no time a decisive factor in any ground offensive. *Vozdushnyy spravochnik*, the yearbook of the IRAC, published a detailed monthly chronicle of air operations for 1914 and 1915. Reading these brief annotated reports, one is struck by the scattered character of air sorties and air battles, which appear to have been occasional and seemingly independent of the larger Russian military activities on the ground. As in the West, the initial air operations were for aerial reconnaissance or artillery spotting. Air battles as such were rare, but not unknown as each side decided to challenge the other for control of the skies.

Most Russian army commanders were slow to exploit fully the potential of the airplane, a consequence of their traditional training and the small number of aircraft then deployed to the front. The Russian army began to realize the utility of the airplane by the summer of 1915, prompting many field commanders to request systematic reconnaissance missions against the enemy. As the war progressed, Russia acquired a large number of modern fighters and bombers from France and the Allies. With these new aircraft the Imperial Russian Air Force expanded and assumed a much more aggressive posture toward the enemy. In 1916, Russia established an advanced school for fighter pilots near Odessa.

Russia did pioneer a number of tactical innovations with the Navy, which had possessed an air arm as far back as 1910. At Sevastopol, the Imperial Russian Navy equipped a special warship to carry seaplanes for attacks against Germany's allies along the Black Sea, in particular Turkey. Russian seaplanes (both Grigorovich and Curtiss types) were transported to sea, off-loaded by crane, and then launched for surprise raids against the enemy. This experiment in 'naval aviation' reflected the capacity of the Russians to seek out new ways to use air power to penetrate the borders of the enemy.

The Russian public idolized and romanticized the military aviator, seeing him as a special kind of hero, a knight of the air, the standard bearer of national honor. Russia did produce a number of its own air heroes. Peter N. Nesterov, who had executed the world's first loop (in a Nieuport IV on 14 August 1913), became Russia's first great hero in the opening weeks of the war. Over Galicia, Nesterov spotted an Austrian aircraft on patrol. Nesterov deliberately rammed the enemy airplane in the ensuing air battle, a tactic that led to his death; after hitting the Austrian airplane Nesterov could not break clear and make a safe landing. Nesterov's death became the occasion for national mourning, a way to eulogize the heroic sacrifices of Russian soldiers against the enemy. The ramming maneuver – the *taran*, if extreme and rarely emulated in the West, was not a suicide move as with the Japanese pilots a generation later; Nesterov in fact aimed to survive the intentional mid-air collision. It is noteworthy that Nesterov's use of the *taran* became a unique Russian air tactic, one that would be repeated in the Second World War by many Soviet pilots.

Nesterov's story and the exploits of many other Russian pilots were lionized in P. Kritskiy's wartime book, *Podvigi russkikh aviatorov* (Heroic Deeds of Russian Pilots), first published in 1915. Alexander A. Kazakov, Russia's 'ace of aces', emerged as one of the war's most accomplished fighter pilots. Another effective Russian fighter pilot was Ye. N. Kruten, who published a manual on air tactics in 1916. Among the wartime aces, only Kruten would be celebrated in the Soviet era as a hero of the Great War. Most Russian aviators were hostile to the Bolsheviks and found themselves ignored in the period after 1917. Boris Sergievsky, Alexander P. de Seversky, and Ivan Smirnoff survived the war as military pilots, only to emigrate to the West for second, perhaps more celebrated careers.

Russia's aviation industrial base grew dramatically during the First World War. War-induced government contracts stimulated a rapid expansion. The Duks factory and the Lebedev factory, among others, were granted government contracts for aircraft production. Typically, such factories assembled or built under license French-designed fighters and bombers. Earlier the Russian Army had purchased Farman biplanes, now obsolete, but still useful as trainers. Through-out the war Russians obtained a series of Nieuport fighters, beginning with the Nieuport 10 model. Allied shipments allowed the Russian air arm also to obtain Voisin bombers, which constituted the mainstay of army bomber aircraft.

Shidlovskiy obtained a government contract for the construction of *Il'ya Muromets* bombers in August 1914. Shidlovskiy placed the talented designer of the *Il'ya Muromets*, Igor I. Sikorsky, in charge of this ambitious project at the RBVZ plant in Petrograd (formerly St Petersburg). The first production run called for ten bombers with subsequent wartime production reaching a total of 76 aircraft. The newly organized strategic bomber squadron, *Eskadra Vozdushykh Korablei* (EVK or Squadron Flying Ships) eagerly accepted these bombers as they came off the production lines in the fall of 1914. Russian interest in large bombers was not restricted to the RBVZ. Across town, V. A. Slesarev worked tirelessly on the ill-fated quest to build the *Svyatagor* – a massive bomber that Slesarev hoped would stay aloft for over 30 hours. The Russian industrial base, if small, was firmly established on the eve of the First World War.

No coverage of Russian military aviation in this period is complete without reference to the remarkable story of the *Il'ya Muromets* bomber and the EVK, or *Eskadra Vozdushnykh Korablei* (Squadron of Flying Ships). Designed as a transport by Sikorsky, the *Il'ya Muromets* quickly found its way to the front as a bomber as early as September 1914. Over 70 of these aircraft would be manufactured for the war effort. The combat role of the EVK proved to be varied and ended only in 1917 when the czarist regime collapsed. M. V. Shidlovskiy, who had been instrumental in bringing Igor Sikorsky's four-engine bomber into reality, became commander of the EVK unit. His appointment as commander was remarkable in many respects because Shidlovskiy had served in the army and came to his wartime duties as an older man, known in Russia as an industrialist, not as a military figure. His assignment represented a slight to the Grand Duke Alexander Mikhailovich who was Russia's nominal wartime air commander.

The Grand Duke Alexander Mikhailovich had joined the ranks of those in the aeronautical community who openly opposed the development of the large multi-engined bombers. These individuals, consisting of many of the most prominent pilots and air leaders, saw little utility in the huge bombers; they argued that the smaller fighters and bombers were the most essential element in the Imperial Russian Air Force. One famous Russian pilot from the prewar years, the aforementioned Rudnev, actually tested the *Il'ya Muromets*, to see if the bomber could achieve the requisite flight performance standards for acceptance by the army. Rudnev's negative evaluation strengthened the Grand Duke's argument that the EVK should not be maintained as an operational air unit. Finally, these objections were overruled and the EVK survived as a special air unit reporting directly to Stavka, the supreme headquarters of the Russian Army in Petrograd. Throughout the war years Shidlovskiy maintained an awkward and strained relationship with the Grand Duke Alexander Mikhailovich.

The operational life of the EVK remains an interesting story, although it has been largely ignored in most accounts of air warfare in the Great War. Looking back, it is obvious that this remarkable squadron operated largely as a reconnaissance air unit. This is also the arena in which the EVK made its most substantial contribution to the war effort. However, one could argue that the *Il'ya Muromets* inaugurated strategic bombing as a facet of air operations in modern warfare. When the EVK launched bombing sorties against East Prussia in 1915, hitting targets in the enemy homeland such as communications centers, rail junctions, and troop concentrations, the era of strategic bombing had begun. It is noteworthy that these raids preceded German Zeppelin attacks on England by several months, and German bomber attacks on London (with Gotha bombers) by over two years. The participation of the EVK in both reconnaissance and bombing missions – even if primitive, occasional, and often without any superintending air theory – confirms the fact that almost all modes of modern air warfare were present in some rudimentary form in the First World War.

One is amazed at the complex organizational structure the EVK possessed during the war. At its peak, the EVK operated in three *otryady* or air units. The EVK maintained its own meteorological unit and a highly sophisticated aerial photography lab. A large and talented ground crew serviced the EVK bombers. They faced enormous problems keeping the Sikorsky bombers airworthy – shortages in spare parts and replacement engines forced all sorts of improvised steps, including the cannibalization of aircraft.

Existing records of EVK combat operations reveal that most raids by the Muromets bombers were launched in the summer months, roughly from June to September. Flights were made in the winter months, almost exclusively for reconnaissance missions, but these were limited. All EVK sorties were influenced by the fact that for the duration of the war, beginning with the German 1915 offensive, the Russian military was on the retreat. The EVK began the war at Yablonna, a base outside Warsaw. This frontline air base was spacious and ideal, but subsequent moves by the EVK – as part of the general Russian retreat – meant more primitive and inadequate air bases. Viewing the entire combat history of the EVK, one is impressed with the ability of Shidlovskiy to maintain the air unit as a viable force in the face of military withdrawal and material shortages. At Vinnitsa in 1917, the last remnants of EVK were demobilized, bringing to an end one of the most experimental air units in the First World War.

THE CZARIST LEGACY

Caught up in the maelstrom of revolution, representatives of Russia's beleaguered aviation community met in Petrograd in May 1917. At the time, the Provisional Government still retained formal authority, even as the Bolsheviks waited for the opportune moment to seize control. At this 'All-Russian Congress' civil and military figures – pilots, mechanics, engineers, professors – met to preserve the institutional basis for aviation in revolutionary Russia. Their debates and plans have been preserved in the short-lived journal, *Vestnik lëtchikov i aviatsionnykh motoristov obnovlennoi Rossii* (Messenger of Pilots and Aviation Mechanics of a Reformed Russia), which appeared in two issues in May and June 1917. The congress delegates adopted a 'non-partisan' posture. They called for 'work, knowledge, and discipline'. Mixed with the appeals for preserving aviation were radical ideas on how best to transform Russian aviation in the new revolutionary epoch: the vision of a new 'Aero-Technical Institute', a vast complex of hangars, aircraft plants, and educational facilities. The rationale for such a grandiose undertaking was to catapult Russia to the forefront of aeronautical technology, to enable Russia to escape 'foreign domination'. These blueprints for the future echoed the vision of the Grand Duke Alexander Mikhailovich for air progress and self-sufficiency.

The call for one vast, integrated institutional structure for aviation anticipated the thrust of Soviet planners in the decades that followed. The Soviet regime exercised control over all technology through a centralized bureaucracy in complete subordination to party discipline. The blueprint of the All-Russian Congress ironically would be adopted in its general outlines by the Soviets once they assumed power, but with few concessions to autonomy for Russia's aeronautical specialists.

By 1918, in fact, the Bolsheviks had initiated a sequence of programs to mobilize the surviving elements of the old Russian air establishment. The ravages of war and revolution had nearly destroyed the institutional basis for Russian aviation. In the July 1918 issue of *Vestnik vozdushnogo flota* (Messenger of the Air Fleet), the new Bolshevik air periodical, a military pilot reported on the state of the old Sevastopol Flying School: 'There are no students. The staff of the school abandoned the facility, leaving it open to pilfering by demobilized soldiers.' That same year the Bolsheviks confiscated the great Duks plant and Lebedev factory in Petrograd. A new 'Red Workers–Peasants Air Fleet' took shape between 1918 and 1922 as the Bolsheviks struggled successfully to consolidate their power. At Kazan in 1918 and at other battles during the Civil War the Red Air Fleet contributed in a modest, if symbolic, way to the war effort.

The Russian aeronautical community faced enormous challenges under Soviet rule. The old network of aero clubs had been shattered, factories were closed or destroyed, the old pre-revolutionary cadres of pilots, mechanics, and engineers were dispersed. Under the Soviets there would be a concerted effort to revitalize this demoralized technological sector. As it turned out, Soviet central planners gave the aviation sector more than a polite bow when it came to budgetary allotments: Soviet budgets routinely emphasized the aviation expansion and under Stalin aviation occupied a central role in national life as an exemplar of modernity.

Nevertheless, during the Soviet era the isolation of the Russian aviation community meant continued technological backwardness in certain spheres, for example aero propulsion, which forced the Soviets to engage in a systematic program of 'technology transfer' from the West – on occasion through licensing agreements or on other occasions through more clandestine means. Viewed from this perspective, one can quickly see that the stillborn czarist aviation community left a mixed legacy, one of indigenous strength and dependence on Western technology. Behind the campaign to make the Soviet Union a world-class air power was the indigenous aviation community, which possessed considerable talent and resources going back to its pre-revolutionary origins. Over time, Soviet aviation acquired a peculiar 'style' of its own, a blend of Western and Russian elements.

In retrospect, one is struck by the continuity between the czarist and communist periods, a reality which has been obscured by the ideological divide between the two regimes: both regimes saw the government as playing a central role, although differing in the amount of freedom and autonomy offered the aviation community; both sought national self-sufficiency as they grappled with chronic weaknesses in the aviation sector; both possessed a strong inclination for the gigantic and spectacular, as an avenue to demonstrate progress and parity with the West; and both encouraged a strong theoretical approach to aeronautics, often at the expense of applied measures. Both eras would possess at certain levels a common style which, on reflection, could be described as distinctively 'Russian'.

The pre-revolutionary Russian aviation community still awaits a detailed history, one that is sensitive to its peculiar character and relationship to the larger, overarching evolution of aviation in Russia in the twentieth century.

From Chaos to the Eve of the Great Patriotic War, 1922–41

REINA PENNINGTON

Following the end of the Civil War in 1932, the new Soviet state was in a period of retrenchment, both politically and militarily. As political struggles determined who would be Lenin's successor, the leadership and direction of the military was also in flux. Despite these uncertainties, aviation was expanded and modernized, at a relatively slow pace during the 1920s, then with increasing emphasis in the 1930s. During the mid-1930s, the Soviet Air Force, or VVS (*Voenno-vozdushnye sily*, which simply means 'military air forces') was making striking progress. By 1941, it was the largest air force in the world, with more than 10,000 aircraft.

Yet when the Germans invaded the Soviet Union in June 1941, the massive VVS was nearly obliterated in the first few days of war. More than 1,200 Soviet aircraft were destroyed on the first day alone; most never got off the ground. Soviet bombers which had to fly without fighter escort during the first days of the war were destroyed so easily by German fighters that the *Luftwaffe*'s Field Marshal Albert Kesselring called it 'infanticide'. No air force before or since endured such severe attrition as the VVS experienced in the summer of 1941.

In examining the Soviet Air Force during the interwar period, it is natural to look for weaknesses that would account for the disaster of June 1941. A number of shortcomings and failures can readily be found: a leadership shaken by frequent changes at the top; an understaffed and inexperienced officer corps, demoralized by the purges within its ranks; a dissociation between theory and practice, leading to the absence of effective operational plans; the failure of Soviet theory to account for the possibility of strategic defense; Stalin's refusal to permit mobilizations or increased alerts in the border areas. Other factors that are more a matter of poor timing rather than failure also help explain the near-destruction of the VVS in 1941: massive reorganization, re-equipping, and retraining that left units in a state of chaos, and the rapid expansion of the VVS which, combined with the effects of the purges, left many units badly under-strength.

However, to look only for weakness is misleading. What was truly remarkable and unique about the VVS was its astonishing recovery in the Second World War. What was it that enabled the VVS to withstand staggering losses, yet rise again to defeat the *Luftwaffe*? When evaluating the status of the VVS in the interwar period, it is important to look not only for the weaknesses that would lead to its initial humiliation in combat, but also to search for the strengths that would account for its incredible resiliency and eventual victory over the *Luftwaffe*.

Many important events in the prewar period affected the Soviet Air Force. The effects of industrialization played an important role in connection with the First and Second Five-Year Plans (1928–32 and 1933–37). The Soviet Air Force was involved in a number of small wars and conflicts: the Spanish Civil War (1936–38), skirmishes in China (1930s), the battles at Lake Khasan (1938) and Khalkin Gol (1939) in the Far East, and the Soviet–Finnish War (1939–40). The experiences of these conflicts led to an extensive reorganization in 1940–41. Purges also directly affected the military throughout this period.

Personalities played an important role. The dynamic Yakov Ivanovich Alksnis (commander of the VVS from 1931–37) was a major influence during this period. Some contemporaries considered him the 'most competent and the most authoritative' of all the Air Force leaders, describing him as a man of 'boundless energy' who was also 'merciless to idlers and those who would not pull their weight'. Alksnis was known for his concern for providing good working conditions for VVS personnel, and for his efforts to protect them during the early stages of the purges.

Alksnis' most important prewar successor, Yakov Vladimirovich Smushkevich (VVS commander in 1939–40), was a war hero in Spain and at Khalkin Gol and had been twice decorated as a Hero of the Soviet Union. Smushkevich 'had a straightforward way about him which made people like him very readily', according to Chief Marshal of Aviation Golovanov. 'You could talk to him about anything without worrying that he would somehow misunderstand you' – a valuable trait in the purge years, when the most innocent remark might be misconstrued. Alksnis and Smushkevich shared a concern for operations more than administration, for practice more than theory. Unfortunately, neither survived Stalin's purges.

THE 1920s

Like Soviet political leadership, the command of Soviet military aviation was unstable in the post-Civil War period. In 1922 A. V. Sergeev handed over command of military aviation to A. A. Znamensky, who in turn was replaced in 1923 by A. P. Rozengolts, and in late 1924, P. I. Baranov became head of the VVS. Some continuity was provided by Vasily Vladimirovich Khripin, a widely-published theoretician who served from 1921 to 1937 as inspector and chief of staff of the Soviet Air Force. The name of the service itself also changed frequently, from RKKVF (*Rabochie-krest'ianskii Krasnyi Voenno-vozdushnyi flot*, or 'Workers and Peasants Red Army Military Air Fleet') to UVVS (*Upravlenie Voenno-vozdushnykh sil*) and then to VVS SSSR (*Voenno-vozdushnye sily SSSR*) in 1924, and finally VVS RKKA (VVS of the Red Army) in 1925.

The composition of the VVS also changed in the 1920s. Most units had previously been composed of mixed aircraft types, but increasingly the organization was shifted toward squadrons, and later regiments, of homogeneous aircraft. For example, in 1927 the VVS comprised 24 aviation squadrons and 40 aviation detachments. The squadrons were organized into fighter, bomber, attack, and reconnaissance units, while the detachments consisted of training and naval aircraft.

TsAGI and the Zhukovsky Academy (see Chapters 1 and 5) continued to be the centers of aviation research and design. Soviet aviation had depended on foreign imports and assistance since before the First World War, and even more so after the ravages of the Civil War. In 1922, some 270 aircraft were ordered from Western countries, primarily Holland. Late in 1922, a contract was signed with the German Junkers corporation to build at least 100 aircraft at the Soviet plant at Fili. Later, the Dutch company Fokker took over the contract.

Soviet aviation benefited greatly from the Five-Year Plans of 1928–37. The Soviets sought to eliminate the dependence on foreign imports that characterized the early days of Russian and Soviet aviation. Theoretically this was possible; of all the natural resources then required for aircraft construction, the Soviet Union lacked only rubber (which could be synthesized). During the First Five-Year Plan (1928–32), the aircraft industry was dramatically expanded, with increases of more than 750 per cent in the labor force and over 1,000 per cent in the number of engineers and technicians. The Soviet aviation industry began to produce native designs; at the same time, the Soviets continued to acquire and study various foreign aircraft. The greatest weaknesses in Soviet aviation production were metallurgy and therefore, engines. The two most prominent designers of the period were Andrei Nikolaevich Tupolev and Nikolai Nikolaevich

Polikarpov. Tupolev specialized in bombers, and in 1928–29 some 120 of his R-3 bombers were produced.

The Soviets stressed ruggedness, reliability and standardization in their aviation designs. Some have interpreted the stress on simplicity and maintainability as an effort to match the ability of Soviet mechanics. Others note the benefits of simplicity as enabling fast and even makeshift repairs under field conditions. Standardization was evident not only among military aircraft, but also in the close relationship between military and civil aviation. The Soviets believed that civil aircraft should be readily adaptable for military purposes in wartime, just as many military aircraft (especially transports) should be convertible to peacetime uses. While improving ease of maintenance, this design philosophy also simplified logistics requirements.

GERMAN–SOVIET CO-OPERATION

After the First World War, the Germans were forbidden by the Treaty of Versailles to have any sort of military aviation. In 1924 the Germans and Soviets agreed to co-operate; the Germans would be allowed to construct military aircraft and train pilots in the Soviet Union clandestinely. In exchange, the Soviets would acquire technical knowledge for their own aviation industry, and some joint training would be conducted. The airfield at Lipetsk, about 300 km south of Moscow, was chosen as the training site.

Flying training began in 1925; 120 German fighter pilots received training in 1925–33, and 100 observer pilots in 1928–31. A test unit that evaluated prototypes of military aircraft and weapons grew from 50–200 German staff during this period All activities concerning German involvement at Lipetsk were conducted in the utmost secrecy; the German operation was sometimes referred to as the 'Fourth Squadron of the Red Air Force'. The planned co-operation between Germans and Soviets at Lipetsk produced little of value, however; in general, the atmosphere was one of mutual suspicion and distrust. By 1933, the Germans withdrew their activities to Germany and left Lipetsk to the Soviets. The 120 Soviet senior officers who went to Germany to attend the secret General Staff course may have received the greatest benefit from the co-operation agreements.

THE 1930S

In the early 1930s, the oldest and largest aviation factory was the Gnome–Rhone plant in Moscow, which had been built by the French before the war. In 1930, Alksnis and Tupolev went to France to study the aviation industry; later, France assisted the Soviets in developing their aluminum industry. The Soviets also borrowed from the United States; for example, the I-16 fighter used a Wright Cyclone engine built under license in the Soviet Union. Another boon to the Soviet aviation industry occurred in 1936, when the Soviets arranged with the American company, Douglas Aircraft, to produce the DC-3 transport under license as the Li-2. The 1930s saw a 'technological revolution' in aviation design, in which constant-speed propellers, high-octane fuels, all-metal construction, flaps, and retractable landing gear were developed.

N. N. Polikarpov was the most prominent aircraft designer of the early and mid-1930s, and his designs epitomized Soviet ideals, and proved that the Soviet Union could become industrially self-sufficient. Ironically, Polikarpov was arrested in 1930 after several of his prototypes crashed; he was charged with sabotage. He continued to work from a prison design bureau, and it was there that his I-3 fighter was designed.[1] More than a thousand of these aircraft were built by 1934; Polikarpov and his team were released from prison as a reward. However, his most successful design was the simple U-2 biplane trainer; several thousand of these aircraft were built beginning in 1930, and even used as night bombers in the Second World War.

The I-15 and I-16 fighters, R-5 reconnaissance aircraft, and U-2/Po-2 trainer/transport were all Polikarpov designs. The I-15 biplane and the I-16 monoplane appeared in 1934–35, and placed Soviet fighters ahead of contemporary European design. The I-16 was the first mass-production low-wing monoplane to be equipped with retractable landing gear, making it the world's fastest fighter when it entered service in 1935. Tupolev, the other leading designer of the 1930s, continued to focus primarily on bombers and transports; his TB-3 was a four-engine bomber, of which more than 800 were built beginning in 1932. Later, he produced the SB-2 fast bomber (production run over 6,600 aircraft after 1935). These Polikarpov and Tupolev designs were the most important aircraft used by the Soviet Air Force in combat in the 1930s.

At the same time the Soviet Union was increasingly 'air-minded'. A stress on the importance of civil aviation for the development of the nation had many benefits for military aviation as well. Because of the vast size of the Soviet Union – some 8.6 million square miles, or more than twice the area of the United States – the Soviets quickly recognized aviation as the best means of communication and transportation. A state airline, Aeroflot, was created in 1932 (at about the same time that Pan American, Air France, and Lufthansa appeared in the West). It was a virtual requirement in Soviet aviation that all aircraft be capable of landing on grass and dirt airfields, minimizing the need for special facilities. While aviation was just one component of Soviet modernization, it served as an exemplar of modernity. The aviation industry was one of the few real successes of the First Five-Year Plan, when aviation production grew from 860 per year in 1930–31 to 2,595 per year in 1933. During the Second Five-Year Plan, the combat strength of the VVS quadrupled, and aviation production grew to 3,578 per year in 1937.

At the same time, the achievements of Soviet airpower were more and more politicized, and Stalin's name came to be closely linked with aviation. Soviet pilots conducted many record-setting flights during the 1930s which were given wide publicity. These aviators were called 'Stalin's falcons' in the Soviet press, and, supposedly, Stalin personally helped them select their routes and equipment. The political benefits to the Stalin regime are clear; historian Kendall Bailes, in his 1976 article on aviation and technology in the Soviet Union (see Acknowledgements and Further Reading), terms it 'the legitimizing function of technology'. The benefit to aviation was that as its glamour increased, so did the resources allocated to support it, as well as the pool of potential young pilots. However, it is also likely that in some ways the focus on record setting and aviation 'stunts' detracted from the real needs of military aviation.

In 1931, at the start of this dynamic period, Yakov Alksnis took over command of the VVS and military aviation became increasingly prominent. In August 1933, Alksnis presided over a showy display of airpower in what would become an annual Air Force Day airshow. Mass formations of bombers and fighters performed for crowds that by 1937 reached nearly a million people. In 1934, a new medal, the Hero of the Soviet Union (HSU), was created as a reward for pilots who rescued a stranded party of scientists from an ice floe in the Chukchi Sea. During the Second World War, the HSU would become the highest military honor.

Another important promoter of aviation was Osoaviakhim (from the Russian acronym for 'The Society for the Promotion of Defense, Aviation, and Chemical Warfare'). This group for young people provided training in basic military skills, like marksmanship and driving. Osoaviakhim also created a massive pilot training program to support the needs of both civil and military aviation, qualifying thousands of young Soviet women and men in gliders and biplanes. In 1936, one of its specific goals was to provide a reserve by training five times more pilots than the VVS required.

MISSIONS AND ROLES

Like other air-minded nations, the Soviets were interested in the theories of Giulio Douhet on strategic bombing. Douhet's 1921 book, *The Command of the Air,* postulated that in the future war might be won entirely by airpower. A pre-emptive strike against enemy cities and industries, he believed, would destroy the enemy's will and ability to conduct war and obviate the need for a ground attack. Douhet believed that bombers alone could achieve air supremacy; he did not foresee the need for fighter escort.

During the late 1920s, several important Soviet theorists including A. N. Lapchinsky and V. V. Khripin stressed the importance of strategic bombing. Soviet theorists never embraced the concept of independent air operations to the exclusion of combined-arms operations, however. Both Lapchinsky and Khripin supported the creation of an independent strategic bomber force in addition to aviation forces that would directly support ground forces. I. V. Timokhovich, a prominent historian of the VVS in the post-Second World War period, found fault with Douhet's theory for 'exaggerating the role of bomber aviation' and 'unjustifiably humbling the value of fighter aviation'.

In their study of the development of VVS theory between the wars, V. V. Anuchin and O. N. Zdorov noted that 'it must be particularly pointed out that from the very outset all the specialists considered the winning of air supremacy to be the main purpose of the air forces.' However, there was no unanimity as to how air supremacy would be achieved. In the 1924 regulations, it was defined in political terms, as 'the moral suppression of the enemy', very much in line with Douhet's views. By 1940, air supremacy was described instead in military terms as control of the air (the free operation of friendly ground and air forces, and the negation of enemy air activities).

The 1936 Provisional Field Service Regulations of the Workers and Peasants Red Army established operational and strategic guidelines for the VVS. The basic war aim was 'annihilation of the enemy'. Consistent with overall Soviet military strategy, the VVS emphasized decisive, offensive battles, rather than a war of attrition. The mission of aviation was 'destruction of those targets which cannot be neutralised by infantry or artillery fire or that of other arms'. This would require aviation to be employed 'on a mass scale'. Specific tasks were listed for ground attack, fighter, and light bomber aviation.

In the 1930s, the Soviets had what many regarded as the only real bomber force in the world. A strategic bomber force was created in 1936, a Special Purpose Aviation Army that comprised most of the heavy bomber assets of the western Soviet Union, nearly 900 aircraft. That same year, V. V. Khripin, the deputy chief of the VVS under

Alksnis, proclaimed to the Eighth Party Congress that the Soviet Air Force was the strongest in the world, and bragged that its combat inventory included 60 per cent bombers. Utilizing Tupolev's TB-3 bomber, which carried a two-ton payload, Khripin and Alksnis had created an offensive bomber force theoretically designed to strike military targets deep in the enemy rear. There were few possible scenarios, however, in which the Soviet bombers would have the range to reach enemy rear areas.

Still, the disputes over air doctrine continued. For example, on 26 July 1937, Tupolev wrote an article for *Industriia* stating that while Soviet aviation sought three goals – speed, distance, and altitude – distance was the most important, especially for heavy bombers. On 18 August, Alksnis published an Aviation Day piece in the same publication, but stressed that speed was absolutely key. Like many fighter pilots, he subscribed to the maxim, 'speed is life'. In reality, the capabilities of both bombers and fighters lagged behind their theoretical potential; the bombers lacked the range to strike crucial targets, while the fighters of the 1930s lacked the speed and maneuverability that was desired.

These disagreements about whether to prioritize bombers or fighters extended beyond aviation circles. As in the *Luftwaffe* at that time, there were divergent opinions within the top military leadership, and between the VVS and political leaders. The shifts in production from fighters to bombers and back to fighters during the 1930s provide proof of this, as do the dozens of articles published in aviation journals of the period.

The Soviets were among the first to experiment with airborne operations, in which parachute-equipped soldiers and their equipment could be dropped by Air Force transports behind enemy lines. Marshal M. N. Tukhachevsky was an energetic proponent of this concept. Airborne units were created on a small scale as early as 1931. In 1935, airborne troops were featured in a widely publicized military exercise near Kiev to which a number of foreign observers had been invited. More than 600 paratroopers were dropped and simulated securing a defensive perimeter; additional troops and equipment were then dropped to reinforce the area. Similar exercises were conducted in 1936 in the Caucasus and Belorussia. Such exercises were experimental, and were not incorporated into general practice until much later.

SPANISH CIVIL WAR 1936–38

The Soviet Air Force was involved in a series of small battles and wars throughout the 1930s. When General Francisco Franco revolted against the Republican government in Spain in mid-1936, Germany and Italy sent aid to Franco's Nationalist forces. The Soviet Union supported the Spanish Republicans through a mostly Communist 'International Brigade' of volunteers. The Soviets provided aircraft and other military equipment beginning in October 1936, and sent military specialists to train Republican troops. A total of around 1,500 aircraft was sent, though only 500 or so were operational at any one time. The bulk of the force consisted of 1,000 fighters (500–600 I-15s and 400–500 I-16s), 200 light bombers (SB-2s), and R-5 reconnaissance aircraft.

Thirty-four-year-old fighter pilot Yakov Smushkevich commanded the Soviet air brigade in Spain for ten months, beginning in the fall of 1936. Smushkevich is described by his biographer, I. Svetlichnyi, as being of 'above average height with an athletic build, an open strong-willed face, curly chestnut hair, an intelligent, penetrating gaze'. He was active both as a pilot and a commander, and 'was able to work 24 hours at a stretch without sleep or rest'. The Soviet members of the International Brigade often used false names in Spain; Smushkevich was known as 'General Douglas'. Marshal of Aviation Pavel Kutakhov stated that Smushkevich commanded a total of 160 Soviet pilots in Spain, but the number was probably higher. These airmen were portrayed as great heroes in the Soviet press, and 31 including Smushkevich received the Hero of the Soviet Union medal.

By 1937, Soviet aircraft comprised 90 per cent of the Republican force. In the spring of 1937, Franco's forces renewed their offensive against Madrid at a time when bad weather would normally prevent flying operations. Smushkevich decided to use the spring thaw and low visibility to his advantage. He concentrated his aircraft on two airfields near Madrid: nearly 100 SB bombers, R-5 reconnaissance aircraft, and I-15 and I-16 fighters were gathered for a mass strike. As Franco's forces, including a motorized division, concentrated for the attack, Smushkevich led the first group of fighters, flying in a dispersed formation through low clouds. His air attack was able to inflict enough damage and cause enough confusion to delay the offensive, buying time for Republican forces to strengthen their defenses around Madrid.

The Soviet-backed Republicans held on to air superiority until late 1937. Then the Germans upgraded the Condor Legion with the new Me-109 fighter and Ju-87 dive bomber, which were a generation ahead of Soviet designs. Rapid innovation in fighter design had made the Soviet I-15 and I-16 aircraft, only three years before considered the best in Europe, obsolete. It is unclear how soon this lesson hit home with Soviet decision makers. In his memoirs, A. S. Yakovlev, a prominent designer of fighter aircraft during the Second World War, commented that the initial successes in Spain led Soviet leaders to be complacent. 'No one was in a hurry to modernize Soviet fighter aircraft', he claimed, while the Germans were actively improving their fighter designs. However, he also wrote that by 1937–38, Stalin saw the Spanish venture as a failure, and reacted with 'dissatisfaction and wrath' against the very 'heroes' he had so recently praised. By mid-1938, the Soviets began withdrawing Air Force personnel and equipment from Spain; the Soviet contingent was gone by the end of the year. While the experiences of this war were studied by the Air Force, other factors prevented their full assimilation. Many of the direct participants were arrested or executed during the purges, and attention was diverted by events in the Far East.

Historians generally agree that both the Soviets and the *Luftwaffe* found that bombing of cities had little effect on the course of the war in Spain, and concluded that grand-strategic bombing was less effective than had been hoped. Both the *Luftwaffe* and the VVS subsequently stressed the importance of fighter aviation, and bombers were given diminishing emphasis in the late 1930s. At about the same time (1937–38), Alksnis, Khripin, Tupolev, and other major advocates of strategic bombing became victims of the purges. Lapchinsky, one of the most important theoreticians of the air offensive, appears to have died a natural death in 1938. Not until after the Second World War did the Soviets once again examine strategic bombing concepts in light of the Anglo-American experience.

THE PURGES

At the same time that Soviet aviation was being tested in the Spanish crucible, the purges began to take a serious toll of VVS and other military leaders. The military purge began in earnest in May 1937, when the chief of the general staff, Tukhachevksy, was arrested for treason. Tukhachevsky and a number of 'conspirators' were tried on 11 June. Alksnis, a long-time associate of Tukhachevsky's, was among the judges. Within only a few hours, the accused were declared guilty; they were shot the next day.

It cannot have been easy for Alksnis to sit in judgement on his long-time friend. Alksnis could not refuse to participate without endangering his family and incurring certain imprisonment, and there was nothing he could do to save Tukhachevsky. If Alksnis failed a test of honor by pronouncing Tukhachevsky guilty of treason when he knew he was innocent, Alksnis was in the company of thousands of other Soviet citizens who faced similar draconian choices during the purges. In any event, it did not save him. Within months, Alksnis, along with most of the other judges from the Tukhachevsky trial, was arrested. Alksnis was tortured and sentenced to penal servitude; on 29 July 1938, he was shot. Alksnis was succeeded as chief of the VVS on 28 November 1937 by A. D. Loktionov, a mediocre and undistinguished officer whose previous commands had been in the ground forces.

Alksnis was hardly the only high-ranking Air Force officer to suffer in the purges. Out of 13,000 officers in 1937, the VVS lost 4,724 in the purge – more than 36 per cent of the officer corps. In June 1937, the deputy chief of the VVS, I. I. Proskurov, was arrested and later executed. Smushkevich was recalled from Spain to replace Proskurov. Even top designer Tupolev was not immune; he was arrested in October 1937 and charged with selling information to Germany. Like Polikarpov before him, Tupolev continued his design work from prison. Not all victims of the purge were executed; some were arrested and released later, while others were only temporarily removed from duty. Still, the impact must have been considerable. Russian historian Dmitri Volkogonov says that silence, paralysis of will, doubt, confusion, and outright disbelief were some of the main reactions in the military as a whole. It seems likely that the pervasive atmosphere of fear and suspicion served to inhibit innovation and the willingness to take risks. In aviation, an inherently risky profession, excessive caution can result in stagnant and unrealistic training.

Ironically, it was just during the height of the purges that a series of striking record-setting flights occurred, including a landing at the North Pole and a nonstop flight from Moscow to the United States. The August 1937 Aviation Day airshow featured air formations spelling out the words 'Lenin', 'Stalin', and 'USSR'. All these events were surrounded by a media frenzy; some believe they were intended, at least in part, to distract the public from the terrors of the purge.

THE FAR EAST 1929–39

The VVS was also directly involved in defending Soviet borders in the Far East, first against the Chinese, and later with the Chinese against Japan. Beginning in 1929, the Special Far Eastern Army (ODVA) had its own air force. In that year, ODVA commander V. K. Blyukher employed 32 aircraft in one operation against the Chinese. Some believe this action influenced the Japanese to take over Manchuria in 1931 before the Soviets got any stronger in the Far East. The Soviets responded by increasing their forces; by 1934, the ODVA controlled nearly one thousand aircraft, and double that number by 1938.

Most of the conflict in the Far East was sparked by Japan's incursions and conquests in China. After late 1937, the Soviets backed the

Chinese against the Japanese. Soviet aircraft and pilots were sent to China, Soviet instructors trained Chinese pilots, and Soviet pilots flew in combat. By 1938, Soviet forces in China were making a significant impact. They sank Japanese warships and destroyed rail traffic, and made attacks against Japanese airfields and fuel depots. One Chinese observer reported that 'Russian pilots are highly regarded by both military and civilians who are connected with aviation in China.' One of the most important engagements of this period occurred in late April 1938, and was planned by Claire Chennault (the American chief flying instructor who later led the Flying Tigers in 1941) in conjunction with the Russians. Combined Russian and Chinese forces were launched to counter a Japanese raid. The Japanese sent 15 light bombers and 24 fighter escorts. The ensuing air battle has been compared to later air engagements during the Battle of Britain. Russian-built planes flown by 30 Chinese and 35 Russian pilots engaged the Japanese and disrupted the attacking force. Somewhere between 17 and 33 of the 39 Japanese aircraft were lost.

An American ambassador reported by late 1939 that the Soviets had sent 1,000 aircraft and 2,000 pilots to China. While this pilot-to-aircraft ratio may seem high, the figures are probably correct, since it is known that the Soviets rotated 'volunteer' pilots to China on a relatively short six-month tour. The Russian pilots reportedly followed an 'iron discipline' while on alert. Chennault recounts that Russian pilots sometimes sat on cockpit alert for up to 12 hours at a stretch, rather than being allowed to relax near the aircraft. He rated the Russians as unpredictable in combat during the early period; sometimes they avoided a fight, but during engagements they fought 'with the teamwork and tenacity of ants, swarming over the Japs and overwhelming them with sheer determination'.

The I-15 and I-16 fighters used by the Russians in China were superior to Japanese aircraft in 1937–38. Chennault was impressed by the ruggedness of the I-16. Although it was less maneuverable than Japanese fighters, he reported that an I-16 under attack could perform a power dive and snap roll that allowed it to evade the attack and even achieve an offensive position against its enemy. Other American observers disparaged the quality of Soviet metals and engines, but praised their ease of maintenance and assembly. The Soviets were most active in China during the first half of 1938, when the Japanese lost 254 aircraft in combat; after that point there were few important air battles. Fourteen pilots who performed 'internationalist duty' in China received the Hero of the Soviet Union medal.

The French air attaché to China reported that his Soviet counterpart outlined a number of conclusions gained during the air battles in China. The Soviet attaché said that they had decided that a proper fighter escort should consist of twice as many fighters as bombers being escorted; that escorts should be dispersed both above the bomber formation and to the most exposed flank; and he noted the advantages of the Japanese use of drop-tanks to extend their operational range. Above all, he stressed the importance of fighter escort for bombing missions. Other lessons learned included the need for better cockpit visibility, improved bombing tactics, and new procedures for cooling aircraft engines. If these conclusions were reported to VVS HQ, they were apparently not incorporated into VVS practice.

The Soviets were also involved in combat in the Far East in two conflicts with the Japanese following their incursion into Manchuria in 1938–39. The first occurred in a disputed area on the Korean–Soviet border; the second, along the boundary of Outer Mongolia and Manchukuo. Flying in the Far East was dramatically affected by terrain and weather. The scarcity of geographic landmarks was aggravated by the dearth of navigational aids; sudden dense fogs and fierce wind and dust storms could make flying extremely hazardous.

In July 1938 a border skirmish instigated by the Japanese near the Korean–Soviet frontier escalated into the 'Battle of Lake Khasan'. During the two-week battle, Pavel Rychagov (who had fought in Spain under Smushkevich) commanded the VVS contingent. Soviet aviation was able to penetrate in depth against relatively light Japanese opposition; however, their weapons were less than effective against entrenched and hardened positions. According to Dmitry Volkogonov, Stalin called Blyukher at one point, demanding to know why an aerial bombardment of the Japanese had not been carried out. Blyukher replied that bad weather had prevented the air attack, however he had ordered Rychagov to launch his aircraft, but he feared that Korean settlements and Soviet units might inadvertently be hit. Stalin reportedly questioned Blyukher's resolve: 'What's a bit of fog to Soviet aviation when it really wants to defend the honour of the Soviet motherland? I'm waiting for your answer.' The air attack, of course, was launched despite the weather.

In late May 1939, a more serious incident erupted in the Far East at Khalkin Gol. The Japanese cavalry, supported by 40 aircraft, attacked a Mongolian border position, and the Soviets became involved in what turned into a small-scale war. Soviet forces eventually included 35 infantry battalions, 20 cavalry squadrons, 500 tanks, and 500 aircraft. General Georgii Konstantinovich Zhukov, of later Great Patriotic War fame, was given overall command; Yakov Smushkevich served as head of the air force contingent.

On 20 August 1939, the Soviets went on the offensive across a 48-mile front. In a three-day operation, the Japanese forces were encircled. Relatively large-scale air attacks were employed; as many as

100 aircraft at a time were used by both the Soviets and the Japanese, with the Soviets employing a total of 150 bombers and 100 fighters. Dense clouds of dust and smoke over the battlefield made it difficult for aircraft to find their targets. Still, Soviet aircraft employed rockets and the 20mm aviation cannon for the first time at Khalkin Gol, to some effect. Aviation was particularly successful in curbing Japanese battlefield reinforcement. Corps Commander Smushkevich received his second Hero of the Soviet Union medal – one of only three pilots at the time to have been twice-decorated.

There were some real successes at Khalkin Gol. Soviet logistics performed impressively; four months of supplies had been transported and stockpiled to support the battle. The performance of combined arms was also generally commendable; Zhukov had insisted on having pilots sent to study the terrain along with ground forces personnel. It was the first real Russian combat test in which aircraft supported tanks and artillery on a large scale, and the Soviet forces showed real tactical flexibility.

Following the experience of the VVS in Spain and the Far East, increased stress was placed on developing and manufacturing new aircraft. In September 1939, the Soviet government issued a decree calling for the construction of nine new aviation plants, and the overhaul of nine existing plants, to be completed by 1941. Certain armaments upgrades also resulted from the experiences in the Far East. The Soviets realized that the ShKAS 7.62 mm (Shpitalny–Komaritsky aviation rapid fire) machine gun was inadequate, and began producing the 12.7mm (the equivalent of the American 50-caliber), and the 23mm aviation cannon.

In late 1939, Yakov Smushkevich replaced Loktionov as VVS commander. Loktionov was sent off to command the Baltic Military District, then was arrested in early 1941 and later executed. Smushkevich had established a shining reputation as a fighter pilot and aviation commander in Spain and the Far East. As deputy commander of the VVS from late 1937, Smushkevich had been actively involved in the development of new aircraft. He even insisted on flight-testing some of the new models himself. During one such test of the R-10 aircraft in April 1938, Smushkevich crash-landed and ended up in hospital with multiple injuries; it took him several months to recover. Despite these qualifications, Smushkevich did not have organizational or theoretical experience; he has even been dismissed as 'essentially nothing more than an aviation brigade commander'. It is true that Smushkevich lacked experience at high levels of command, but as a seasoned combat pilot and authentic war hero, he was a distinct improvement over Loktionov.

Smushkevich took over command at a difficult time. By the end of

1 Imperial Russian Air Force at war, c. 1915. G. V. Yankovskiy (far left), a prewar Russian air hero, stands with Russian airmen next to a Sikorsky S-12 biplane. Courtesy of the National Air and Space Museum.

2 Czar Nicholas II inspects Igor Sikorsky's four-engine airplane, called *The Grand*, in 1913. This became the proto-type for the *Il'ya Muromets* bomber in the First World War. Courtesy of the National Air and Space Museum.

3 An American-designed Curtiss flying boat at Sevastopol in the Crimea, *c.* 1913. Courtesy of the National Air and Space Museum.

4 The Soviet Air Force employed the British De Havilland DH-9 during the interwar years. DH-9 aircraft are shown here at a Moscow air show in 1930. Courtesy of the National Air and Space Museum.

5 The Soviets engaged in a systematic campaign to break world records for long-distance flight, as shown here in the 1929 Moscow–New York trek of the 'Land of the Soviets', a TB-1 (ANT-4) bomber design. Courtesy of the National Air and Space Museum.

6 As late as 1942, the Soviet Air Force was making use of obsolete I-153 biplanes, such as these two fighters equipped with rockets for ground-attack missions in Northern Russia. Courtesy of the National Air and Space Museum.

7 Andrei N. Tupolev (far left) stands with crew of the ANT-25 *Stalin Route*, which made the first transpolar flight in June 1937, a nonstop trek from Moscow to Vancouver, Washington. (Left to right) Alexander Belyakov (navigator), Valery Chkalov (command pilot), and Georgiy Baidukov (co-pilot). Courtesy of the National Air and Space Museum.

8 The all-purpose Polikarpov PO-2 (U-2) entered service in 1927. During the Second World War it served as a night bomber, liaison aircraft for partisan warfare, air supply, and trainer. Courtesy of the National Air and Space Museum.

9 Soviet poster portraying one of 'Stalin's Falcons', the name given to the Soviet Union's air heroes of the interwar years. Courtesy of the National Air and Space Museum.

10 Artem I. Mikoyan, famed Soviet aircraft designer. Courtesy of the National Air and Space Museum.

11 The highly maneuverable I-16 fighter at a Moscow airfield in 1939. The I-16 entered service in the early 1930s and flew in the Spanish Civil War, 1937–39, but would have limited use at the start of the Second World War. Courtesy of the National Air and Space Museum.

12 The Soviet Air Force four-engine bomber, the Pe-8, flew V. Molotov to Bolling Air Force Base, Washington DC, in 1942. Courtesy of the National Air and Space Museum.

13 A. A. Novikov
served as Soviet Air
Commander 1942–46.
His decisive wartime
leadership played a
key role in the rebirth
of the Soviet Air
Force in the long
struggle with the
Luftwaffe. Courtesy of
the National Air and
Space Museum.

14 Soviet airman holds a bomb inscribed 'Gift to Hitler'. Courtesy of the National
Air and Space Museum.

15 The Il-4 medium bomber performed diverse missions on the Eastern Front; shown here with an Il-4 specially equipped with a torpedo for support of the Soviet Navy. Courtesy of the National Air and Space Museum.

16 Soviet airman re-enact a successful air battle against the *Luftwaffe* with an Il-4 bomber in the background. Courtesy of the National Air and Space Museum.

17 A Soviet ground crew services a SB-2 bomber in the Second World War. Courtesy of the National Air and Space Museum.

18 MiG-3 fighter interceptors in winter camouflage paint in 1941. Courtesy of the National Air and Space Museum.

19 Yak-3 fighter flown by French volunteers, the Normandie-Nieman regiment. Courtesy of the National Air and Space Museum.

20 Lavochkin La-7 radial-engine fighter flown by the Soviet Air Force in the Second World War. Courtesy of the National Air and Space Museum.

21 The Ilyushin Il-2 *Shturmovik* ground-attack aircraft played a major role in Soviet offensive operations after 1942, attacking German armored units and troop concentrations with great success. Courtesy of the National Air and Space Museum.

22 A Pe-2 with two-man crew with inscription on the fuselage: 'Forward to the West!' Courtesy of the National Air and Space Museum.

23 The Tu-2 performed numerous tasks as part of the Soviet Air Force's tactical operations in the Second World War. Courtesy of the National Air and Space Museum.

1939, more than three-quarters of senior officers in the VVS had been arrested, executed, or relieved of duty. The atmosphere of fear and suspicion was pervasive. One Soviet historian believes that 'particularly tangible harm was done to our Air Forces' during the purges, especially during the repressions of 1940–41. The purges appear to have been more severe in the VVS than in the ground forces. While about 30 per cent of purged army officers eventually returned to duty, less than 16 per cent of Air Force officers (892 out of 5,616) had been reinstated by the end of 1939.

At the same time that experienced leaders were falling victim to the purge, the officer corps of the VVS was being rapidly expanded – from 13,000 in 1937 to 60,000 in 1940. New regiments and divisions were created to accommodate the influx of personnel and equipment. The new personnel had to be recruited and trained, and existing pilots and technicians had to be retrained in the new aircraft and weapons. Between the purges of senior officers, the influx of untrained junior personnel, and ongoing training and retraining, the VVS had few proficient pilots or combat-ready units by the end of the 1930s. It may have been the largest air force in the world at that point, but it was also one of the least prepared.

THE SOVIET–FINNISH WAR 1939–40

The Soviet Air Force did not rest on the laurels of Khalkin Gol for long; the halo earned in that battle was tarnished only a few months later in Finland. From November 1939 until March 1940, numerically and technologically superior Soviet forces fought the Finns and finally, at great cost, forced their surrender. As in the Far East, the war was sparked by border disputes; the Soviets sought territorial concessions to improve their strategic position around Leningrad.

The VVS should have prevailed by sheer weight of numbers: when the war began, the Soviets had at their disposal some 900 aircraft against 200 obsolete Finnish aircraft. Even so, the Soviets suffered heavy losses and eventually doubled their initial commitment of aircraft. Perhaps the Soviets were overconfident after their victories in the Far East; in any event, they badly underestimated their enemy. A total of 684 Soviet aircraft were lost (240 in air combat) against only 62 Finnish losses.

A major weakness in Air Force training became evident during the Finnish War. In December 1939, foul winter weather played havoc with Soviet pilots who had little training for instrument flying. Weather conditions were much better in January and February, which contributed to the improved success of Soviet aviation missions. The

potential ability of Soviet pilots to perform in bad weather was evident
from the fact that one special regiment staffed with highly experienced
crews trained in instrument and night flying was able to operate on 63
out of 70 days during the war; many other units only managed six to
eight operational days in the same period. The VVS also concluded
that aircraft landing gear would no longer be replaced with temporary
skis in winter; although it made the aircraft easier to take off and land,
it reduced their performance in combat.

Aleksandr Golovanov, later commander of Long Range Aviation
during the Second World War, flew in Finland as a civil aviation pilot.
He confirmed that 'the weather was exceptionally bad during the
Finnish campaign'. But he also noted that bad weather offered certain
protections for transport aircraft, since it 'virtually excluded the possibility of encountering enemy fighter planes'. Few Soviet pilots were
able to take advantage of this, however. 'The Finnish campaign made
apparent the serious deficiencies in our pilots' training for flights
under difficult conditions and for the use of radio communications.'

Though the Finnish War was technically a Soviet victory, Stalin
realized that the Soviet military had performed wretchedly. There
was a subsequent shakeup in the top leadership; Marshal Semyon
Konstantinovich Timoshenko replaced Voroshilov as commissar of
defense in May 1940, and in January 1941 Zhukov took over as chief of
the general staff. Among the indirect casualties of the Finnish war
was VVS chief Smushkevich, who was blamed for the failures of the
VVS. It did not help that the Soviet embassy in Helsinki had been
mistakenly bombed in a night raid by the VVS during the campaign.
In his memoirs, Golovanov says:

> it was only many years later that I learned that Generals Smushkevich
> and Arzhanukin had written a report after the Finnish War in which
> they analyzed the military action taken and the improper utilization of
> our bombing capabilities – which were not administered under a single
> agency but were spread around under various bodies and commanders.
> The report had also referred to our bomber crews' poor preparation for
> flying under difficult meteorological conditions.

Apparently, Smushkevich and other top air force leaders were aware of
the weaknesses in Soviet military aviation. They had to shoulder the
blame and punishment for those problems, but the obvious solutions –
to unify aviation resources and conduct intense training in instrument
flying – were not implemented.

On 15 August 1940 Smushkevich was removed from command
and appointed Inspector General of the VVS; in December, he was
demoted to deputy chief of staff. In 1941, Smushkevich was arrested
and later executed. On 28 August 1940, 29-year-old P. V. Rychagov,

who had been a lieutenant only three years earlier, took over for eight months as chief of the VVS. The period when Rychagov was head of the VVS was characterized by one contemporary as a time when 'there was no strong man in charge and a kind of anarchy existed'. He was transferred in April 1941 to the General Staff Academy and then arrested. Rychagov was later executed for treason in June 1941 for 'allowing' the destruction of so many aircraft by the *Luftwaffe* during Operation Barbarossa. Rychagov's deputy, Lt.-Gen. of Aviation P. Zhigarev, was then designated the new VVS commander (the position was then called the Chief of the Air Force Main Directorate).

The Soviets appear to have been more prone to react to their failures than to examine their successes in combat in the 1930s. Little study was devoted to analyzing why the VVS was effective during the first part of the Spanish Civil War or in the Far East. This was characteristic of the Soviet military as a whole. M. V. Zakharov, the assistant chief of the Soviet General Staff in 1938, noted that 'responsible members of the General Staff' studied the Khalkin Gol experience carefully, implying that other members were less responsible. A history of the M. V. Frunze Military Academy for senior officers indicated, for example, that the experience in the Far East was neglected. After the initial failures in Finland, however, a team was sent to Finland in January 1940 to study the situation 'on the spot' with the goal of a 'rapid introduction of combat experience' into the school's instruction. This resulted in Order No. 120 in May 1940 from the Commissar of Defense, enumerating weaknesses in training and leadership that existed when the war in Finland began.

RESTRUCTURING OF THE VVS, 1940–41

The experience of the wars of the 1930s led to a massive restructuring of Soviet military forces in 1940–41. M. V. Zakharov says that a preliminary Third Five-Year Plan for reorganizing the military had already been drafted in 1938, calling for a radical restructuring of the VVS to increase the maneuverability of various units. The size of aviation squadrons was to be reduced to 12–15 aircraft each; 4–5 squadrons would comprise an air brigade; and air corps of two to four air brigades would be organized. The overall ratio of aircraft types was to be 55.4 per cent bombers, 34.7 per cent fighters, and 9.9 per cent reconnaissance by 1942. This plan was not immediately implemented, but was a forerunner to the 1940–41 reorganization.

In September–October 1940, Marshal Timoshenko personally visited the western military districts, putting units on alert and observing training. His inspection revealed serious shortcomings, especially in operational training. The conclusion was that 'the basic components of the combat force were below par, and this applied equally to the air force'. The restructuring was also partly instigated by technological improvements. A number of new aircraft designs appeared in the late 1930s and went into series production. New designers like Yakovlev, Mikoyan–Gurevich, Lavochkin, and Petlyakov created impressive new prototypes. The new Pe-2 dive bomber, the Ilyushin series of long-range bombers and attack aircraft, and three new fighters (the MiG-3, LaGG-3, and Yak-1) all came into service. All these aircraft were a great improvement on previous designs.

In January 1940, the energetic and capable Aleksei Ivanovich Shakhurin replaced M. M. Kaganovich in the post of People's Commissar of the Aviation Industry; the young designer A. S. Yakovlev was named as his assistant for experimental construction. They made significant changes in the organization of the aviation industry, creating new design bureaux and independent design teams. By the end of 1940, all the old fighters had been taken out of series production. By the start of the war, 2,739 of the new series aircraft had been produced; 1,540 were in the western military districts in June 1941. However, this still meant that the VVS was only 20 per cent equipped with the new model aircraft. For example, there were still 1,762 I-16 and 1,549 I-153 obsolescent fighter aircraft in the western MDs alone. Because of pilot shortages, many of the older fighters were in a sort of temporary storage status until they could be flown to rear areas.

Sweeping changes were instituted by a decree of 25 February 1941, 'On the Reorganization of the Red Army Air Force.' This decree called for an increase in aviation strength, improvements in training, reorganization of the Rear Services, and reconstruction and expansion of the airfield network. The basic organizational structure of the VVS was revamped. In mid-1940, the aviation regiment consisting of three to four squadrons (approximately 60 aircraft) became the basic unit of the VVS. Some regiments were integral (all one type of aircraft) while others were mixed. Aviation divisions comprising three to five regiments were created, replacing the old brigades. Divisions were assigned to military districts, armies, or grouped into air corps under independent VVS control. The air corps included either long-range bombers or fighters used for air defense. Moreover, the VVS was given permission to form 106 new regiments; most were slated to receive the new generation aircraft. Only 19 of these new regiments had actually been formed by 22 June 1941.

On the larger scale, the VVS was reorganized into five elements. Front Aviation (*VVS fronta*) included those units assigned to military districts (which would become fronts in wartime). Army Aviation

(VVS armii) included those VVS units assigned directly to army formations. Corps aviation comprised mainly tactical liaison aircraft assigned directly to ground forces corps. An air reserve was formed, attached to the High Command. Last was Long-Range Bomber Aviation, consisting of five air corps and three independent divisions. The impetus for the new structure was the Soviet experience at Khalkin Gol and in Finland; both those conflicts revealed the need for close co-operation between air and ground forces. It was believed that dispersing aviation units under the control of various Army groupings would facilitate co-operation. This structure immediately proved inadequate during the first days of the war; centralization and rapid concentration and redeployment of aviation at the front level and above proved far more valuable than giving army commanders direct control over the bulk of aviation. M. N. Kozhevnikov, in his 1977 study of the VVS in the Second World War, states that the error was one of scale; dispersal of forces might be effective in small-scale wars, but in a large war, concentration of forces and centralized control acquired primary importance.

Another important part of the restructuring was in VVS Rear Services, which included logistics, airfield support, and maintenance. In April 1941, the Rear Services were reorganized on a regional basis, to provide greater centralization. It had been discovered in the Soviet–Finnish War that the existing system, in which rear service units were directly subordinate to aviation units, had limited maneuverability and flexibility. Each new airbase region would support three to four air divisions. Each division was supported by three to four airfield maintenance battalions (BAOs), depending on the number of regiments assigned. Rear Services was responsible for providing fuel, munitions, housing and food, and maintenance. The new Rear Services structure was supposed to be complete by August 1941, but when the war began it was still in upheaval. Only half the BAOs in the Western Special Military District were operational when the war started, and in the Kiev MD, the rear services had not even begun restructuring.

Rear services problems were exacerbated by the large territorial gains of 1939–40 that resulted from the Nazi–Soviet Friendship Pact. The western border of the Soviet Union was suddenly extended to include the Baltic states, a large portion of Poland, and annexations in Finland and eastern Europe. Soviet strategy required that aircraft be based very near the border, in preparation for immediate retaliation against any enemy invasion. More than 200 new landing strips and airfields were scheduled for renovation or construction. The plan was to have three airfields (primary, auxiliary, and grass-strip) for each regiment. In the spring of 1941, many airfields were closed while new concrete runways were being constructed. While the work was underway, many Soviet aviation units were jammed onto existing primary fields. Some airfields in the Western and Kiev MDs were packed with more than 100 aircraft at a time. Soviet aircraft in the western districts were often located on overcrowded airfields when the war began, preventing any sort of effective dispersal or camouflage, and leading to disaster.

New equipment and the expansion of the VVS required wholesale retraining of flight and maintenance personnel. Conscription had been introduced in December 1940 to try to provide a bigger pool of recruits. At the same time, training courses were cut short so that crews could be produced more rapidly; this decreased the overall quality of training and level of experience in operational regiments. Only four months before the war began, a new training system for the VVS was announced that affected all levels of personnel, from new pilots learning to fly to the command personnel. There were appalling deficiencies in flight training; many crews simply did not fly often enough to attain basic proficiency, much less to acquire sophisticated skills in their new aircraft. During the first quarter of 1941, for example, pilots of the Baltic Military District undergoing retraining averaged only 15½ hours of flying time (less than one hour per week), while in the Western MD the average was 9 hours, and in the Kiev MD, only 4 hours. By 1 May 1941, only 72 per cent of Pe-2 pilots had been retrained; about 80 per cent of MiG-3 pilots and only 32 per cent of LaGG-3 pilots had completed retraining. The training system at all levels was in flux.

Some of the deficiencies in flight training can be directly attributed to the legacy of the purges. An inspection in April 1941 in the Western MD revealed severe shortcomings in the 12th Bomber Aviation Division. Transition training to new aircraft was proceeding far too slowly; fear of flying accidents was cited as the main reason. Flight safety was always a concern under the best of conditions, as it was common during the purge years to accuse anyone involved in an aircraft accident of being a 'wrecker' or saboteur. Almost immediately after the Soviet–Finnish War, in April 1940, the VVS held a conference for command personnel to discuss training and safety. The conference apparently led to little improvement.

Despite the fear of flying accidents, there was an alarming increase in the accident rate in 1941, probably due to the transition of so many units to new aircraft types. Stalin reportedly castigated his VVS commander for this, but Rychagov replied, 'There are accidents and there will be more of them because you are driving pilots to their graves.' Rychagov probably regretted his retort; in any event, he was removed from command soon afterwards. The pervasive fear of independent action in the VVS was only exacerbated by Stalin's policy of placation toward Hitler. For example, there was an increasing number of German overflights in the spring and summer of 1941; the Soviets claim 324 documented cases of airspace violations by German aircraft in January–June 1941. Fighter units were sometimes allowed to intercept and attempt to force the German aircraft to land, but they were prohibited from firing on intruders or acting in a provocative manner.

The combined effect of the purges and the rapid expansion of forces meant that there was a serious shortage of trained personnel at all levels, but especially in the mid- and upper-levels of command. Even the main staff of the VVS was not prepared for 24-hour operations; this would be a serious problem during the first days of the war. The personnel the VVS did possess were often new and inexperienced. In the summer of 1941, 75 per cent of military officers as a whole had been in their posts for less than a year. The situation was even worse in the VVS, where 91 per cent of commanders of major units had held their positions for less than six months.

Theoretical matters were given high-level attention by the VVS. In December 1940, Rychagov spoke at the Supreme Military Council of the Red Army, and defined the main mission of the Air Force as gaining air superiority. This mission would be achieved, he claimed, by destroying enemy aircraft both on their main airfields and in air combat, with simultaneous strikes against frontline airfields, repair facilities, and fuel stores. This first and most important mission of the VVS was consistent with the mainstream of Soviet aviation theory since Lapchinsky. The theory may have been sound, but no concrete plan was outlined to implement it. Soviet historians admit that while everyone agreed on the importance of air supremacy, 'the leadership was unable to arrive at a common point of view on a number of issues concerning the operational employment of the VVS in war'. The participants at the conference exaggerated their own experiences in Spain, while ignoring German tactics in Poland and France, which had already revealed German plans for large-scale surprise attacks against airfields. One of the major weaknesses of Soviet military theory was the failure to account for the possibility of a surprise attack by the enemy. Soviet theory simply did not consider the possibility that a large-scale strategic defense might be required in the first phase of war.

Amazingly, on the eve of the Second World War, many in the Soviet Air Force found reason for optimism. Golovanov recalled that he attended a New Year's party on 31 December 1940 at the Flier's House in Moscow. 'Everyone was in a good mood', Golovanov recalled. 'It seemed as though the recent period of worry and struggle was over ... everyone had a tale to tell, everyone wanted to share something amazing and unexpected that had happened to him during

1939 and 1940, those two troubled years.' He also noted a strong
tendency to gloss over the failures in the Soviet–Finnish war. 'The
events in Finland were somewhat disappointing ... one wanted to
believe that the main reason for [the] hardships had been the bitterly
cold winter that year with deep snow and usually foggy, icy bad
weather.' Golovanov admitted there was a strong feeling that war
might soon erupt with Germany, but many hoped for time to recover
their health and morale.

Golovanov was trying to get approval to conduct an around-the-
world long-distance flight. He wrote that at the New Year's party, he
mentioned this idea to Smushkevich, then serving as Inspector
General of the VVS. Smushkevich told him that more important
matters were at hand. 'Have you given any thought to our aviation, to
its combat readiness during the battle at Khalkin Gol and the Finnish
campaign?' Smushkevich asked him. Smushkevich went on to talk
about the successes in Spain, which he attributed partly to good
weather conditions. Flying at night and during bad weather, he said,
was the 'Achilles' heel' of the VVS. Smushkevich told Golovanov that
this weakness had been recognized in Spain and again in Finland, yet
little had been done to correct the problem. Smushkevich urged him to
write to Stalin and suggest that the VVS develop a special program to
train Soviet pilots in blind flying and using navigation instruments.
Golovanov was reluctant to go directly to Stalin, and asked
Smushkevich why he himself, a far better-known figure in aviation
who held a position of power, didn't address the matter. According to
Golovanov, Smushkevich 'answered that he was not in a position to
do this right now, and that a report from him on the subject would
probably receive little serious attention'.

Golovanov did write to Stalin. As the chief pilot of the Special
Forces Squadron of civil aviation, Golovanov was well known; his
work consisted of 'unexpected flights to different destinations in all
parts of the country' under all weather conditions. He was called in to
see Stalin, who approved his plan. The 212th Detached Long-Range
Bombing Regiment was formed, with Golovanov as its commander.
A combined staff of civil aviation pilots and VVS navigators and
gunner/radio operators worked to organize training programs.
According to Golovanov, Rychagov, then VVS commander, was not
pleased, and told him to 'give up your little plan before it's too late.
There's nothing to be gained by it, no matter what.'

Thus, operational training for flying in bad weather and at night
finally began to get attention. Golovanov concluded that while the
VVS had a solid introductory training program for nighttime and
blind flying, there was little or no ongoing training, which was vital to
maintain pilot skills. Training on radio and navigation instruments

was inadequate at all levels. Although a program was developed to train aircrews, and a new manual was written on navigation, the VVS was slow to incorporate these changes into actual training. There were also scheduled improvements in aircraft instruments. In March 1941, the VVS undertook a program to improve navigational systems, both those on board aircraft and on the ground, in order to improve instrument flying capabilities for bad weather. But it was too little, too late; the war began a few months later.

CONCLUSIONS

The disasters that would be suffered by the VVS during the first period of the Great Patriotic War derived in part from the chaotic effects of restructuring and re-equipping. A wide mix of aircraft types due to the transition to new models of aircraft created many problems for the VVS. Few pilots had completed transition training to the new aircraft when the war broke out, and their inexperience contributed to their poor performance. Moreover, Soviet pilots had apparently received little training on recognizing and identifying their own new aircraft. There would be many reports of fratricide early in the war; Soviet fighter pilots, gunners in bombers, and ground troops alike shot down their own aircraft in the mêlée of the initial German attack. Other problems included poor communications with ground forces and the failure of fighters to protect the ground forces from enemy attacks.

The Soviets lacked solid operational plans to support theoretical concepts. During the interwar period, the Soviets had several opportunities to battle-test their personnel and equipment and to evaluate the capabilities and tactics of their potential enemies: the Germans, the Finns, and the Japanese. They had the chance to develop their tactics and to test concepts like strategic bombing and fighter escort. But many of these concepts never made the leap from theory to practice. One reason for the destruction of so many Soviet bombers during the first days of the war was the total lack of fighter escorts.

The airfield expansion plan and rebasing scheme would prove a disastrous failure. The extreme forward basing of the bulk of VVS aircraft in the Western MDs meant that many airfields would be overrun within hours of the Nazi advance. An order was issued on 19 June to begin camouflaging airfields and dispersing aircraft, but it came far too late to prevent the catastrophe of 22 June. The massive restructuring of 1940–41 proved only partially successful. The new Rear Services organization worked extremely well, but was only beginning to be implemented when the war broke out. The structure of operational forces proved completely inadequate. The dispersal of aviation units between fronts and army units would quickly prove ineffective; Novikov would solve this problem in the spring of 1942 with the creation of the Air Armies.

Both the Soviet and German air forces entered the Great Patriotic War with some degree of recent combat experience. The Soviets had flown in Spain, the Far East, and Finland, while the *Luftwaffe* had tested its mettle in Spain, Poland, France, and the Battle of Britain. But the Soviets failed to take to heart the lessons of those engagements until it was almost too late. Moreover, many of the battle-seasoned Soviet pilots fell victim to the purges; those who remained were a minority in a sea of young and inexperienced recruits who had been rushed through crash-courses to fill out new units. The Germans entered the war with a much larger percentage of experienced pilots. Ironically, the situations of the two air forces were largely reversed by 1943, when Soviet pilots grew increasingly skilled, while the *Luftwaffe* was forced to make up its losses with ever younger and greener personnel.

There were serious problems in the Soviet Air Force by 1941. The rapid changes in VVS leadership were, at the very least, disruptive and unsettling; at worst, they directly contributed to the disastrous defeats of June 1941. If capable and experienced leaders like Alksnis and Smushkevich had been allowed to retain command, they might have provided greater continuity and direction. The purge of the officer corps was far more devastating to the VVS than to the other services. The purges not only affected leadership, as in the other services, but the equivalent of the rank and file: most of the actual combat personnel in aviation were officers. The purges also removed many aviators who had accumulated combat experience during the wars of the 1930s, leaving the VVS to begin the war with an excessive number of young and inexperienced pilots.

The structure of the VVS would be proven deficient in wartime. Scattering air force resources too widely prevented their effective use, and violated classic rules of war like concentration of forces and mobility. Tentative moves toward centralization, such as the restructuring of the Rear Services, proved effective once implemented.

The operational and strategic concepts that would later make the Air Armies effective were delineated by Soviet theoreticians before the war. The work done by theorists was reinforced by the Soviets' own experience in Spain, the Far East, and Finland. In both theory and the test of battle, the VVS had learned the importance of unity of command over aviation resources, the need for fighter escort for bombers, flexibility in fighter tactics and formations, the necessity for sophisticated training in instrument flying, and the need for improved navigation skills. It was largely the legacy of the purges that led to a disastrous reluctance to change VVS practice to reflect theory and the lessons learned in previous wars. Still, the foundations for change had been laid. The Soviet Air Force could never have recovered from its near-destruction in 1941 to defeat the *Luftwaffe* without relying on the hidden resources developed in the 1930s.

NOTES

1. During this period, Soviet aircraft were designated according to function: I for fighters (*istrebitel'*), DI for two-seat fighters (*dvukhmestnii istrebitel'*), TB for heavy bombers (*tiazhelyi bombardirovshchik*), SB for fast bombers (*skorostnoi bombardirovshchik*), and R for reconnaissance (*razvedchik*). Some aircraft began to be designated according to their designer's name (ANT for A. N. Tupolev), a practice that became standard during the Second World War (when, for example, Polikarpov's U-2 was redesignated the Po-2).

Soviet Frontal Aviation during the Great Patriotic War, 1941–45

JOHN T. GREENWOOD

In a short overview, it is impossible in even the grossest detail to cover the air war that the Soviet Air Forces fought against Germany from 22 June 1941 to 9 May 1945. This chapter provides some perspectives that will add a number of important pieces to the larger mosaic of Russian and Soviet air power that this book has crafted. It will show how some Soviet airmen molded the organization, equipment, and tactics to implement prewar concepts as well as to introduce changes based on costly experience gained against the experienced *Luftwaffe* pilots over the battlefields. The focus will be on the role of *frontovaya aviatsiya* (Frontal or Tactical Aviation) in attaining the final Soviet air victory in the Great Patriotic War. The emphasis on frontal air operations remained an integral part of the Soviet tactical, operational, and strategic doctrine and planning until the very end of the Soviet Union in 1991 and continues to permeate the new Russian Air Force of the post-Soviet era. As such, the Soviet wartime experience merits our attention today as much as it did during the height of the post-1945 era of Soviet military power.

THE CENTRAL ROLE OF FRONTAL AVIATION

Following the poor showing of the *Voenno-vozdushniye sily* (Air Forces or VVS) in the Finnish campaign of 1939–40, the People's Commissariat of Defense, the General Staff, and the Air Forces leadership thoroughly reorganized the VVS and began an ambitious re-equipment program geared to its perceived future combat employment and existing Soviet doctrine. Prewar military doctrine had delineated three basic air missions for the VVS:

1. Achieving tactical, operational, and strategic air supremacy;

2. Supporting the ground forces and the navy in their offensive and defensive operations;
3. Air reconnaissance and intelligence.

Soviet military doctrine clearly placed the VVS's overriding wartime role in the tactical and operational spheres of the offensive operations for which the Red Army was being groomed. When the reforms of 1940–41 were completed, *frontovaya aviatsiya*, those air forces devoted directly to tactical and operational missions in support of the ground forces, accounted for the vast majority of the VVS's units, aircraft, and rear service resources. Frontal Aviation then consisted of all air assets assigned to *fronty* (army groups) as elements of the *VVS fronta* (Front VVS) that provided air support to the entire *front* and its armies during wartime, to *VVS armii* (Army VVS) for the direct battlefield support of each ground army of a *front*, and to *voyskovaya aviatsiya* (Army Aviation) that consisted of *korpusnye aviaeskadril'y* (Corps Aviation) assigned to individual ground corps. Fighters, ground-attack aircraft, tactical bombers, and liaison planes made up frontal aviation's imposing array of air assets.

Frontal Aviation's most important task was always the struggle for air supremacy because the ground support and reconnaissance missions depended upon its outcome. Two principal methods were used to attain air supremacy – air-to-air combat and the destruction of aircraft on enemy airfields. Because of the overriding importance of the air superiority mission, *istrebitel'naya aviatsiya* (fighter aviation) always had the priority role in Soviet air warfare. Fighter aviation's primary mission was to gain air control through air combat against the enemy's fighter force so that the other air arms could successfully operate against the enemy on the battlefield and in the immediate rear areas. Frontal fighters also escorted the *shturmovaya aviatsiya* (ground-attack aviation) and *bombardirovochnaya aviatsiya* (frontal day bomber) units that carried the full burden of close air support operations.

Owing to the heavy emphasis placed on it, Frontal Aviation accounted for 61.5 per cent of all Soviet military, naval, and civilian aviation strength when the war began (see Table 3.1), and would vary from 59.7–66.1 per cent during the war. In June 1941, fighters represented 59 per cent of Frontal Aviation (see Table 3.2) while ground-attack aircraft accounted for only 4.5 per cent and day bombers 31 per cent. These proportions varied significantly during the following years as attrition, aircraft production, and operational requirements constantly reshaped the force structure, organization, and composition of *frontovaya aviatsiya*.

Table 3.1

SOVIET AIR STRENGTH DURING THE GREAT PATRIOTIC WAR:
PERCENTAGE OF TOTAL AIR STRENGTH BY COMMAND

Command	22 June 1941	19 November 1942	1 July 1943	1 January 1944	1 January 1945	9 May 1945
Frontal aviation	61.5	59.7	65.8	63.2	66.1	65.5
Long-range aviation	11.5	5.3	5.5	6.9	6.4	8.0
Air defense (IA/PVO)	10.2	17.4	15.8	16.3	15.0	15.4
Naval aviation	12.5	8.0	6.5	7.1	8.2	7.6
Civil aviation	4.3	9.6	6.4	6.5	4.3	3.5

Source: V. Chernetskiy, *Voenno-istoricheskiy Zhurnal* (Journal of Military History [*VIZh*]), 25, 1 (January 1983).

Table 3.2

SOVIET AIR STRENGTH DURING THE GREAT PATRIOTIC WAR:
PERCENTAGE OF FRONTAL AVIATION BY TYPE

Type	22 June 1941	19 November 1942	1 July 1943	1 January 1944	1 January 1945	9 May 1945
Fighter	59.0	30.8	41.8	42.5	41.7	46.0
Ground-attack	4.5	31.2	31.9	26.1	30.0	26.3
Bomber	31.0	34.7	22.7	31.4	23.0	22.0
Recon-naissance	5.5	3.3	3.6	No figures	5.3	5.7

Source: V. Chernetskiy, *Voenno-istoricheskiy Zhurnal* (Journal of Military History [*VIZh*]), 25, 1 (January 1983).

Long-range bombers, which were grouped into the *Dal'ne-bombardirovochnaya aviatsiya* (DBA) (Long-Range Bomber Aviation) during the 1940 reorganization, represented 11.5 per cent of all aviation assets in June 1941. Although for use primarily against strategic targets in the enemy's rear areas and not formally a part of Frontal Aviation, the DBA also had operational and tactical missions. The DBA, redesignated *Aviatsiya dal'nego deistviia* (ADD or Long-Range

Aviation) early in 1942 and the 18th Air Army in late 1944, accounted for from 5.3 per cent to 8 per cent of the VVS during the war. While the VVS maintained a capacity for independent strategic air operations throughout the war, tactical and operational requirements for co-operation with the ground forces always took precedence in the Soviet concept of air warfare.

FROM CATASTROPHE TO VICTORY IN THE AIR: AN
OVERVIEW OF THE VVS IN THE GREAT PATRIOTIC WAR

By noon on Sunday 22 June 1941, the *Luftwaffe* had destroyed nearly 890 Soviet aircraft – 668 of them on the ground – in the most devastating and successful surprise attack in aviation and military history. The early months of Operation Barbarossa exposed many serious weaknesses in the VVS despite the high priorities of 1940–41. In their initial onslaught, the Germans shattered the VVS's command structure in the west, destroyed its base and logistical system, and scattered its ground support and air crew personnel. Frontal Aviation's devastating early losses – the Russians acknowledge the loss of 3,985 aircraft by 9 July 1941 while the Germans claimed 6,857 through 12 July – swiftly showed its inadequacies and the urgent need for change. The large tactical air units formed in 1940 to exercise greater control of the air regiments and to provide more support to the ground forces combined with an ongoing reorganization of the VVS's basing and logistics structure directly translated into full-blown disaster.

The Finnish war had revealed some serious shortcomings in the *Tyl VVS* (VVS Rear Services) – logistics, airfields, and maintenance. In April 1941, the Central Committee and Council of People's Commissars reorganized the VVS Rear Services on a territorial area principle, to improve the VVS's combat readiness and maneuverability. New *raiony aviatsionnogo bazirovaniya* (RAB, or air base regions) were created that could support three or four air divisions with one air base allotted per division. Each RAB had an airfield engineer battalion, a signals company, and mobile aviation repair shops. Depending on the number of regiments in a division, each airfield would have three or four *batal'ony aerodromnogo obslu-zhivaniya* (BAO, or airfield maintenance battalions), each of which included airfield maintenance companies, a signals company, a motor transport company, and various services. One BAO could support a single regiment of twin-engined aircraft or two regiments of single-engined planes. Airfields and BAOs sustained air operations and the operational and base personnel, while stockpiling and providing munitions, petroleum-oil lubricants (POL), supplies, spares, and maintenance facilities and services. Both the RABs and BAOs were established along territorial lines and were not assigned to specific air units. The critical western military districts were divided into 36 RABs that were drawn along the existing *oblast'* (regional) boundaries. These RABs became the main rear organization for the VVS elements in each military district and were responsible for supporting the air units and developing the airfields and stockpiling all needed supplies. Two logistical pipelines flowed into the structure: the VVS Rear Services provided weapons, ammunition, and aviation technical equipment and the military district or front rear services provided POL, food, uniforms, and other items of combined arms supply.

Fully two-thirds of the airfield construction and reconstruction sites were in the western Soviet Union, and especially the Western and Kiev Special Military Districts, where the airfield network was poorly developed and lacked sufficient dispersal fields. The Russian seizure of additional territory from Poland, Rumania, and Finland and the incorporation of the Baltic States in 1939–40 only enlarged the basing problems. Well-developed defense lines, such as the Stalin Line, and the established base and logistical system were left behind as units were relocated to hasily prepared positions and bases in the newly annexed areas. Over 200 airfields and landing strips were rushed into construction and renovation in 1940–41 as units deployed to protect the Soviet Union's new western border. Tactical air units were quickly assigned to these airfields even before they were completed and the new BAOs were being established to support them. Although the new rear services reorganization was slated for completion in August 1941, only 31 BAOs had been organized and 32 more were in the process of formation in the Western Special Military District (largely Belorussia) and the Kiev Special Military District had not even begun to revise its base structure when the Germans attacked in June 1941. Moreover, during 1940–41 special emphasis was given to stockpiling supplies of aviation *matériel*, including munitions, at the airfields and aviation dumps in the western border districts. As a result, 70 per cent of the bomb supply in the European Soviet Union was concentrated in the west, but scant consideration was given to what would happen in the case of a German surprise attack and how these huge stocks would be evacuated in such an event.

Air regiments of 60 aircraft, and air divisions of over 100 aircraft that were crammed onto the inadequate airfields in the western military districts with little room for dispersal and less opportunity for camouflage, made extremely inviting targets for the *Luftwaffe* attackers who took full advantage of their opportunity. The swift German advance meant that the numerous VVS airfields and dumps were quickly overrun and captured or were destroyed by withdrawing units. In either case, as of 1 August 1941 the VVS of the North-western, Western, and Southwestern Fronts had expended only 18 per cent of their bombs supplies in combat operations – the rest had been lost to the enemy or destroyed. Thus, the loss of significant quantities of its air munitions severely limited the offensive striking of the VVS's remaining front line tactical units for months to come. In addition, the dispersion of tactical air units down to the army and corps levels that was intended to provide air support had in fact prevented centralized command and control, the exploitation of opportunities for massed use of air power, and the concentration of air strength on the most critical sectors of the front. Lastly, the VVS found itself waging a life-or-death defensive air struggle that it was not trained or equipped to fight due to its prewar preoccupation with offensive air operations.

Despite the desperate situation of 1941, the *Stavka Verkhovnogo Glavnokomandovaniia* (Staff of the Supreme High Command, the *Stavka* or SVGK) and VVS headquarters began to create both the ground and air reserves that would be necessary for later defensive and offensive operations as well as the doctrine and organization needed to conduct defensive air operations. In August, the DBA and the air forces of the military districts not yet engaged in active military operations relinquished their air units to form the air reserves of the *rezerv* of the Supreme High Command (RVGK) that could be maneuvered rapidly and massed quickly on critical sectors of the front. The *Stavka*'s careful husbanding of its scanty air resources and withholding of reserve air units from all but the most critical frontal operations allowed it to concentrate and use its rebuilt air power with maximum effectiveness. It would take the VVS many long months to recover fully and replace the staggering loss of a total of 21,200 aircraft – 10,300 combat aircraft, 7,600 non-combat losses of combat aircraft, and 3,300 support aircraft – between 22 June and 31 December 1941 (see Table 3.5). Volume 4 of the German official history, *Das Deutsche Reich und der Zweite Weltkrieg* (Germany and the Second World War), that covers the attack on the Soviet Union through 1942, lists the *Luftwaffe*'s total tally of Russian aircraft destroyed by the end of 1941 as 20,392 – only 808 fewer than the Russian figure – at the cost of 2,505 German combat aircraft lost and 1,895 damaged.

The *Stavka* and VVS leadership learned many expensive but very crucial lessons from the early fighting. Air power was employed most effectively in defensive operations when central control and direction existed. Although at first successful, the massed use of air units in the winter offensive of 1941–42 could not be sustained throughout ensuing operations. At Stalin's urging, the *Stavka* had largely squandered its reserves in too many operations along the entire front rather than concentrating them on a few major sectors. This dispersion of air assets prevented the planning and conduct of effective, large-scale air operations, while improved, air–ground co-operation and fighter operations still left much to be desired. With over 50 per cent of frontal aviation assets still allotted directly to *VVS armii* units, which the commander of the *VVS fronta* did not control, the centralized planning and unified employment of Frontal Aviation in large-scale operations was virtually impossible.

The experience of the summer of 1941 also led to changes in the VVS Rear Services. Air bases as separate organizations were abolished in August 1941 and their assets turned over to the RABs which were then assigned to the operational control of the commander of the *VVS armii*. This change was intended to establish a mobile and maneuverable rear service unit that was tied neither to a certain territory nor tactical air unit. This restructuring established the basis for the subsequent growth and refinement of the VVS's logistical and airfield operations structure during the war.

Despite the obvious shortcomings, Soviet airmen believed that the air operations at Moscow and during the winter offensive that followed had confirmed the basic tenets of Soviet air doctrine. Air power had been massed on the major attack zones in an offensive. Air reserves were maneuvered and extensively used. VVS commanders and staffs had worked closely with the ground forces to achieve close co-ordination of Frontal Aviation and ground units in combined-arms operations. Both DBA and *Istrebitel'naia Aviatsiia/Protivovozdushnaia Oborona (IA/PVO)* (Fighter Aviation/Air Defense) units had been used in co-operation with Frontal Aviation.

Late in January 1942, General Pavel Zhigarev, then commander of the VVS, issued a directive highlighting the basic problem:

> The use of aviation of the fronts, considering its limited quantity, at present is being incorrectly carried out. Instead of the effective and massed use of aviation on the main sectors against primary objectives and enemy groupings which are impeding the successful execution of the front's tasks, the commanders of the front air forces have scattered the equipment and efforts of aviation against numerous objectives in all sectors of the front ... Massed air actions in the interests of the designated operations are carried out indecisively on the part of the commanders of the air forces of the fronts or are totally absent.

Zhigarev's memo prompted the *Stavka* and VVS to initiate major changes in the existing air organization to align it more closely with current operational concepts and requirements. The first task was to restore the unity of command of aviation assets and to create large air units that could be used more efficiently and effectively in air operations. On 11 April 1942, J. V. Stalin appointed General A. A. Novikov, then deputy VVS commander, as commander of the VVS to make the required changes.

An outstanding air commander with extensive experience on the Northwest Front and at Leningrad who would remain the VVS commander for the rest of the war, Novikov was a strong advocate of large operational air units in centralized air operations. Under *Stavka* leadership, he astutely directed the VVS's reorganization and rehabilitation, its buildup for offensive air operations, the fleshing out of the *Stavka* reserve, and the VVS's eventual seizure of air supremacy. Novikov's appointment initiated a critical period of continuity and development in the leadership of the VVS as well as in its sustained growth. Under his wartime command, S. A. Krasovskiy, K. A. Vershinin, S. I. Rudenko, S. A. Khudyakov, N. S. Skripko, V. A. Sudets, and numerous others emerged as battle-tested combat air commanders who later formed the core of the Air Forces' leadership in the postwar era.

Drawing on prewar air concepts outlined by the prominent air theorist, A. M. Lapchinskiy, as well as on his own wartime experiences, Novikov moved quickly to combine front and army air units into large *vozdushnye armii* (air armies) that provided the required flexibility and operational efficiency. The new centralized organizational structure allowed the Soviet command to use Frontal Aviation more effectively in the struggle for air supremacy, to concentrate and organize its strength on the main sectors in support of ground force operations, and to conduct co-ordinated, independent air operations by multiple air armies against the enemy air forces.

As a result of this reorganization, an air army was eventually assigned to each active *front*, with its commander acting as the air advisor to the *front* commander. The first air armies were formed on the crucial strategic fronts in the west and southwest during May 1942. By November 13 air armies were operational on the German front and all tactical aviation (*VVS armii*) had been removed from the direct control of the ground armies. During the war a total of 18 air armies were operational at various times, four of them in the Far East.

Basic to the reorganization of the spring of 1942 was the complete restructuring of most of the Frontal Aviation in the new air armies into integral air divisions of one type of aircraft – fighter, ground-attack, or bomber. The change to integral air regiments and divisions was also a significant step in the VVS's organizational and tactical development. These integral air divisions simplified operational, logistical, training, maintenance, and command problems. The attack units were now gathered into divisions which, by the end of 1942, were further grouped into air corps rather than scattered among the combined-arms armies. Thus, their impact and effectiveness on and over the battlefield were greatly enhanced. After 1942, only some attack air divisions retained a mixed force structure with one fighter and two or three ground-attack regiments.

Balanced and flexible air combat organizations of fighter, bomber, attack, and reconnaissance units, the air armies were capable of fulfilling all of the VVS's principal air missions. Their strength and exact composition could be altered swiftly and easily to suit the shifting requirements of frontal operations. The organizational changes of 1942 set the stage for the steady improvements in tactical employment and aircrew quality that paralleled the production of more and better combat aircraft after 1943. Combined, these advances produced powerful and effective air forces which were neatly tailored to the tactical, operational, and strategic requirements of the Soviet–German war.

Novikov could also now begin to build up the Air Corps of the *Stavka* Reserve for operational use from the rapidly increasing output of combat aircraft. As with its ground forces, the *Stavka* could mass these reserve air units with its operational frontal air armies to attain superiority over the Germans on the critical sectors of the front for major multi-*front* strategic operations.

During the war, the air corps and divisions of the *Stavka* Reserve represented at times as much as 63 per cent of the VVS's total frontal air forces, although it generally fluctuated at 40–50 per cent. Thirty air corps (seven bomber, 11 attack, and 12 fighter) and 27 independent air divisions of the *Stavka* Reserve were formed and provided significant reinforcements for the frontal air armies. The 19 reserve air corps used at Kursk in July 1943 represented 61 per cent of all operational aircraft engaged. At Berlin in April 1945 the 17 reserve air corps accounted for 51 per cent of VVS aircraft in action. The creation of these air reserves allowed the *Stavka* to mass powerful air forces for strategic defensive and offensive operations after 1942. These forces gave the VVS the air superiority it enjoyed after 1943 and the ground forces the support they required in the decisive battles of 1943–45.

Safe from German interference after the industrial dislocations and arduous relocations of 1941–42, the plants of the Soviet aircraft industry equipped the air armies and reserves to seize air supremacy in 1943 and retain it in 1944–45. In 1942, the total aircraft output reached 25,436, of which 21,577 were military aircraft. New and improved aircraft, such as the Lavochkin La-5, Yakovlev Yak-1M, Yak-7B, and Yak-9, and the Ilyushin Il-2m3, entered full production and combat before the end of 1942. Production increased to 34,884 in 1943 and 40,241 in 1944. These large yearly increases allowed the VVS to replace its continuing heavy combat losses (7,800 alone in 1942 and

11,200 more in 1943), to add entirely new air units, and to reorganize and strengthen existing regiments with more squadrons and aircraft. Moreover, these aircraft were better armed with heavier machine guns and cannon that meant improved weight of fire per second for both fighter (2.1 times) and attack aircraft (3.1 times).

With more combat aircraft pouring into frontal units, the air armies, which numbered 200–1,000 aircraft in 1942–43, grew to an average strength of 1,500 in 1943–44 and to as many as 2,500 to 3,000 aircraft in 1944–45. The growth of the VVS's frontal air armies was particularly impressive in 1944 and in large measure explains the Soviet dominance in the air and the great successes in ground operations. From 1 January 1944–1 January 1945, the number of air armies deployed against the Germans decreased from 11 to 10 as the front shortened, but the number of air corps increased by 32, climbing from 89 to 121. Actual frontal VVS combat air strength grew from 5,775 to 11,530 – or by nearly 100 per cent. Of these, fighters increased by 109 per cent to 5,184 (45 per cent of the frontal forces), attack aircraft increased by 107 per cent to 3,845 (33 per cent), and bombers grew by 75 per cent to 1,857 (16 per cent) (see Table 3.3).

Table 3.3

GROWTH OF VVS FRONTAL COMBAT STRENGTH, 1944–45

Date	Air Armies	Air Corps	Air Divns	Fighter	Ground Attack	Bomber	Misc	Total
1/1/44	11	19	89	2,478	1,858	1,063	376	5,775
1/6/44	12	19	93	3,997	2,845	1,511	445	8,798
1/1/45	10	29	121	5,184	3,845	1,857	644	11,530
Change	−1	+10	+32	+2,706	+1,987	+794	+268	+5,755

Source: Col Ye. Simakov, *'Boyevoy i chislenniy sostav VVS v tretyem periode voyny'* (Combat Strength of the VVS in the Third Period of the War), *VIZh*, 17, 7 (July 1975), pp. 75–7.

Air commanders translated these large *matériel* increases into victories in the skies over the battlefields. In 1943, 70–75 per cent of air armies' strength was massed to support the major offensive operations and in 1944–45 this increased to 90–95 per cent. Although the operational sectors were never more than 15–25 per cent of the entire Soviet–German front, the VVS concentrated a very large proportion of frontal and ADD air power in these operations – 60 per cent at Kursk, 50 per cent for the Belorussian offensive in the summer of 1944, 43 per cent for the Vistula–Oder operation (January–February 1945), and 48 per cent at Berlin.

An in-depth look at the organization and equipment of an operational air army of 1944 will more clearly show many of the strengths that these gross numbers cannot adequately relate. The 2nd Air Army (S. A. Krasovskiy, commanding) of Marsal I. S. Konev's 1st Ukrainian Front in the L'vov–Sandomir operation of 13 July–29 August 1944 is an excellent example of many features of the fully matured VVS and its air armies. The flexibility created by the reorganization of 1942, the importance of the *Stavka* reserve, and the ability of the VVS headquarters to maneuver units among the various air armies was shown in the initial build-up of the 2nd Air Army for offensive operations. In June, the 2nd Air Army had two air divisions and two independent air regiments in addition to five air corps (one bomber, two attack, and two fighter) from the *Stavka* Reserve that were moved to forward bases only shortly before the operation's start. On 26 June, the *Stavka* transferred four air corps and three independent air divisions, along with their supporting aviation technical units, from the 5th and 17th Air Armies. These tranfers brought the 2nd Air Army to the strength outline shown in Table 3.4.

Table 3.4

2ND AIR ARMY IN THE L'VOV–SANDOMIR OPERATION
(13 JULY–29 AUGUST 1944)

Command	Air Corps	Air Divisions	Aircraft (Type)	Aircraft (Assigned/ Operational)	Total Aircraft (Assigned/ Operational)
Fighter	3	9	La-5 & Yak-1,3,7,9	844/783	844/783
Attack	4	9 Attack 4 Fighter	Il-2 Yak-1,7,9	1,043/1,016 575/530	1,618/1,546
Bomber	2	6 Day 1 Night	Pe-2 Po-2	572/543 105/104	677/647
Reconnaissance		3 Air Regts	Pe-2 Il-2, Yak-7,9	68/54 34/22	102/76
Total	9	29			3,241/3,052

Source: 'L'vovsko-Sandormirskaya nastupatel'naya operatsiya 1-go Ukrainskogo fronta v tsifrakh (13.7–29.8. 1944g.)' (The L'vov–Sandomir Operation of the 1st Ukrainian Front (13 July–29 August 1944)), *ViZh*, 11, 8 (August 1969), p. 59.

The total strength of Krasovskiy's air army at the beginning of the offensive on 13 July 1944 was 3,241 aircraft assigned and 3,052 operational – for an operational availability rate of 94 per cent. By composition, the attack divisions of 2nd Air Army accounted for roughly one-half of the assigned (1,618/3,241) and operational aircraft (1,546/3,052), with the 1,016 Il-2s representing 33 per cent of the operational aircraft and the escort fighters 17 per cent. Air superiority fighters accounted for 26 per cent of the operational aircraft, while total fighter strength (including fighters in the attack corps) totaled 1,313 or 43 per cent of operational strength. Day bombers accounted for 21 per cent (647/3,052) and night bombers 3 per cent (104/3,052) – a total of 24 per cent. Thus, the total force assigned to battlefield support – attack, bomber, and escort – accounted for 74 per cent of the 2nd Air Army.

Even today, however, full information is still not available on VVS operational readiness and sortie rates during offensive operations. The Soviet offensives of 1944–45 often involved swift and deep penetrations at high rates of advance, that required air units and their supporting airfield maintenance and aviation technical battalions to redeploy forward quickly several times to occupy former German airfields, or to build new airfields to keep the aircraft within operating range of the ground forces. These conditions severely strained the logistical system that had to supply the VVS combat units with fuels, ammunition, equipment, and spare parts, service and maintain the aircraft, and operate the airfields. To maintain unit operational capabilities at the highest possible levels during the course of ground offensives, pilots and aircrews flew aircraft individually from the rear area reserve air corps and divisions to the frontal units to replace combat losses. Operational wear and tear, enemy action, and logistical difficulties eventually meant diminished air support from fewer aircraft until the ground offensive operations finally came to a halt. Only then could the front line air units and their airfields once again be brought up to fully operational status and readied for the next major operation.

FIGHTER AVIATION: THE KEY TO SUCCESS

Prewar Soviet air doctrine had stressed the role of fighter aviation in gaining air superiority so that all other VVS missions could be effectively performed and the ground forces could operate freely. Soviet doctrine emphasized that victory in modern war was achieved only through the combined efforts of all branches of the armed forces and all arms and services, so the struggle for air supremacy was intricately linked to all other combat actions. Moreover, air supremacy was seen as a precondition for successful ground, air, and naval operations.

Fighter aviation had three principal wartime missions:

1. Air cover for the ground forces over the battlefield and rear areas.

VVS fighters flew 47 per cent of their 1,400,000 wartime sorties for this purpose, although during the early years when the *Luftwaffe* generally controlled the air as many as 60–70 per cent of the missions were to provide air cover. The principal tactics used were the combat air patrol, which accounted for 90.3 per cent of all air cover sorties, and *svobodnaya okhota* (free hunting) and sweeps;

2. Support to other air components, primarily the escorting of attack and bomber forces, which accounted for 37 per cent of the sorties;

3. Operations against ground targets, such as trains, field emplacements, and troops, were mainly conducted during 1944–45 and accounted for 16 per cent of the total.

When the Germans attacked in June 1941 Soviet air tactics were outdated, stereotyped, and rigid. Flyers lacked initiative and flexibility, but were stubborn opponents once engaged. Whenever Soviet flyers met the Germans, they adopted either the defensive *krug samoletov* (circle of aircraft) or headed for their own lines flying in *zmeyka* (snake or 'S') formation on the deck in an effort to draw the Germans into Soviet anti-aircraft guns. Soviet fighter pilots were hesitant to attack escorted German bombers and were equally inept at escorting their own bombers and attack aircraft.

In the prewar tables of organization for the fighter arm, the principal tactical unit was a large regiment of 63 aircraft in four squadrons and three planes for the regimental headquarters. The three-plane *zveno* (flight; plural *zvena*) was the basic tactical air element, and each squadron of 15 aircraft had five three-plane *zvena*. The regiments were assigned to either fighter air divisions or, along with bomber or ground-attack regiments, to composite divisions which were then the highest mixed or composite tactical air units. Soviet fighters generally continued to use the inadequate three-aircraft *zveno* during the early months of the war, and suffered heavily at the hands of the tactically more advanced and experienced German fighter pilots. Early in August 1941, the Commissariat of Defense reduced the fighter air regiments to three squadrons of nine aircraft in three three-plane flights. Within two weeks, on 20 August, that strength was slashed to 20 aircraft in two squadrons of nine aircraft each in three *zvena* in addition to the regimental commander and his wingman.

The successful German tactical organization of the fighter *Rotte*, *Schwarm*, and *Staffel* was eventually modified to fit the VVS and adopted as the basic Soviet tactical air organization. As with the Germans, the *para* (pair) of two aircraft, one attacker and one defender, was the basic element upon which the tactical fighter structure was built. Two *para*, one offensive and one defensive, came to make up the *zveno*, and three or four *para* formed a *gruppa* (group) of six to eight aircraft. The *gruppa*, which became the most common formation for fighter and ground-attack aircraft after 1942, usually flew with each successive flight of four aircraft echeloned higher and in back of its predecessor. The adoption of the German's 'four-finger' organization, with squadrons in two flights of two pairs, gave the fighters greater tactical flexibility in attack and defense. The new *para-zveno* structure of two four-plane *zvena* and a *para* of the commander and his wingman perfectly fitted the reduced size of the fighter regiments and their 10-plane squadrons of 1941–42. First used at Moscow in the winter of 1941, the four-plane flight was formally adopted as the standard fighter organization in September 1942 and became the basic structural building block of the Soviet tactical air organization. From 1942 on, the *zveno*, *para*, and *gruppa* were the basic VVS fighter combat elements and have remained so ever since.

Soviet air doctrine stressed the role of the fighter arm in gaining air superiority so that all other VVS missions would be effective. Soviet leadership devoted considerable effort to turning the tactical fighter arm into an élite force capable of regaining the air superiority so abruptly lost in the summer of 1941. Fighter production was given emphasis, while new aircraft were designed and built in ever-increasing numbers. Despite continuing relatively high losses, the VVS fighter force was gradually rebuilt during 1942 with modern aircraft as the more heavily-gunned, faster, and more maneuverable Yak-1M, Yak-7B, La-5, and Yak-9 appeared in increasing numbers at the front in 1942–43. With better aircraft to fly, the Soviet fighter pilots grew more confident and aggressive in aerial combat. This new attitude was clearly reflected in their growing effectiveness in combat air patrols and in penetration, counter-air, escort, and even ground-attack operations. The Soviet fighter pilots increasingly used *svobodnaya okhota* (free-hunting) tactics on patrols with *okhotniki* (hunters) operating in pairs on wide-ranging combat air patrols and fighter sweeps. Such sorties became the primary fighter tactic after Stalingrad.

A major reason for this new effectiveness was the introduction of new fighter tactics that Captain Aleksandr I. Pokryshkin, then a squadron commander in the 55th Fighter Air Regiment of the 4th Air Army, largely developed during the heavy fighting over the Kuban early in 1943. Later a three-time Hero of the Soviet Union and second leading Soviet and Allied fighter ace with 59 victories, Pokryshkin's innovations played an important role in breaking the hold of outdated horizontal maneuvering and introducing vertical tactics that best took advantage of the qualities of the new Soviet aircraft.

Air operations in the skies over the Kuban Peninsula in southern Russia very significantly influenced the development of VVS tactics and fighter employment. Fighter tactics were improving steadily largely because of the assignment of better fighters to the frontal units, the adoption of the *para-zveno* as the standard combat formation, and the presence of a number of superior air tacticians, such as Pokryshkin, Dmitriy (50 victories and fourth leading Soviet ace) and Boris Glinka, G. E. Rechkalov (57 victories and third leading Soviet ace), Vadim I. Fadeyev, and others in the 4th Air Army. Fighter employment in the Kuban areas changed after commanders discovered that the pilots tended to engage German fighters and not the bombers. Interceptors were therefore directed to attack bombers before they reached the front lines. To do this, a greater number of fighters were sent over the battlefield to protect the ground forces from enemy air attacks. Formations were always echeloned upward because altitude carried the advantage in aerial combat.

Pokryshkin was one of the first to use altitude staggering in an *etazherka* (a three-tiered stepladder, also known as the 'Kuban stepladder') that placed patrolling air groups in mutually supporting low, middle, and high tiers (altitudes). The low and middle groups engaged enemy bombers and provided air cover for the ground forces while the high group contained enemy fighters and provided high cover to the other two groups. The low and middle groups could thus fully focus on attacking the enemy because the high group covered them from surprise attacks from above. From his Kuban experience, Pokryshkin learned that sudden, swift attacks were the key to success and survival in the air. His simple and direct tactical formula of 'altitude–speed–maneuver–fire!' soon spread throughout the VVS and became the Soviet fighter pilots' formula for aerial victory.

In addition to Pokryshkin's growing tactical influence, another significant reason for the increased effectiveness of the Soviet fighter force was the introduction of advanced radio, radio navigation, and ground radars in 1942–43. The lack of airborne radio equipment had seriously handicapped the entire VVS during the early period of the war. Starting in October 1942, every other fighter carried a radio transmitter-receiver, and from 1943 on every new fighter came equipped with full radio gear. In addition, radio navigation aids, such as the fixed-loop radio compass and radio direction finders, meant aircraft could more easily return to their home airfields after missions.

During the latter part of 1943, the introduction of the mobile *Redut*-2 and the fixed *Pegmatit* (RUS-2) aircraft detection radars, with maximum effective ranges of 120 and 150 kilometers respectively, provided the air armies with advanced warning when German aircraft were 15 minutes from the front lines. With this warning time and the new radio control capabilities, airborne Soviet fighters could now effectively be directed for interception or a *zasad* (ambush). Rather than being committed to combat air patrols or sweeps over the battlefield, VVS fighters were now more often held on ground alert at airfields close to the front from which they could respond swiftly and directly to the threats as they materialized. The vastly improved Soviet air defenses meant that the German attack and bomber units found it much more difficult to operate effectively during daylight over the battlefield without suffering heavy losses. The German ground forces, which depended so heavily on the *Luftwaffe* to supplement its modest organic artillery support, were thus increasingly denied a critical source of fire support at the exact time they most needed it and could least replace it.

The increased number of modern aircraft reaching the frontal units in 1943 prompted another revision in the basic fighter tables of organization that October. A third squadron of 12 aircraft was added to each regiment to increase its operational strength to 36 and four administrative aircraft for a total of 40 combat aircraft per regiment. As similar increases in the ground-attack and tactical bomber regiments did for them, this growth greatly enhanced the combat strength and staying power of the fighter force. Moreover, this new squadron-regiment organization was perfectly tailored to the *para-zveno* tactical structure because each squadron could put three four-plane *zvena* in the air. A regiment could sortie any combination of up to nine flights which fit the new *etazherka* tactics of three mutually supporting, altitude-echeloned fighter groups.

After 1942, the combination of improved basic flying training under combat veterans, including the reintroduction of aerobatics, a realistic combat training program in frontal air training regiments and in the Air Corps of *Stavka* Reserve, revised tactics based on a careful analysis of German tactics and Soviet weaknesses and on Pokryshkin's tactical innovations, and increased combat experience produced better individual tactical skills and abilities among the Soviet pilots. The resurgence of the fighter arm and VVS in general also can be at least partly ascribed to vastly improved morale. Considerable credit for this must be given to the Communist Party and its organs, especially the *Komsomol* (Young Communist League). Membership of the Party and *Komsomol* reached 85 per cent in some air units, and was over 90 per cent for pilots and navigators.

By late 1944, the individual Russian deficiencies mattered little due to the overwhelming Soviet numerical superiority and the general deterioration of the *Luftwaffe*. A measure of the weakening German air threat was the decline in actual monthly air combats from 1,100 in 1943 to 806 in 1944. The new Yak-9U, Yak-3, and La-7 fighters that were coming on line in 1944 were at least equal to if not better than the

German fighters in speed and maneuverability, especially below 10,000 feet where the air war in the east was largely fought. The Yaks and Lavochkins were best at the lower altitudes and this gave the Soviets the edge in aerial combat.

The narrow edge that the VVS forged in the Kuban and at Kursk in 1943 was constantly widened until by mid-1944 the fighter arm had finally achieved its top priority mission – the seizure of strategic air supremacy. The struggle for air superiority over the *Luftwaffe* was always the overriding mission of the Soviet fighter force. Air-to-air combat was the primary means for achieving air superiority and accounted for 77 per cent of all German aircraft destroyed during the war, although the VVS increasingly used attacks on German airfields to weaken the opposition prior to and during major ground operations later in the war. However, the Soviet fighter force paid a heavy price for its victory – 20,700 combat losses, which represented nearly 28 per cent of the 74,300 fighters put into service by the VVS during the war years. Including non-combat losses, Soviet fighter losses totaled 46,800 aircraft or 62.9 per cent of the total fighter aircraft strength.

STRIKE AVIATION: CLOSE AIR SUPPORT, SOVIET STYLE

Ground-attack aircraft and frontal day bombers, together forming 'strike aviation', carried the burden of the Soviet tactical and operational close air support mission throughout the war. From Stalingrad to Berlin, strike aviation's strength varied between 53 per cent and 66 per cent of Frontal Aviation. Only a mere 4.5 per cent of all Frontal Aviation in June 1941, ground-attack aviation quickly rose to over 31 per cent by November 1942 due to the vastly increased demands of the ground forces for air support, and remained between 25 and 33 per cent until 1945. Direct battlefield operations accounted for over 80 per cent of attack sorties and for 82 per cent of frontal bomber missions. Ground-attack aviation received an emphasis second only to that of the fighters. As the attack units increased significantly in number, strength, and effectiveness, so too did their corresponding impact on the battlefield.

The VVS began restructuring its ground-attack air arm in January 1940 when it established the ground-attack air regiment at five squadrons with a total of 61 combat aircraft, five trainers, and one liaison aircraft. Each squadron had three combat *zvena* of three aircraft each, which was then the standard combat element for tactical bombers and attack units, plus one reserve flight for a total of 12 aircraft. Regiments were either independent or assigned to composite air divisions that also included a fighter regiment. As with the bomber force, these large organizational establishments were unwieldy, inflexible, difficulty to maneuver, and on their congested airfields were exceptionally vulnerable targets.

On 7 August 1941 the *Gosudarstvennyi Komitet Oborony* (State Defense Committee) ordered the change to an attack air regiment of three squadrons and 33 aircraft. Basically, the new regiments had three squadrons, each of which had either 10 obsolescent Polikarpov I-153 or I-15 *bis* biplanes, or I-16 low-wing, monoplane fighters, or 10 new single-seat Ilyushin Il-2 *shturmovik* (ground-attack aircraft) that began reaching the units in numbers during July–August 1941, a *zveno* of the ineffective two-seat Sukhoi Su-2 attack aircraft, and a fighter squadron of ten aircraft. This new structure was modified again on 20 August to reduce the ground-attack regiment to two squadrons of nine aircraft or three combat *zvena* each plus the regimental command's aircraft for a total of 20 per regiment. The continuing heavy losses, limited number of aircraft then available to satisfy the huge demand for air support from the ground combat units, and the desperate need to provide air support all along the front had prompted this change.

With the VVS reorganization of 1942, ground-attack air divisions of two attack regiments and one fighter regiment were created to control the tactical air support operations under the newly organized air army. Later that year, in September, ground-attack air corps were organized as the first operational-tactical command control organization to over-see ground-support operations under the overall control of the air army commander.

During the early months of the war, attack aircraft flew at low altitudes to achieve surprise, minimize losses to flak, and keep German fighters from rear hemisphere attacks. These low-level attacks also had some significant shortcomings: such a short time-over-target and flying at such extremely low altitudes fully occupied the pilots' attention and distracted them from maintaining formations and finding targets, which limited the effectiveness of the Il-2s. These tactics also prevented full use of the Il-2's firepower. In December 1941, crews of the 1st Reserve Air Brigade held a conference at which pilots recommended that diving attacks be made from medium to low altitudes (800–1,200 meters) from the *zamknutoi krug* (closed circle) formation. Such tactics would make locating targets easier and improve the accuracy of strafing and bombing. So many VVS ground-attack unit commanders opted to stay with the old low-level tactics that the Commissariat of Defense finally issued a directive on 17 June 1942 that required attacks to be made from medium altitudes. Once implemented, these tactical innovations immediately produced positive results.

The success of the medium altitude attack tactics also meant that the maneuverability of the attack groups had to be increased. The continued use of the three-plane flight in their basic *klin* ('V' or wedge) formations did not suit the newly introduced tactics. Thus, the ground-attack arm followed the example of the fighter force and switched to the more flexible *para-zveno* structure of two two-plane pairs. The addition of the rear-seat gunner in the Il-2m3 and airborne radios greatly aided this tactical change by providing greater defensive firepower and improved in-flight co-ordination between the aircraft in the pair and within the flight. The *Nastavleniye po boyevym deistviyam shturmovoy aviatsii (NShA-44)* (The Regulation on Combat Operations of Attack Aviation), issued in 1944, completed the tactical-organizational development of attack aviation and established the two-plane *para* as the basic 'battle formation of attack aircraft.'

The Soviet Union built over 36,000 Il-2 *shturmoviki* (attack aircraft) – the best, most durable, and most numerous ground-attack aircraft of the Second World War. The 'Ilies', as they were affectionately called, carried a wide variety of ordnance from cannon to rockets and bombs, and received outstanding assistance from the Petlyakov Pe-2 and Tupolev Tu-2 twin-engined frontal bombers. During 1942–43, the squadrons were gradually built up once again to 12 aircraft each and by October 1943 a third squadron was added to form the basic air regiment of three 12-plane squadrons and a total of 40 aircraft. The Il-2 was constantly upgraded in performance and armament and was only slowly giving way to the new Il-10 by late 1944. With its vitals heavily armored, the rugged *shturmovik* could take tremendous punishment, while its rear gunner and 12.7 mm machine gun often provided a lethal defense that claimed many German fighter pilots bent on an easy kill. Colonel Dieter Hrabak, one of the *Luftwaffe*'s aces with 125 victories, said: 'I never saw any aircraft that could absorb battle damage and still fly as did the Il-2.' Not only did they fly, but the *shturmoviki* handed out so much punishment that the German soldiers called them the *Die schwarze Tod* (The Black Death).

Attack aviation's unchanging primary mission was direct support of army ground operations. Once Soviet forces initiated offensive operations, attack aviation became the predominant element in the air war, clearing the way for and supporting the tank and mechanized armies upon which the final decision rested. By 1944–45, however, the variety of targets had expanded beyond the immediate battle area to include such rear area objectives as rail and road traffic, troops, command and communications centers, supply depots, and *Luftwaffe* airfields. Groups of specially selected attack aircraft now ranged far into the German rear areas in support of breakthrough operations to attack fixed and floating bridges, road junctions, and convoys to create 'plugs' on the German withdrawal routes to bottle up retreating forces so that other VVS units could make concentrated follow-on attacks.

Attack aviation showed definite tactical refinement and maturity from 1943 on. Attack units operated either in support of (*aviatsion-naya podderzhka*) or under assignment to (*pridacha*) ground force units. By far the most standard form for tactical air employment, support stressed centralized control of air units in support of tank, mechanized, and combined-arms units. When operating under assignment, attack units were attached directly to specific ground formations, usually tank and mechanized units.

Ground-attack operations were constantly modified to prevent stereotyped tactics which might give the advantage to the German air and ground defenses. They were also improved to account for new tactics, weapons, German reactions, terrain, weather, and the units level of combat experience. Depending upon the targets, defenses, and battlefield conditions, attack aviation flights, groups, squadrons, and larger units used an increasingly wide variety of sophisticated tactics in their operations after 1943:

1. *Klin* (wedge or 'V') of flight elements (two pairs, four aircraft) for flight on route and attacks against large targets offered the greatest density of offensive and defensive fire and the shortest time-over-target.
2. Flight-sized elements in *peleng* (echelon) formations were usually employed for low-level operations and striking long, narrow targets because this tactic provided complete freedom of maneuver and better aimed fire and bombing.
3. 'Broken file' flight-sized elements in *kolonna* (trail) for nap-of-the-earth flights and sequential attacks.
4. The *zamknutoi krug* (continuous orbit or 'wagon wheel') was used for striking most targets with maximum effect and afforded the best defensive fire over target areas while 'Figure 8s' were used in heavily defended areas.
5. For long targets, such as road convoys and trains protected by flak but not fighters, the *zmeyka* ('S') was used while pairs employed the *nozhnitsy* (scissors) for dispersed, long columns.

The attack squadron of 12 aircraft was the principal combat unit that had sufficient striking power and flexibility for command and control in the air. The four-aircraft flight of two pairs was the smallest tactical combat and fire unit, but the six to eight aircraft group and especially the *vosmerka* (eight-plane) formation possessed the greatest flexibility and maneuverability and was also the smallest group that had sufficient defensive firepower to operate without fighter escort. A fully equipped operational squadron of 1944–45 could put two groups of six aircraft in the air. Beginning in 1943, formations of 24 to 60 aircraft drawn from two or more co-operating regiments in a ground-attack air division were often used over the battlefield in waves of 6, 8, 12, 18, and more aircraft which continuously attacked enemy positions, personnel, and equipment. Most ground-attack missions were flown in concentrated formations or as waves of attackers. Soviet airmen mostly used wave tactics because in that way they could best support the ground forces with continuous attacks.

Coming in low and often from unexpected directions to maximize surprise, attack units struck quickly and returned to their own territory before German fighters or flak could respond. Where enemy defenses were weak, attack aircraft formed *zamknutoi krug* (wagon wheels) and loitered over their targets, repeatedly attacking until all of their ordnance was expended and the targets destroyed. When flak was present, special flights and groups were assigned for suppression because German flak downed as many *shturmoviki* during the war as the prowling fighters did. A favorite Soviet tactic was to attack from the German rear, usually the west, in the afternoon so that the aircraft came directly out of the sun at rooftop altitude against the blinded defenders. The primary point targets for the Il-2s were German tanks and assault guns on or approaching the battlefield, artillery emplacements, and especially antitank gun positions. During the war, Il-2s achieved a kill ratio of 15 per cent in their attacks on small groups of German tanks.

While attack aviation's emphasis was on the immediate battlefield area, from mid-1943 on the Il-2s were increasingly used in 'hunting' operations into the German rear areas. At first only designated 'hunters' or 'rovers' were sent to attack any rail or road traffic 'targets of opportunity' they could find or against specified depots, marshaling yards, units, etc. Eventually ranging as far as 150 kilometers into German rear areas during 1944–45, Il-2s played an ever larger role away from the immediate battlefield.

The development of improved air-to-air and air-to-ground radio communications and radio networks as well as better visual signals and marking techniques for ground forces made the attack arm more effective and responsive to ground force requirements. Air liaison officers were attached to ground units down to the regimental level to co-ordinate operations and direct attack elements to the most critical targets. Air army, attack air corps, and air division command posts were established close to the headquarters of the supported ground units, and auxiliary air command posts and checkpoints proliferated from 1943 on to control tactical support on the battlefield. 'In the clear' voice communications were often used to prevent confusion and make sure that the most important and time-sensitive targets were clearly identified and struck. Despite these improvements, poor tactical discipline in the air often hampered these direction efforts.

Recent Russian figures indicate that the total wartime losses of ground-attack aircraft in combat were approximately 12,400, or 36.8 per cent of the 33,700 aircraft devoted to this mission during the war. When non-combat losses are added, total losses from all causes reach 23,600 aircraft – fully 70 per cent of all air attack aircraft that the VVS received during the war.

Frontal bombers, largely the twin-engined Pe-2s and small numbers of the new twin-engined Tupolev Tu-2s after 1943, provided the heavier, massed striking power often required in ground support operations. Early in the war, these units were often sacrificed to stop advancing German columns and to buy precious time for the ground forces. Slowly, the poorly trained bomber regiments moved from the entirely horizontal bombing tactics of 1941 to using more precise dive-bombing techniques for as much as 49 per cent of their sorties in 1944. With this change, accuracy against point targets improved by 100 per cent. Tactical bombers flew both battlefield support and attack missions against targets in the German rear areas. Battlefield support strikes were by far the most important, accounting for 91 per cent of bomber sorties in 1943, 78 per cent in 1944, 87 per cent in 1945.

The prewar organization of bomber aviation – both short-range or tactical and long-range or strategic – concentrated aircraft in very large bomber air divisions of from two–three to four–six air regiments, each with five squadrons of 12 aircraft or a total of 60 aircraft and two staff aircraft. This meant that potentially each bomber division had a total strength of from 186–372 aircraft. The regiment was the principal tactical unit that could perform missions individually either as a divisional element or independently and was built on the three-bomber *zveno* that remained the smallest bomber combat element throughout the war. On 15 July 1941 the VGK concluded that bomber divisions with three or more regiments were 'very cumbersome, unwieldy, and unsuited for highly fluid maneuver combat'. Moreover, they were very inviting targets because it was difficult to disperse such units of large aircraft on an airfield.

The new organization provided air regiments of 30 aircraft (two squadrons of 15) and divisions of two regiments – a total of 60 aircraft. The VVS command directed that composite regiments of three squadrons would have 32 aircraft in two bomber squadrons of five *zvena* and a spare and ten aircraft in a fighter squadron with two bombers for regimental headquarters. As it also did with fighters and attacks regiments, the Commissariat of Defense soon cut this organization down to 20 aircraft on 20 August 1941. The regiment was reduced to two squadrons of nine aircraft each and two regimental air-craft because of the huge losses and need to spread newer aircraft along the front to the maximum extent possible. With growing aircraft production and dropping losses, late in 1943 the organizational structure of the bomber force finally stabilized at a regiment of three ten-plane squadrons (three *zvena* and a spare) and two regimental staff aircraft. Three regiments were assigned to each bomber division which now had up to 100 aircraft. This increased strength meant that the tactical bomber force could maintain much greater combat effectiveness over extended periods of service at the front.

Frontal bombers, like their ground-attack comrades, suffered heavily during the war. Wartime combat losses were approximately 10,000 aircraft and total losses were 17,900 – thus, combat losses represented 36.2 per cent of the 27,600 tactical bombers employed during the war, and total losses accounted for 64.8 per cent of all tactical bombers.

The pilots and crews of strike aviation contributed significantly to the final Soviet victory. Among all VVS branches the ground-attack and tactical bomber forces paid dearly for their success because they suffered the highest ratio of losses to sorties flown. One Il-2 was lost for every 62 sorties flown, while the Pe-2s and Tu-2s lost one aircraft for every 96 sorties. A total of 1,450,000 sorties, some 46 per cent of all VVS wartime sorties, were flown for tactical and operational support of the ground forces, so the magnitude of strike aviation's sacrifice and contribution is obvious.

THE AIR OFFENSIVE

Soviet military theoreticians developed the concept of deep operations during the 1920s and 1930s. They proposed the use of modern weaponry, especially tanks, artillery, airborne forces, and aircraft, to smash enemy attacks and then to destroy him in depth. Penetration and exploitation operations would convert tactical success into operational and strategic victory. Aviation had an important role to play in these offensive operations because it was the most powerful and maneuverable means for supporting and protecting the ground forces and for the long-range destruction of objectives in operational depth – the enemy's rear. Once the Soviet armed forces went over to the offensive after November 1942, these concepts were implemented with deadly effectiveness.

The VVS played an increasingly larger role in all of the major strategic operations after Stalingrad. In these it always had four primary missions to plan and execute:

1. The struggle for air supremacy.
2. Destruction of enemy personnel and equipment on the battlefield and in operational depth.
3. Destruction of enemy regrouping efforts, movement of reinforcements and reserves, and blunting of counterattacks and counteroffensives.
4. Air reconnaissance and intelligence operations.

Based on wartime experience in such operations, the VVS flew an average of 35–45 per cent of its sorties for air supremacy; 40–50 per cent for ground support; 4–12 per cent against enemy regrouping, reinforcements, and counterattacks; and 2–13 per cent for air reconnaissance. As the size of VVS grew, so did the proportion of forces allocated to the various missions and the impact of the VVS. The number of combat aircraft available increased from 1,327 at Stalingrad to 7,496 at Berlin, and the daily average of frontal and ADD combat sorties jumped from 500 at Stalingrad, to 2,600 at Kursk, and finally to 4,157 at Berlin.

The large operational air forces and reserves created in the reorganization of 1942 allowed the development of a new and more effective form of operational air employment – the *aviatsionnoye nastupleniye* (air offensive). The revised Soviet infantry field service regulations issued on 9 November 1942 specified that 'the air offensive is introduced with the aim of providing for the continuous support of infantry by the massed, effective fire of aviation during the entire period of the offensive.' Designed to assist the main infantry and tank offensives, the air offensive was first to provide *aviatsionnaya podgotovka* (air preparation) for a planned breakthrough and then to support the actual attack and subsequent operations of infantry and tanks in the depth of the enemy defenses. In his excellent work on the wartime operational art of VVS, Colonel I. V. Timokhovich wrote 'The main essence of the air offensive was the continuous support of the ground forces by the massed operations of aviation throughout the entire depth of an offensive operation.'

The initial air offensive was planned for the opening of the counteroffensive at Stalingrad on 19 November 1942, but bad weather conditions prevented its proper execution. The major offensive operations of 1943 allowed the VVS to put the theory of the air offensive into practice. After blunting the last great German summer offensive of the war at Kursk in early July, the *Stavka* immediately launched offensives north of the Kursk salient at Orel and south of it on the Belgorod–Kharkov axis. These offensive operations, which lasted from 12 July through 25 August, were significant milestones in the development of Soviet air power. For the first time the VVS began a strategic offensive with a definite superiority of over 3:1 in fighters. Also, for the first time, this advantage cleared the way for the air armies to conduct a complete air offensive in both the preparatory and support phases.

Air cover and support for the tank armies that carried the burden of the major Soviet offensives after 1943 were critical to the overall success. The VVS was the most mobile, flexible, and powerful means for supporting tank armies during deep operations. Once fighters had assured control and cover, air support fell principally to the attack and frontal bomber units. Covered by one or two fighter divisions, one or two attack divisions usually worked in depth with specific tank armies during the breakthrough and exploitation phases in 1943 and this number rose to two, three, and even four divisions in 1944–45.

Front and air army commanders and their staffs carefully worked out plans for air support which was an integral part of each front's offensive operations. Air officers from VVS headquarters or the *Stavka*, often Chief Marshal of Aviation A. A. Novikov himself, acted as the *Stavka* Representative to assure the co-ordination and planning of air operations when more than one front and air army were involved in strategic operations. With the authority of the *Stavka* and intimate knowledge of its plans, these air officers played important roles in arranging logistical support, maneuvering units, training and preparation, providing reserve air units, defining air objectives, and assisting each air army with detailed operational planning.

The preparatory phase of the air offensive could be either preliminary or direct. Both were intended to crush German defenses in the line of the planned Soviet advance, with the only distinctions being the duration and timing of the attacks and nature of the defenses. Air strength for these operations grew from a regiment to a division and finally to several divisions by 1944–45. Usually lasting one to three days, preliminary preparation, or 'softening up', was used where heavily fortified tactical defense zones had to be penetrated and strong opposition destroyed. In addition to frontal bombers and attack aircraft, ADD strategic bombers were often used against targets in the tactical zones. In fact, 40 per cent of the ADD's total wartime combat sorties were completed in direct support of the ground forces in operations such as these.

Direct air preparation came immediately prior to the ground attack and could last for anywhere from 20 minutes to 2.5 hours. With the greater number of aircraft available, preparation was generally reduced to 20–30 minutes in 1944–45 while the density of operations and their effect were vastly increased. From an average bomb density of 17–20 tons per square kilometer in 1943, attack weight grew to 60–80 tons in 1944 and then 100+ tons in 1945. Massed bomber

strikes, followed by massed employment of attack aviation in groups
and successive waves, coincided with artillery preparation and were
directed against artillery batteries, tanks, reserves, and any strong
points beyond Soviet artillery range that could retard the advance.
Targets were carefully divided between air and artillery based on
detailed study of repeated aerial reconnaissance and photography. For
example, in the Vistula–Oder Operation of January 1945, air reconnaissance identified 2,803 point targets for artillery and air attack.
Such support was carried over into the actual ground attack and after
the insertion of the second echelon of the offensive.

Once the breakthrough began, the VVS shifted to air support
operations to cover the infantry, armored, and mechanized forces.
This involved the co-ordinated and combined employment of all VVS
arms, fighters for cover and escort, attack for close support, and
bombers for support and deeper interdiction. Mainly, however, the
emphasis fell upon the attack and frontal bomber units co-operating
with the ground forces. Covered by fighters, attack aircraft in groups
of 4–18 planes were continuously over the battlefield during the day.
The massed employment of air forces in the direction of the advance
and the rapid maneuvering of units to the most important sectors in
each phase of the offensive characterized such air support operations.
During the breakthrough alone, fighters averaged 2.5–3 sorties daily,
attack aircraft 2–2.5, day bombers 1.5–2, and night bombers 2–2.5
sorties. Overall, as much as 50–70 per cent of frontal aviation's available forces conducted extensive operations in the first critical days of
the offensive. In large operations such as Belorussia, Lvov–Sandomir,
Vistula–Oder, and Berlin, as many as 1,500–2,000 aircraft were used in
these narrow attack zones. Although the breakthrough period was only
6–9 per cent of total duration of an offensive operation, 20–25 per cent
of all combat sorties were flown during these few days.

After the penetration was achieved, the most important phase of air
support was the commitment to battle of the tank and mechanized
forces. These mobile groups were usually committed on relatively
narrow fronts, highly concentrated and in dense formations – all of
which made them vulnerable to German air action and to flank attacks
or obstruction by defenders. During this critical time, fighters protected the armored columns while ADD and frontal bombers, attack,
and fighter units struck German airfields, reinforcements, counterattacks, and defenses. Once in the open, the mobile groups operated as
much as 70 miles ahead of the combined-arms armies and thus
depended heavily on the air forces for fire support, cover, reconnaissance, and resupply of fuel, food, and ammunition. When the mobile
groups outran their supplies during the Belorussian offensive of
June–August 1944, the VVS airlifted 1,182 tons of fuel, 1,240 tons of

Table 3.5

SOVIET AIRCRAFT LOSSES (COMBAT/TOTAL) BY AIRCRAFT TYPE IN THE GREAT PATRIOTIC WAR: 1941–45

Type	1941	1942	1943	1944	1945
Attack	600/1,100	1,800/2,600	3,900/7,200	4,100/8,900	2,000/3,800
Bombers	4,600/7,200	1,600/2,500	1,700/3,600	1,500/3,200	600/1,400
Fighter	5,100/9,600	4,400/7,000	5,600/11,700	4,100/12,700	1,500/5,800
Total combat	10,300/17,900	7,800/12,100	11,200/22,500	9,700/24,800	4,100/11,000
Non-tactical	300/3,300	1,300/2,600	500/4,200	700/5,700	200/2,300
All aircraft	10,600/21,200	9,100/14,700	11,700/26,700	10,400/26,700	4,300/13,300

Source: F. G. Krivosheev (ed.) et al., *Grif sekretnosti snyat: Poteri vooruzhennykh sil SSSR v voinakh, boevykh, deistviiakh i voennykh konfliktakh* (Secret Stamp Removed: The Losses of the Armed Forces of the USSR in Wars, Combat Operations, and Military Conflicts) (Moscow: Voennoye Izdatel'stvo, 1993), pp. 359–60.

Table 3.6

SOVIET AIRCRAFT LOSSES AS A PERCENTAGE OF AIRCRAFT RESOURCES BY TYPE

Type	Combat	Total	Total resources	Combat losses as % of total resources	Total losses as % of total resources
Attack	12,400	23,600	33,700	36.8	70.0
Bomber	10,000	17,900	27,000	36.2	66.3
Fighter	20,700	46,000	74,300	27.9	61.9
Total combat	43,100	88,300	135,600	31.8	65.1
Non-tactical	3,000	18,100	35,000	8.6	51.7
All aircraft	46,100	106,400	170,600	27.0	62.3

Source: F.G. Krivosheev et al. (eds), *Grif sekretnosti snyat: Poteri vooruzhennykh sil SSSR v voinakh, boevykh, deistviiakh i voennykh konfliktakh* (Secret Stamp Removed: The Losses of the Armed Forces of the USSR in Wars, Combat Operations, and Military Conflicts) (Moscow: Voennoye Izdatel'stvo, 1993), pp. 359–60.

ammunition, and over 1,000 tons of technical equipment and spares for tanks. In 1943–45, the VVS also often provided the strength that fast-moving mobile groups lacked to contain and destroy encircled German units after their initial envelopment.

SOVIET AIR LOSSES

One of the major questions that has lingered for 50 years about the VVS participation in the Great Patriotic War has been the matter of combat and total aircraft losses. In 1993, the Ministry of Defense published a long classified study of Soviet and Russian personnel and equipment losses from 1918 through Afghanistan. For one of the first times, Soviet air losses were officially catalogued – if in less than exactly precise numbers. The figures in Tables 3.5 and 3.6 speak for themselves, and need little additional elaboration beyond what has already been said.

During 1942, the VVS lost 6,178 pilots from operational combat squadrons. This represented about 24 per cent of the combat flight personnel in the VVS (approximately 25,750). In 1943, as a result of the sustained air operations in the Kuban, at Kursk, and in other offensive operations, VVS combat pilot losses rose to 8,255, or 39.2 per cent of the 21,000 flight personnel then assigned to the operational air armies.

From 1 January 1943 through 1945, VVS personnel losses totaled 48,993 killed, missing, wounded, captured (0.29 per cent of total casualties), of which key officer personnel (comanders, staff, and flight personnel) accounted for 35,385, or approximately 3.8 per cent of the more than 1 million officer losses suffered by the Red Army. Commanders and staffs of air corps, divisions, and regiments suffered 464 killed or lost, while a total of 1,946 squadron commanders and staff officers were lost, and 2,110 were missing, wounded, or captured. A total of 2,792 flight section commanders were killed out of a total of 5,379 casualties, and a total of 18,464 pilots were lost – 7,855 of them killed. Such loss figures for officers are certainly miniscule compared to the huge losses of the ground forces which had 296,744 platoon, 90,210 company, 14,547 divisional battalion, and 2,545 regimental commanders killed alone.

The VVS's low casualty totals are accounted for at least partly by the nature of the air war in the East – most Soviet aircraft were one or two-seaters (fighters and ground-attack aircraft) and most Soviet air losses took place over Soviet-occupied territory during the first three years. Thus, much as in the *Luftwaffe*'s air offensive against Britain in 1940, the force on the offensive lost aircraft and crews over hostile territory where many of the crews were captured and lost for war. The defenders, when they lost aircraft, lost them over friendly territory, and the crews were frequently quickly recovered and returned to their units where they soon took to the air again.

SUMMARY

Unfortunately, the foregoing account has touched only briefly on the rich experience of Soviet tactical air power during the Great Patriotic War. The Soviet frontal air armies played a critical role in the defeat of the *Luftwaffe* and the eventual Soviet and Allied victory over Nazi Germany. The VVS flew 3,124,000 combat sorties during the war, of which just over 1,400,000 were flown for air supremacy and 1,450,000 for support of the ground forces. Recovering from its devastating early defeats, the VVS rebuilt itself into a formidable fighting force. Wartime experience demonstrated the inherent simplicity, flexibility, and adaptability of Soviet air power and the leadership skills and foresight not only of those who commanded it but also of those who provided the aircraft, engines, armament, airfields, and so on. By the end of the war, the Soviet Air Forces were the best honed and most effective tactical air power in the world. The hard-won knowledge and experience gained during the Great Patriotic War significantly shaped the doctrine and organization of the postwar Soviet Air Forces just as certainly as it molded the leaders who would direct the Air Forces throughout the entire Cold War era.

Aviation and the Transformation of Combined-Arms Warfare, 1941–45

TOM ALISON AND VON HARDESTY

The Soviet Air Force (VVS) deployed 7,500 combat aircraft for the final assault on Berlin in May 1945, by which time the Soviet Union possessed the largest tactical air arm in the world. At the core of this air juggernaut was *Shturmovaya aviatsia* or Ground Attack Aviation. Operating as a 'Flying artillery', Soviet Il-2 aircraft had flown at the forward edge during the war, providing close air support for the huge ground offensives that defeated Germany in the east.

The wartime priority given to tactical aviation is mirrored in the fact that Soviet aviation factories manufactured 36,163 Il-2 Shturmovik aircraft as opposed to a mere 78 Pe-8 four-engined long-range bombers. Designed by Sergei V. Ilyushin, the ubiquitous Il-2 Shturmovik flew highly disciplined air sorties in close synchronization with the army in all its defensive and offensives modes of operation throughout the war. The Il-2 or 'Ilyusha' became emblematic of the Soviet Air Force in the Second World War.

Prewar Soviet air doctrine did not necessarily reflect the wartime stress on tactical aviation. In fact, Soviet air strategists and war planners had placed an inordinate emphasis on fighters, bombers, and special aircraft designed for record-breaking flights. Coming to power in 1927, Stalin stressed the expansion of his aviation sector, giving aviation figures – in particular his pilots or 'Stalin's Falcons' – heroic stature in the new revolutionary society. Flying 'faster, farther, and higher' committed Soviet aviation to an intense and expensive competition with the West for world records. Consequently, the Soviet aviation industrial sector gave priority to the design of aircraft that would establish records or provide evidence of dramatic strides in aeronautics. In the military sphere fast fighters and large long-range bombers enabled Soviet air power to evoke images of strength and parity with the West in the interwar years.

Stalin oversaw an aviation establishment in the 1930s where aerial spectaculars replaced any concrete program of military preparedness.

This preference for image over national security came at a dangerous time. As the threat from a rearming Nazi Germany became manifest, the Soviet Union would be slow to shift its air policy. Soviet fighters in the early 1930s were fast and highly maneuverable as in the case of the I-16 type monoplane. But by the end of the decade these same fighters were obsolete. The ANT-25 long distance aircraft had made two successful transpolar flights in 1937. The design work on the ANT-25, as with other special projects, compelled Tupolev and his design bureau to ignore other, more practical military projects. Perhaps the most spectacular Soviet airplane of that era was the red-winged, eight-engined *Maxim Gorky*, another Tupolev project which reflected the fascination of Russian aircraft designers with gigantic aircraft. The Tupolev-designed TB-1 and TB-3 which became operational in the early 1930s mirrored the interest of Soviet air leaders in long-range bombers. On the eve of the war, these obsolete machines no longer possessed value as strategic bombers.

Parallel to these misguided design priorities was the contradictory nature of Soviet air doctrine. Jan Alksnis, the Soviet air commander in the mid-1930s, had been an apostle of Guilio Douhet, the Italian air theorist who argued that the bomber would be the strategic weapon of the future. This same sentiment was shared by the army high command. Because of the 'Douhetists' at the highest levels of the Soviet military there was minimal interest in the development of tactical aviation.

During the 1930s, the Soviet Air Force found itself caught up in a sequence of clandestine air operations. These skirmishes would test Soviet aviation technology in actual combat conditions, exposing the relative obsolescence of Soviet aircraft and the ineffectual nature of Soviet air tactics. These events cast a long shadow on the fate of tactical aviation. When the Spanish Civil War erupted in 1936, the Soviet Union quickly mobilized arms shipments for the Republican Government. A dramatic aspect of Soviet aid came with the shipment of late model fighters (I-15s and I-16s) and bombers (the fast SB-2 medium bomber and the R-5 adapted for both bombing and reconnaissance). Along with the aircraft the Soviets sent 'volunteers' (pilots and mechanics) to serve in a special air group. Throughout most of 1937, Soviet air intervention allowed the Republican air force to assert air superiority, especially in the defense of Madrid against the advancing Nationalists under Franco. Soviet participation in the air war over Spain still remains a shadowy episode. While no precise figures are available, perhaps as many as 1,400 aircraft were deployed and over 700 pilots were sent to defend the Republic.

Among the pilots was Yakov V. Smushkevich, who like many of his air force comrades fell victim to the purges, being summarily executed for his involvement in the Spanish Civil war. The intervention, at first, went well. But Franco's Nationalists eventually gained control of the skies with the active assistance of Nazi Germany and Fascist Italy. Modern German aircraft, in particular the Messerschmitt Bf-109E, turned the tide. Most of the Soviet aircraft deployed in Spain were fighters (around 1,000 aircraft). They were quickly outclassed by German and Italian models. German pilots flying in the 'volunteer' Kondor Legion experimented with new air tactics, which only enhanced the technical superiority of their fighters. When the Soviets withdrew their air group in August 1938, many technical and tactical shortcomings of the VVS had been revealed. Subsequent air skirmishes in the Far East against Japan at Khalkin Gol produced mixed results, but the overall weaknesses of the VVS had been revealed. At the time of the Winter War with Finland in 1939–40 these failings would be exposed to the world, suggesting that the lethal character of the prewar Soviet Air Force had been wildly exaggerated.

The air débâcle in Spain no doubt played a role in the decision of Stalin to purge the Soviet air establishment, a systematic repression of aviation leaders that would devastate both the civilian and military sectors. Jan Alksnis, along with many high-ranking air commanders such as Smushkevich, were arrested and executed. Andrei N. Tupolev, the most renowned of Soviet aircraft designers, was arrested and escaped death only by being placed in a special design bureau run by the secret police. A whole new generation of aircraft designers arose to replace Tupolev, the most prominent being Alexander S. Yakovlev, who had criticized the stress on large aircraft which had served only as a means for aerial spectaculars, not national defense, at a time when Fascism threatened the Soviet Union.

Yakovlev's critique dovetailed with a series of events that prompted Stalin to revise his aviation establishment in 1940. This dramatic reorganization brought a whole new generation of aircraft designers to prominence together with a deliberate campaign to build a new ground-attack aviation. In January 1940, Stalin appointed A. I. Shakurin to head the aviation industrial sector. Shakurin had built a reputation in the high-risk purge era for toughminded efficiency. With Tupolev and many older designers in prison and prewar emphasis on large aircraft discredited, Stalin moved Yakovlev to the top (during the war he served as deputy commissar for the aviation industry); behind Yakovlev came a younger group of designers such as P. O. Sukhoi, S. A. Lavochkin, A. I. Mikoyan, M. I. Gurevich, and S. V. Ilyushin. The shift signalled a new drive for modernization and the emphasis on a fewer number of designs each capable of technological refinement over time.

By 1941, the Soviet aviation industry had delivered the first installment of the new generation of aircraft. More than 1,900 MiG-3, LaGG-3, and Yak-1 fighters reached the VVS in the six months prior to the German invasion. In addition to this dramatic effort to revitalize fighter aviation, the VVS received a major augmentation of tactical aircraft, a total of 458 Pe-2 bombers and 249 Il-2 Shturmoviks. Missing in the 1941 production figures were the large bombers, a strong indicator that the VVS had made the shift to tactical aviation in a decisive fashion. For the rest of the war the VVS would be committed to combined-arms operations with ground-attack aviation as a vital component in all defensive and offensive operations.

THE QUEST FOR A NEW GROUND-ATTACK AIRCRAFT

Several Soviet design bureaux received orders to develop a multi-purpose monoplane that could fill the reconnaissance and ground-attack roles. This competition, given the code name of the 'Ivanov Program', mirrored the high priority given to the production of a specially designed warplane for low-level ground-attack operations. Polikarpov's 'Ivanov' design was one such attempt, along with other designs such as I. G. Neman's R-10/KhAI-5, Kocherigin's R-9, and D. P. Grigorovich's DG-58R. All of these prototypes failed to win approval for series production because they failed to meet the fundamental requirement for an armor-plated aircraft specifically designed for the ground attack, a role now again recognized. Interestingly, Ilyushin did not participate directly in the Ivanov program.

In the race to build a new ground-attack aircraft the Sukhoi Su-2 became the chief rival to the Ilyushin Il-2. The Su-2, in fact, became operational on the eve of the Second World War. Designed as a short-range bomber, the Su-2 was a low-wing monoplane with a crew of two. The cockpits were armor-plated, and the Su-2 could be configured to carry bombs, rockets and up to four ShKAS machine guns. Initially moderately successful, 800 were ordered and most were in service when the war began. Slow and lacking maneuverability, the Su-2 later proved to be ineffectual against *Luftwaffe* fighters. As a consequence, the Su-2 was withdrawn and production canceled.

The Sukhoi design bureau made one further effort to design a ground-attack aircraft. A single-seat aircraft with heavy armament, the Su-6 prototype appeared to be a major challenge to the Ilyushin Il-2 design. Sukhoi chose as a powerplant the new 18-cylinder, 2,000 hp, Shvetsov ASh-71 radial engine, which was a constant source of problems during the flight-test phase and caused the testing to be postponed several times. The Su-6 had better performance than the Il-2 and carried a heavier armament load, but was not put into production because of the unreliable engine. While opinions vary considerably as to how the Su-6 would have compared with Ilyushin's Il-2 design under the rigorous combat conditions to which it would have been subjected, in actuality it was too late to even compete. Il-2 production was already underway and, owing to the urgent requirement for the aircraft, it was decided not to disrupt that process.

In January 1938, Sergei Ilyushin had submitted his proposal for a revolutionary ground-attack aircraft. Ilyushin's design, the Il-2 Shturmovik, eventually attracted favorable attention. Rather than adapt a prewar design to meet the requirements for a rugged, armor-plated ground-attack airplane, Ilyushin fashioned a radical new design. His design was not a hybrid aircraft, one that modified an old design to meet certain tactical requirements. The Il-2 possessed an innovative fuselage, one that incorporated an armor-plated shell. The design of the protective shell or 'Bronekorpus', as an intregal part of the fuselage, gave the Il-2 its legendary ability to withstand anti-aircraft fire in low-level sorties flown against enemy tanks and artillery. Ilyushin took a keen interest in all aspects of contemporary aircraft technology. He exploited fully new advances in Soviet aero propulsion technology. He adapted for his Il-2 aircraft the powerful, liquid-cooled aircraft engines designed by Alexander Mikulin at the TsIAM (Central Institute of Aero Engine Construction). He was also aware of the newly developed armor-plate metallurgy and the production techniques for double-curvature armor.

What made Ilyushin's Il-2 design so revolutionary was the concept of the aforementioned 'bronekorpus', a bathtub-shaped shell that formed the engine bay and forward fuselage of the aircraft. Constructed of 5 mm to 12 mm-thick armor plate, the 'bronekorpus' enclosed and protected the engine, radiators, fuel tanks, and the crew, which consisted of a pilot and gunner. The 'bronekorpus' became an integral stress-bearing part of the aircraft structure eliminating the need for a standard aircraft tubular steel alloy framework. The sides and bottom of the forward fuselage to the rear of the cockpit were also armor-protected. The wings, and horizontal stabilizer were of standard duralumin construction, the rear fuselage and vertical stabilizer were made of plywood. The movable flight control surfaces were a fabric covered metal framework. The aircraft made its first flight on 2 October 1938.

Factory flight tests of the first Ilyushin prototype proved to be less than satisfactory. Over time, the Ilyushin design would be perfected, a process that would incorporate certain wartime field modifications which had been fashioned at the front. The Il-2 aircraft underwent a metamorphosis that led ultimately to the Il-2m3, arguably the optimal combat version of the Il-2 Shturmovik. Owing to the austere conditions that existed and, at times, the urgent requirements of the front-line regiments, many of the changes were first made by technicians and engineers at the front. Ilyushin and his design team would spend many days with operational units obtaining first-hand information from the pilots of his Il-2.

Based on this feedback Ilyushin had already begun a redesign of the aircraft. This first redesign included changes that were requested by the Soviet Air Force. One significant modification that the VVS had requested, removal of the gunner, was anticipated but was particularly unpalatable to Ilyushin. The VVS position was that the ground-attack aircraft would be supported by Red Air Force fighters providing cover from attack by enemy fighters. Ilyushin did not have a high level of confidence in the capability of Soviet fighters to do this. The issue was finally settled by the Politburo, which accepted the VVS position. The first Il-2 to enter combat would be a single-seat aircraft.

The requirement for immediate production was urgent. The war had begun in Europe and the Soviet Union needed a ground assault aircraft desperately. Some idea of the large impetus that had been given to the Il-2 program can be gained by the fact that the first production aircraft flew from the assembly line two days before the completion of the State acceptance trials with the prototype. The first Soviet Air Force unit to be equipped with the new Il-2, the 4th ShAP (4th Ground Attack Aviation Regiment), began to receive its aircraft in late May 1941.

GROUND-ATTACK OPERATIONS IN THE SECOND WORLD WAR

At the start of the war, as noted above, the VVS had deployed only 249 Il-2 Shturmoviks. The larger program of integrating new fighter and ground-attack designs into frontal aviation was about 20 per cent complete. The plight of the Fourth Ground Attack Aviation Regiment is indicative of the problems faced by VVS commanders. As most of the air regiment's pilots had only made a few orientation flights in the new Il-2 aircraft, the unit was barely operational. One veteran, V. B Yemel'yanenko, recalled, 'we still had not flown in formation, and no one had fired the cannon and machine guns on the proving ground ... no one had any idea how to bomb accurately'.

As Operation Barbarossa unfolded, Shturmovik pilots were quickly given their baptism of fire, often with high attrition in aircraft. Still, the Il-2s acquitted themselves well in this uneven air war. Simple piloting techniques, powerful armament, an apparent invulnerability to ground fire, and to some extent even to fire from small-caliber anti-aircraft cannon made the Il-2 an effective weapon against enemy ground troops, especially tanks and mechanized infantry.

From the moment it entered the combat arena the Il-2 was recognized as a devastating ground assault weapon. One of the main reasons that the early results were at all favorable was due to the sheer tenacity of the pilots who pressed home their attacks, even in the face of lethal flak barrages. During the early days of the war the German tanks and motorized infantry used their high level of mobility to strike primarily against poorly defended areas. The vast distances on the Eastern Front spread the Soviet defenses from a thin to a nonexistent level, which the German blitzkrieg used to great early advantage. Under these conditions the role of aviation grew, specifically the role of the Il-2 with its ground-assault capabilities. The Shturmoviks struck the German mechanized infantry and artillery positions, as well as fortified lines, but they were most potent attacking groups of German tanks at collection points and when they were stopped for refueling. Soviet pilots reported that rocket attacks against the tanks were particularly effective and that they appeared to demoralize the Germans. Many of the tank crews did not survive Shturmovik rocket attacks. The ones that did had abandoned their tanks.

At the same time, it was also noted that losses suffered by the Il-2 units were extremely high. While the Il-2's rugged construction enabled it to withstand numerous strikes from light and heavy machine guns and even 20 mm cannon shell hitting at oblique angles, it could only operate effectively in the absence of *Luftwaffe* fighters. The German pilots discovered early that the Il-2, with its low speed and inability to make rapid evasive maneuvers, could most easily be dealt with from behind and above. This was the design weakness that Ilyushin had recognized from the beginning. A rear defense capability was urgently needed.

From early in the conflict the design office had been receiving requests from front-line pilots to install either a machine gun, remotely controlled by the pilot, or provide for a rear gunner with a defensive machine gun. These requests had also reached the headquarters of the Supreme Command. It was with this background, the high attrition rates, and advancing German troops, that a special conference was held in February 1942, attended by combat-experienced Shturmovik pilots, test pilots and design personnel. Stalin and Ilyushin were also in attendance. Ilyushin had been incensed by the VVS decision to change the Shturmovik from a two-seat aircraft to the single-seat version and on several occasions had protested about the change to the Central Committee. His position had been vindicated, but at great cost. Stalin is said to have commented to Ilyushin that he now realized that the VVS had been wrong to change the requirement.

He acquiesced to Ilyushin saying, 'Do what you like but do not stop production – only send them to the front as two-seaters!'

Ilyushin had already been working on the two-seat ground-attack aircraft for some time. As a result of the conference he ordered his design team to begin preparations to convert the factory production of the Il-2 to a two-seat model called the Il-2m. While Stalin's directive forbade modifications that would in any way slow production, Ilyushin made changes that would make the Il-2m a much more effective ground-attack aircraft. The major modification, of course, was the addition of the rear gunner position again, as it had been in Ilyushins' initial version of the Shturmovik.

In the two-seat Il-2 the gunner, unlike the pilot, was only partially protected by the armor-plated fuselage and canopy. The canopy was open in the rear quadrant and the gunner's head and upper body were essentially unprotected. The machine gun was placed on a pivot mount and, because of the cabin armor, had a limited range of fire. The semi-circular turret screen created a large amount of drag and lowered the Il-2's performance capability significantly. In an attempt to maintain the flight range portion of the aircraft's capability, and to replace some of the fuel capacity lost when the fuel tank originally behind the pilot was replaced with the gunner, the capacity of the lower fuel tank under the pilot's seat was increased. At the same time the two-wing cannon were converted to the armor-piercing VYa 23 mm cannon with a muzzle velocity that would ensure it could penetrate armor up to 25 mm thick. A further modification of the rear machine-gun turret increased the gunner's firing angles to 45 degrees up, 35 degrees to the side and 12 degrees downward, increasing the gunner's offensive capability.

Some units, however did not wait for the Ilyushin design bureau to add the rear gun on the factory assembly line. Field conversions to the two-seat Il-2 were the first into the battle. These modifications were carried out by engineers of several front-line regiments. The various conversions differed from regiment to regiment, but they were effective. Mechanics and fitters flew as gunners, sometimes sitting on the ammunition box. These field conversions continued even after the first rear-gun-equipped Il-2s were delivered directly from the factory to the front in the fall of 1942. The factory-produced Il-2m made its combat debut over the central front on 30 October 1942. During the first operational sorties the Shturmoviki gunners were credited with the destruction of ten *Luftwaffe* fighters.

The addition of the gunner in the Il-2m instilled a higher level of confidence in the Shturmovik pilots. They felt that the gunner's presence ensured they would not be attacked unexpectedly by enemy fighters from the rear hemisphere. The gunner would also inform the pilot of anti-aircraft fire from the rear, giving the pilot an opportunity to maneuver against these attacks. This increased the Il-2's combat efficiency, made it possible for the pilot to approach the target to bomb and strafe more accurately, and reduced aircraft losses. But the *Wehrmacht* was the first to admit that the Shturmovik pilots were both aggressive and courageous. They were less inflexible than other VVS pilots and flew their aircraft recklessly and tirelessly.

During this time Ilyushin and his brigade of engineers were constantly looking for ways to improve the Il-2 performance and lethality. The changes Ilyushin incorporated included the results of TsAGI wind tunnel research that showed that a 15-degree sweep back applied to the outer panel of the Il-2 wing moved the center of lift aft which compensated for the aft center of gravity, the cause of the worst handling characteristics for which the aircraft was known. The result would be called the Il-2m3 and would become the most effective and most produced of all the Shturmoviks. It is generally agreed that the Il-2m3 was the definitive Shturmovik and was destined to become the most respected aircraft on the Eastern Front.

While the *Luftwaffe* was by no means a spent force, it no longer maintained the overall air superiority that had rendered the Il-2 operations as hazardous as it had during the first two years of the war. This only intensified the devastation of the Il-2 attacks on the German troops, especially for the members of the Panzer Corps. To them it was known as 'Black Death'. Even German fighter pilots, such as the famous *Luftwaffe* ace Guenther Rall, express a level of respect not only for the Il-2 but for the courage and tenacity of the pilots and gunners. In many cases the *Luftwaffe* fighter pilots attacking the formations of Il-2s would expend their complete ammunition load with what appeared to be many direct hits on the Il-2 only to have it continue flying its ground-attack course.

The Americans and British were also aware of the emerging Soviet ground-attack capability. America's famous First World War ace, Eddie Rickenbacker, when shown a demonstration of the Shturmovik's abilities, was recorded as having the view that it was the best aircraft of its type in the world; that his country had never produced anything in the same class of machine; that as the only truly armored aircraft in the world it should form part of the equipment of every army and every air force. None of the Soviet Union's Western allies was equipped with the Il-2. However, in the only recorded occasion in which US and Soviet aircraft flew together on a combat mission, two squadrons of Il-2s from the 951st ShAP (Ground Attack Regiment) rendezvoused with four US P-38 Lightnings on 9 May 1945 over St Polten, Austria and together they strafed and destroyed a German armored column on the road.

The Soviet ground-attack tactics that were in use at the beginning of the war reflected ideas of interwar air theorists such as A. N. Lapchinskiy, who had advocated groups of five or six aircraft employed in a 'V' formation. Lapchinskiy refers to this as the 'crane formation'. The attacker is protected from above by friendly fighters. This was still the VVS tactical plan when the decision was made to produce the Shturmovik as a single-seat aircraft instead of the two-seat version that Ilyushin had proposed. This flawed decision resulted in the massive attrition rates suffered by the VVS ground-attack forces early in the war.

The Soviet fighter pilots of the early war period were generally not up to the task of protecting the ground assault forces. Their tactics were outdated, they lacked training and combat experience. Most importantly, the obsolete aircraft they were flying could not match their *Luftwaffe* adversaries. As a result of these factors, when confronted by German fighters, the VVS fighters either tended to spend more time covering each other rather than the ground assault aircraft that they were there to protect, or worse, they simply headed back toward their own lines leaving the ground-attack aircraft to fend for themselves. The *Luftwaffe* fighter pilots quickly realized this situation and would divide their flights; part going high to attack the fighters and part going low to attack the vulnerable ground-assault aircraft without having to be concerned about Soviet fighters entering the fight.

At the same time, tactical use of the Il-2 had been a subject of intense study within the VVS. In an attempt to protect themselves from German fighters, the Il-2 pilots were beginning to develop their own specific tactics. They would strike ground targets from a so-called free circle with 150–500 m (492–1,640 feet) between aircraft. If German fighters appeared, the formation would begin a series of serpentine turns while remaining in their line formation or would turn into a closed defensive circle in which the leading aircraft was covered by the fire of the weapons of the aircraft behind him.

One critical source of information on the effectiveness of wartime ground-attack air units is the *Sbornik Materialov* reports. These wartime reports on VVS air operations were prepared for the Stavka, and for years remained classified in the Soviet Ministry of Defense archives. Now released, these reports provide an accurate picture of Soviet air operations and, in particular, tactical aviation. Both the strengths and the weaknesses of VVS operations are mirrored in the reports. Most important, the reader can see how the wartime VVS under A. A. Novikov made a systematic effort at adaptation, to assure that technique and technology always possessed a practical application. There was a ruthless embrace of the goal of effectiveness.

For the Il-2, the reports reveal that flight discipline was recognized as an important survival tactic, as important as the potential offensive capabilities of the Il-2. In one example from an October 1942 report from the 6th Guards ShAP (Ground Attack Air Regiment), a flight of three Il-2s that were attacking a railroad station were, in turn, attacked by three German Bf-109 fighters. By maintaining formation discipline the Il-2s were able to down one of the Bf-109 fighters while all the Soviets returned to their base safely. Examples were also cited where formation discipline broke down, and, as a result, the Il-2s became vulnerable to enemy attack. On 25 August 1942 nine Il-2s from the 212th Ground Attack Division (ShAD) were attacked by five Bf-109s. Formation integrity was broken by three of the Il-2s and one was destroyed.

Another aspect of the VVS operation that needed to improve was the command, control, and communication both between the Il-2 units and the ground commanders and between the airborne aircraft. These shortfalls resulted in missed opportunities to stop German troop movements, poor defensive order when attacked by *Luftwaffe* fighters, and in one documented report, recently declassified, caused disorientation in the Shturmovik flights which then struck Soviet troops by accident.

By the fall of 1942 the standard ground-attack formation had shifted to a line formation, with the Il-2s attacking their targets either in flights of two (*para*) or four (*zveno*)at very low altitudes; almost always below 300 m (984 feet). The Shturmoviki quickly demonstrated flexibility in their tactics, along with a tenacity against withering German rifle and machine-gun fire, always a threat at low altitudes. The Il-2s primarily used three basic types of approaches during their attacks. In open terrain and against targets such as vehicles, gun emplacements, or attacking infantry the aircraft would make its attack at altitudes of five to ten meters (15–30 ft) above the ground, releasing bombs and firing machine guns, cannon, and rockets in a nearly horizontal trajectory. Targets such as buildings or hardened sites, such as pill-boxes, would be bombed and strafed using the conventional dive-bombing pass; which, in the case of the Il-2, was a steep 30 to 40-degree dive. The third, and possibly best-known tactic, was the 'circle of death'. The Shturmoviki would cross the front line to one side of the target area, then circle and attack from the rear, in trail formation or line astern, in a shallow dive. The *zveno* would continue to circle, usually with no more than 600 m (1,968 feet) between aircraft, each protecting the tail of the one in front, and repeatedly attack the target until their ammunition was completely expended.

At the Battle of Kursk in 1943 Soviet ground-attack operations achieved a new maturity. Attacking the German panzer divisions in groups of 30–40 or more aircraft, the Il-2s became a formidable weapon against German armored units and artillery. Typically, Il-2s were armed with the new Nudelman-Suranov NS-37, a 37mm anti-tank cannon, and the new PTAB, a small, effective anti-tank bomb. The Il-2s scored results that established the Shturmovik as the premier ground-attack aircraft in history. The Ninth Panzer Division lost 70 tanks in 20 minutes; the Third Panzer Division suffered nearly 2,000 casualties and lost 270 tanks in a continuous attack that lasted two hours; and after an attack that lasted almost four hours, the 17th Panzer Division faced virtual extinction as an effective unit, losing 240 tanks out of a total of 300. Further examples in VVS reports cited the effectiveness of the PTAB. They reveal that on 15 July four Il-2s from the 614 ShAD on the Briansk front attacked 25 enemy tanks, destroying seven and damaging another four. On 16 July, 23 Il-2s from the 610 ShAD in Podmaslovo district destroyed 17 tanks and 40 other vehicles.

The VVS had also realized that under certain conditions the Il-2, especially at low altitudes and primarily in the horizonal plane, had an offensive capability against *Luftwaffe* fighters. In March 1942 a flight of six Guards ShAP Il-2s were attacking enemy troops when they encountered German Ju-52s being escorted by nine Bf-109s. In the ensuing air battle the Il-2s reported downing one Ju-52 with no VVS losses.

With the Il-2s appearing in the ground-attack regiments in greater numbers after 1943, and with the introduction of the two-seat Il-2m, the Shturmoviki also began using another tactic known as '*svobodnaya okhota*' or free hunt. The '*okhotniki*' or 'Free Hunters' usually in a two-aircraft *para* or small four-aircraft *zveno*, but occasionally even as a single aircraft, used free-hunt tactics. Roaming the frontal areas of the battlefields, searching for targets of opportunity, the Il-2s were again extremely effective, especially in bad weather that grounded the German fighters.

Range, when measured in terms of a combat radius, was one of the continuing shortfalls in the Il-2's capabilities and was one of the ground commanders' few complaints regarding their beloved 'Ilyusha'. The Il-2s were normally based 60–70 km (37–43 miles) to the rear of the front line. This was mainly out of respect for the *Luftwaffe*'s talent for surprise counter-air capability. During their combat missions the Il-2s rarely penetrated more than 30–40 km (19–25 miles) beyond the front line. Considering the fact that the primary mission of the Il-2 was direct support of the ground forces in attack corridors, the distance from the front line takes on less importance. But that reduced capability translates directly to time on station – an important factor when supporting ground troops. The

Il-2's missions were rarely longer than 20 minutes in duration. While the Soviet infantrymen and artillerymen loved to see the Shturmovik overhead, they would even more have loved them to loiter longer!

WARTIME AIRCRAFT PRODUCTION AND ATTRITION

Soviet wartime aircraft production capacity for the Il-2 proved to be remarkable, given the modest base with which the Soviets began the war. For example, on 30 September 1941, when the German 'Operation Typhoon' began – the drive on Moscow – the Soviets could only muster five Il-2-equipped Shturmovik regiments to oppose the advance. The rapid advance of the German forces made necessary the evacuation of Soviet aircraft factories along with most of the other industrial centers to safer areas east of the Ural Mountains. In October 1941, Ilyushin's design office was evacuated from Moscow. Production stopped completely during the move of the factories and extremely high attrition rates reduced the available aircraft drastically.

One of the most impressive feats, of the many the Soviets accomplished during the Great Patriotic War, was the speed with which they were able to evacuate their industrial base completely to a position east of the Urals, out of range of German bombers, and begin the manufacturing process again. Production was slow at first and a distinct drop in the quality of the aircraft being produced by the factories was also evident. The engineers in the aviation regiments often found themselves making field level repairs and modifications.

But the time spent during the evacuation was not wasted. Further development studies were carried out by Ilyushin's design bureau with the goal of simplifying and speeding production techniques. In the case of the Shturmovik the manufacturing efficiency of the aircraft's design contributed greatly to this effort because the entire process by which airframe assemblies and parts were produced could be broken down into comparatively small operations which did not require highly skilled assemblers. This made it possible to expand the variety of jobs each could perform. Less than two months after the evacuation, Il-2s were once more reaching the front, albeit in small numbers.

But Il-2s were urgently needed in battle and the new factory assembly lines' production of one aircraft a day could not meet the demand. In December 1941, Stalin sent the famous 'Our Red Army needs Il-2 airplanes like air, like bread' telegram. This has, over time, been described as a Stalin tribute to the Shturmovik. In fact, when the complete text of the telegram is seen, it becomes clear that the message, while a tribute to Ilyushin's design and to the Il-2 as an effective attack aircraft, was really a 'last warning' – typical of Stalin's method of motivation. The telegram, sent to GAZ 18 managers Shenkman and Tretyakov, read:

> You have deceived our country and our Red Army. You still are not producing Il-2s. Our Red Army now needs Il-2 airplanes like air, like bread. Shenkman gives one Il-2 a day, while Tret'yakov gives one MiG-3 every two days. This is a mockery of the country, the Red Army ... Please do not exhaust the government's patience; I demand that more Il-2s be produced, I am warning you for the last time. Stalin.

Stalin's exhortations and the extensive recruitment of unskilled workers using Ilyushin's simplified manufacturing processes began to show results. By the end of 1941, a total of 1,293 Il-2s had been completed. By the end of January 1942, GAZ 18 workers were producing seven a day.

By 1943, Soviet aviation plants had delivered over 11,000 Il-2s to the VVS with peak monthly production exceeding 1,000 aircraft. The early months of 1944 saw the rate of Il-2s coming from the production lines exceed 1.5 per hour – almost 40 aircraft per day. There were now approximately 130 Shturmovik regiments, to which 30 more would be added by the year's end. New pilots and gunners were pouring from specialized Shturmovik training schools.

The high production numbers were required to offset the massive attrition suffered by the Shturmovik regiments. These attrition rates were due initially to several factors. The first was associated with the inherent danger of attacking heavily armed forces at very low altitudes and remained prevalent throughout the war. Added to the equation was the initial lack of protection afforded by the gunner. While this was rectified early, the Germans were very adept at adjusting their attacks and the attrition rates stayed quite high. At this point in the conflict it was not unusual for an Il-2 pilot to survive a half-dozen or more gunners.

The high attrition suffered by the Shturmoviki is attested to by the fact that the Il-2 pilots were awarded the Hero of the Soviet Union award after only ten sorties. This award, the highest the Soviet Union could bestow, normally was awarded for 100 sorties. As an example of the attrition rates suffered by the early Shturmoviki; in April 1942, the 6th GShAP (Guards Ground Attack Aviation Regiment) on the Kalinin Front, possessed no more than three serviceable Il-2s.

With the *Sbornik Materialov* records (see Table 4.1 and Table 4.2) accurate Soviet figures for actual attrition rates have become available, and these confirm the German figures accurately. The German estimates placed Il-2 losses in 1943 at 6,900, and in 1944 at 7,300 aircraft. Despite the huge number of Il-2s produced, at no time were there more than 6,000 in the VVS inventory. Few aircraft logged more than 100 flying hours before battle damage necessitated replacement. One Shturmovik regiment is on record as having written off its entire complement of aircraft twice, as a result of combined combat losses, battle damage, accidents, and other causes, in the short period of two and one-half weeks of intensive operations.

Table 4.1

ATTRITION: BATTLE OF KURSK, 5–18 JULY 1943

Date	Air Battles	Enemy Losses			2nd AA Losses			
		fighters	bombers	total	fighters	Shturmov	bombers	total
5.7	81	71	83	154	36	27	15	78
6.7	64	40	65	105	23	22	0	45
7.7	74	44	78	122	24	13	0	37
8.7	65	54	52	106	24	16	1	41
9.7	62	49	22	71	16	15	1	32
10–14.7	152	112	93	205	N/A	N/A	N/A	N/A
15–18.7	43	45	27	72	49	75	14	138
Total	541	415	420	835	172	168	31	371

Table 4.2

ATTRITION: OFFENSIVE AT RIVER MIUS (NEAR ROSTOV), 18–23 AUGUST 1943

Date	Air Battles	Enemy Losses			VVS Losses			
		fighters	bombers	total	fighters	Shturmov	bombers	total
18.8	19	5	9	14	3	8	1	12
19.8	39	16	6	22	6	8	0	14
20.8	40	21	17	38	14	11	3	28
21.8	27	8	8	16	10	5	1	16
22.8	25	13	7	20	3	3	0	6
23.8	16	5	7	12	2	2	0	4
Total	166	68	54	122	38	37	5	80

Source: The above charts were cited in *Sbornik materialov, po Izucheniyu opyta voiny* (Moscow: Voyenizdat, 1944), pp. 11 and 13.

On the other hand, the Il-2 was an extremely rugged aircraft. It could take a lot of punishment and still bring its crew home safely. A study of Ground Attack Aviation Regiments of the 3rd Air Army revealed that 50 per cent of the Il-2s returned from a mission with some battle damage, but only 2.8 per cent of them were actually lost.

Six per cent of the crippled Il-2s made crash landings, but almost 90 per cent of them could be repaired at the front.

CONCLUSION

The Second World War, or 'The Great Patriotic War' for the Soviets, saw a profound transformation of the Soviet Air Force, which flowed from the wartime commitment to ground-attack aviation, as a vital part of combined-arms operations. To achieve victory over Nazi Germany, the Soviets launched massive offensives which dictated the co-operation of air, ground, and, on occasion, naval forces. At the cutting edge of these air operations, the Soviets deployed the ubiquitous Il-2 Shturmovik, often in air division strength, to destroy enemy tanks and artillery. The Soviet Air Force made effective use of the 'Shturmoviki' as part of a larger air arm committed to tactical operations.

In retrospect, the outcome of the war on the Eastern Front was shaped in a profound way by ground-attack aviation. Sergei Ilyushin's design, the legendary Il-2, became emblematic of the Soviet Air Force. While close air support of ground troops was part of the Order of Battle for all the armies engaged in the conflict, there is no Western counterpart for the Il-2 or the air tactics associated with its wartime history. The designation 'Shturmovik', a generic term meaning ground-attack aircraft, was applied to the Il-2, in much the same way as the term 'Stuka', meaning dive-bomber, was applied to the Junkers 87 aircraft. In addition, the two aircraft also tend to be thought of as having similar missions and as having similar capabilities. Several other equally well known aircraft of the time, with the same ground-attack mission, have also received similar comparisons from some military aviation historians. These include: the Republic P-47 Thunderbolt, the Douglas A-20 Havoc, the North American B-25 Mitchell, and Glenn L. Martin's B-26 Marauder. However, none of these distinguished veterans of the Second World War aerial campaigns had the level of design focus or the ultimate impact on the outcome of the war for their respective national constituencies as did Ilyushin's Il-2. Only in more recent times has the legacy of Ilyushin and his Il-2 found a new incarnation in the American ground-attack aircraft, the A-10.

Ilyushin went on to design many more aircraft, becoming well known for his large commercial and transport designs. But the Il-2 remained the most memorable of all the Ilyushin aircraft. There is no question that the Il-2 Shturmovik, Ilyushin's 'Flying Tank', the Soviet infantryman's 'Ilyusha', and the German artillery and tank force's 'Black Death', earned its legendary reputation fairly and is richly deserving of its place in military aviation history. This simple, austere, but enormously rugged aircraft, probably the best known of all Soviet military machines of its period, and certainly the most effective ground-assault aircraft in history, was largely responsible for the Soviet victory on the Eastern Front. Produced in great numbers, under tremendous duress, by motivated but largely unskilled workers, it was flown by brave and daring pilots and gunners, who were well aware that their life expectancy could very well be measured only in minutes and hours. It is superbly fitting that the surrender ultimatum which was delivered to the German Army Group South Ukraine was dropped by an Il-2. It was accepted.

Looking back, it is remarkable that the Soviet Air Force triumphed over the *Luftwaffe*. Few in 1941 would have predicted such a victory. The wartime emphasis on ground-attack aviation transformed Soviet air doctrine, even as it strengthened the traditional ties of the air forces to the army. The great irony arose in 1945, at the time of supreme victory: the Soviet Air Force with all of its tactical air might was ill-equipped to fight the next war – one that would call for a strategic air force with long-range bombers. Still, one cannot understand the larger history of Soviet air power without reference to the years 1941–45 when tactical aviation prevailed.

Russian and Soviet Naval Aviation, 1908–96

CHRISTOPHER C. LOVETT

INTRODUCTION

When the public envisions naval air power, they rightly see aircraft carriers on the horizon, both commanding the air above the waves and projecting power from distant shores. Although this model has served Western naval powers well for most of the twentieth century and reflects the prime tenets of Alfred Thayer Mahan, the Russian experience requires more thought.

Unlike the United States, Great Britain, and Japan, the Russians have been imprisoned by their geography. Russia is virtually land-locked and its access to the sea is limited to the Pacific and the far North, with the exception of critical choke points in the Baltic and Black Seas. The political leadership in Russia, realizing the nature of the threat to Russia's security and their own industrial limitations, have opted to spend the state's precious economic resources on the Army, making the Navy, for the most part, handmaiden to its sister service.

Russian naval theorists often sought to convince their own political leaders of the necessity to alter strategic plans, especially in regard to the role of the Navy in defense matters. Russia's political masters saw the Navy as secondary to the Army. This rather narrow view of the Navy, combined with the inability to modernize, led Russia to disaster beginning with the Crimean War. As a consequence, the Navy was downsized, and by the eve of the war with Japan in 1904, Russia had lost her place as a major naval power in the eyes of the world. The combination of downsizing and shortsightedness ultimately led Russia to repeated tragedy, when the Navy was expected to protect Russian national interests in the Far East. Once again, the Russian Navy suffered humiliation at the hands of a more technologically advanced foe.

But this was not solely a problem for Imperial Russia. During the Soviet era, the Soviet naval air force became a prisoner to the state's technological failings. Still, the Russians and Soviets managed to make do. They found a way to compensate for their technological weaknesses. Continuity and change has linked the history of naval aviation in both the Russian and the Soviet Union eras; continuity of maintaining the state's interests, while reflecting the latest technological trends in aviation in order to assert those objectives. But as Imperial Russia found the path to modernization treacherous, so did the Soviets in attempting to preserve the Soviet Union's security in the atomic age.

RUSSIAN NAVAL AIR DEVELOPMENTS BEFORE THE
FIRST WORLD WAR

The birth of Russian naval aviation took place in a period of revolutionary technological innovation in Europe and the United States. The Russians were motivated by the same factors that drove the other naval powers to seek an air component for their naval forces. Unfortunately, the Russians lacked the industrial base to provide the machines required to fill that need, forcing St Petersburg to seek foreign concerns to supply the required aircraft until Russian industry could fill that void.

The Russo-Japanese War (1904–05) demonstrated the need for a reconnaissance element assigned to Russian surface and ground forces. If the Russian Army, for instance, had placed observation balloons on the Korean front, unauthorized troop withdrawals might not have occurred, with the loss of critical weaponry from the battlefield. The Navy confronted a similar dilemma.

As the Baltic Fleet sailed to the Far East to restore Russian strength in the region, lookouts believed that they had sighted Japanese balloons on the horizon in October 1904. The Russians were so concerned they altered their refueling schedule in Danish waters. Unfortunately, this was not the only case in which the Russians panicked after inaccurate sightings of Japanese forces. By not furnishing the Baltic Fleet with a credible reconnaissance capacity, the Russians made a similar blunder near Madagascar, when lookouts thought that they had spotted Japanese torpedo boats shadowing the fleet. As a result of not having a credible aerial observation capability, the Russians played into the strength of Admiral Togo, the Japanese fleet commander, at Tsushima on 27 May 1905.

Before the disaster at Tsushima, the Russians experimented with a converted liner, renamed the *Rus'*, purchased from the north German Lloyd Line, as a gift to the Russian government from Count

Sturganov. The goal was to modify her so that she could handle a complement of four kite-balloons or a combination of balloons and kites in order to gather intelligence at sea. Unfortunately, the vessel was not ready for deployment when the Baltic Fleet sailed to the Far East.

After the Russo-Japanese and the Balkan Wars, the Naval Ministry believed that aircraft could play an important role in the future. So in 1907, the Russian Navy, like those of the other naval powers, embarked on the creation of a naval air arm. In the years immediately prior to the First World War, the Russians began primitive combined-arms operations with fixed-wing aircraft and surface forces in the Black Sea. But due to the lack of an indigenous aircraft industry, aircraft had to be purchased abroad in order to meet the Navy's requirements.

Even when the decision was made to invest in fixed-wing aviation, elements within the Navy still sought to utilize lighter-than-airships rather than propeller-driven aircraft. The opponents of propeller-driven machines believed that the current state of aviation development could not provide a machine with a sustained flying time equal to that of lighter-than-airships, despite the cost differential, as Britain's Royal Navy was showing. In their view, the airplane was the weapon of the future.

The Naval Ministry was aware of aviation developments abroad. The success of Eugene Ely in 1911, for instance, in flying an aircraft from the deck of the USS *Pennsylvania* impressed the Russians. Still, Russian officers believed that Ely's use of a catapult was too cumbersome for the Russian Navy and perhaps impractical at that time. The Naval Ministry decided to send a representative abroad to evaluate aircraft models, especially seaplanes, in France and Britain and to discuss possible purchases. After evaluating a Donnet–Lévesque model constructed by Franco-British Aviation, the Naval Ministry placed an order for 12 of these seaplanes.

The Russians ordered American models from the Glenn Curtiss Company before the First World War. Yet despite the purchase of foreign aircraft, the total never reached more than 50 planes. That number was favorable in relation to the naval inventories in Britain and the United States. But the Russian Navy lagged well behind the Russian Army, which had 250 planes at the time. Unfortunately, war broke out before all the Russian naval orders could be filled.

RUSSIAN NAVAL AVIATION DURING THE FIRST
WORLD WAR

Following the commencement of hostilities with the Central Powers, the Russian naval air arm was given the dual mission of providing aerial reconnaissance along the coastal sectors and sea lanes for the navy. In order to implement those assignments, each fleet was allotted an air division comprised of two air brigades. Each brigade had two squadrons of nine flights. To support their air assets, each brigade had three air stations and six airfields to handle the fleet's aircraft. The naval air stations, often located at key sites along the coastal frontier, allowed the Russian naval command the flexibility to shift aircraft or units from one sector to another. Likewise, the Baltic and Black Sea fleets had a seaplane tender to repair each fleet's seaplanes.

The Naval Ministry did not authorize the design of a viable seaplane tender for the Russian Navy before the outbreak of the war. As a consequence, the Black Sea fleet had to make do with a series of conversions, which were called 'hydrocruisers'. The Black Sea fleet received two – the *Imperator Nikolai* and the *Alexandr I*. The armaments for each vessel included 6 x 12 mm and 4 x 75 mm guns. In 1916, Admiral A. V. Kolchak created a strike force composed of three pre-Dreadnought-class battleships and the two seaplane tenders. Kolchak used this formation to conduct an early aerial interdiction campaign along the Turkish coast. The Russians sank a 7,000-ton Turkish transport and disrupted enemy communications in the process.

Much of the credit for the success of the Russian Navy in the Baltic resulted from the pioneering efforts of Admiral N. O. Essen, who fostered the integration of aviation into combined-arms operations in his command until his untimely death in 1915. Often outgunned by a superior enemy, Essen registered a number of triumphs during the war, most notably an intelligence coup when he captured the German naval codes from the grounded German cruiser *Magdeburg* in 1914. Reconnaissance and aerial interdiction remained the primary duties for Baltic Fleet's aviators during the war.

The most singular achievement of Russian aviation during the old regime remained the design and development of Igor Sikorsky's *Il'ya Muromets*. Earlier, the Navy had conducted test flights with aircraft carrying an assortment of aerial armaments, including rockets, 82mm recoilless and 37 mm cannon. Alexander P. de Seversky, in later years a primary champion of air power, tested Sikorsky's designs. With the addition of the four-engine *Il'ya Muromets* to the inventory, the naval air arm's combat power increased dramatically. Yet the government's inability to supply spare parts and aircraft limited

Russian naval aviation's combat effectiveness as the Revolution loomed on the horizon.

THE CIVIL WAR

It is generally acknowledged that the mounting losses at the front and the collapsing Russian economy in the rear hastened the Russian Revolution. As the czarist regime staggered from crisis to crisis in 1916–17, tensions increased between officers and men, often along class lines. In the Russian Navy the division was between officers and rankings; in naval air units the clash occurred between pilots and mechanics. From the Bolshevik seizure of power in October 1917 to the signing of the Treaty of Brest-Litovsk in 1918, the Bolsheviks sought to establish a fresh air force, but the new regime first had to find pilots. To locate aviators, the Bolsheviks required all pilots to register with the new government.

Those naval aviators who served the new regime fought admirably for the Bolsheviks. Naval pilots flew seaplanes in support of the Red Army against General Deniken and Admiral Kolchak. In those engagements, naval pilots supplied air cover, reconnaissance, and conducted air strikes on suspected White positions for the Red Army.

The crowning achievement for naval aviators came with the capture of Kazan on 9 September 1918, when the naval air arm provided the only air cover the Reds expected during the battle. Two pilots as identified from former Soviet sources, S. Stoliarskii and I. A. Svinarev, logged 40–43 flying hours during the battle, and hit enemy gun emplacements from altitudes of 300–500 meters.

In the battle to retake Tsaritsyn in the summer of 1919, the Bolsheviks sent additional reinforcements to the front, and Lenin personally ordered the consolidation of the Volga–Caspian and Astrakhan–Caspian Flotillas, with additional forces sent from the Baltic. During the Tsaritsyn counteroffensive, the Bolsheviks conducted night air operations with all of the ensuing dangers of distinguishing the landing site and the water's surface. To increase those difficulties, the aircraft did not have luminous instrument panels at the time.

In the Civil War, naval aviation supported the new regime by furnishing the Red Army with effective close air support, including the North Dvina–Archangelsk sector as well as assisting in solidifying the Bolshevik position at Samara in August 1918, following the Czech rebellion. Likewise, Red naval aviators played an integral role in the air defense of Petrograd and conducted offensive operations against the British naval and land targets in that sector on 20 June 1919. Yet despite the naval air arms achievements in the Civil War, its fate had been sealed with the unification of army and naval air arms into the Red Air Force in 1920.

KRONSTADT, NEP, AND THE REBIRTH OF THE NAVY

After the Civil War, many officials raised questions involving naval expenditures in light of the state's financial wellbeing. Despite those interests, the Navy's surface forces were refurbished between 1921 and 1928. For example, the Baltic Fleet increased its strength from two battleships, one cruiser, eight destroyers, and nine submarines in 1924, to three battleships, five cruisers, 24 destroyers, and 18 submarines by 1928, the last phase of naval construction before the start of the first Five-Year Plan.

The strains of the Civil War and war Communism greatly affected the peasants who filled the ranks of the Red Navy. During the early days of the Revolution, Baltic Fleet sailors served as the shock troops, but the replacements saw the pain and suffering in their villages firsthand and began to sympathize with the workers in Petrograd. And going further, the sailors at Kronstadt made 15 political demands, that threatened the very fabric of the Bolshevik dictatorship. Lenin could not allow this to continue, and Trotsky's Red Army crushed the rebels. Lenin wanted not only to scuttle the Baltic Fleet, but to break up the Navy as well. Only Trotsky disagreed, and he kept Lenin from following his instincts.

THE 1920S

After Kronstadt, and the decision to maintain the Navy, a debate arose over its composition and mission. During the pre-1914 period, the Mahanian approach, which emphasized the command of the sea, dominated Russian naval thought. During the early 1920s, Professors B. Gervais and M. Petrov with the faculties at the Frunze Naval Academy and the Voroshilov War College advocated a continuation of this policy, modifying the naval mission according to the development of improved weapon systems. This view, better known as the 'Old School', continued the predominance of battleships and cruisers in the Soviet Navy, but overlooked the importance of both aircraft and submarines as viable options.

But by the late 1920s, others in the Navy actively opposed the 'Old School' approach. The opponents emphasized that the submarine had supplanted the battleship as the principal strike weapon of a modern

fleet. Going further they argued that both modern aircraft and submarines made traditional naval blockades obsolete against contemporary naval forces. The 'Young School', as they were called,
advanced the view that a balanced fleet could be achieved by combining light surface forces with submarines and aircraft. The major
naval powers did not co-operate with the theorists of the 'Young
School', since they did not reduce their naval forces to a level comparable to that of the Soviets. Still, the 'Young School' believed that
the Civil War combined with Communist Party ideology provided the
basis for all future military planning. Ultimately, Stalin's economic
plans, as reflected by the First Five-Year Plan, settled the struggle
between the two schools.

During the early 1920s, the Soviets divided the Red Air Force into a
land and naval component. As late as 1 October 1923, 36 aircraft were
designated for naval use and were utilized according to their particular
mission. Earlier, the Soviets had organized a Black Sea Naval Aviation
Command in the spring at Nikolaev, but moved the facility to
Sevastopol. The 'Air Fleet of the Two Southern Seas' had 20 aircraft
at their disposal, most of which were of obsolete foreign design. Yet
naval aviators received the assignment to maintain the security of
Soviet air space as well as to maintain their efficiency in the air.

The modern Soviet aviation industry emerged from the crisis
atmosphere of the 1920s. As a result, the Soviets managed to replace
their obsolete foreign models with up-to-date aircraft constructed in
the Soviet Union. Much of the credit for the revival of Soviet aviation
fortunes went to A. N. Tupolev, who had pioneered in 1923 an all-
metal aircraft prototype, the R-1, a reconnaissance biplane similar to
the De Havilland. As the R-3 it became the workhorse of the Red Air
Force for the rest of the decade.

IN THE SHADOW OF WAR

The 1930s were years of innovation for the Red Army and Navy.
Under the direction of M. N. Tukhachevskii, A. A. Svechin, and
V. K. Triandifilov, the Red Army developed a doctrine encompassing
shock, mobility, and firepower in order to destroy an adversary in
depth. The major components of this new doctrine included armor,
mechanized infantry, and artillery, as well as air power, the latter
becoming an integral component within Soviet tactical doctrine on
land and at sea.

The 'Young School' dominated Soviet naval thought and relied
upon a combined-arms approach to naval operations by emphasizing
surface forces, submarines, and aviation in order to challenge potential

naval adversaries. The Naval Regulations of 1937 directed the Navy to support the Red Army in defending the coastal zones as well as conducting separate operations at sea. The adoption of the Naval Regulations came at a time when the Navy, as well as the Army, suffered the same fate as did much of the Soviet Union during the Stalin years. During 1937 much of the Young School had been purged, which set the stage for Admiral V. M. Orlov's own execution on about 28 July 1938. The purge reached the staffs of the Baltic, Black Sea, Northern, and Pacific fleets, and the Caspian flotilla. All of the fleet commanders, with the exception of the commander of the Pacific fleet, were executed during 1938.

The purges not only marked a break with the 'Young School', but also a return to traditional naval autonomy from the War Commissariat on 30 December 1937. Even Stalin's loyal henchman, V. M. Molotov, told the Supreme Soviet in January 1938 that the 'Soviet State must have a sea-going and ocean-going fleet, consistent with its interest worthy of our great task.' With its independence from the Army, the Navy received additional deliveries of surface combatants; more important, the purges forced the Navy to accept direction from above, when Stalin was planning a shift in Soviet foreign policy.

From 1929 through 1939, the Soviet Union faced national security threats in both Asia and Europe. Even with Hitler's rise in January 1933 and the Civil War in Spain, the most profound menace to Moscow came from Imperial Japan. Naval aviation's role was to protect Soviet air space and territory from the Japanese. Both air force and naval pilots flew missions during the border skirmishes with Japanese forces at Changkufeng in August 1938 and, a year later, at Khalkin Gol in August 1939, which administered a decisive defeat upon the proud Japanese Kwantung Army.

From the mid-1930s to the outbreak of the Second World War, the Soviets had witnessed the German, Italian, and Japanese application of air power in Spain, Ethiopia, and China. Yet the German mastery of combined-arms operations in Poland, and later in France, demonstrated the advances the *Luftwaffe* and *Wehrmacht* had made in doctrinal development in the tactical application of air power. Those lessons were not lost on the Soviets, who applied their own air doctrine against the Finns during the Winter War of 1939–40.

Before the war began, the Soviets had a clear numerical superiority over the Finns, who could muster 162 front-line aircraft; a few were modern models, such as Brewster Buffaloes, Fokker D-XXIs, and Bristol Blenheims. In time, the Soviets gained air superiority and naval aviators supported landing operations by conducting close air-ground support and reconnaissance missions for the Navy and Army, but at a heavy price in pilots and aircraft.

Weather, as well as the fierce resistance of the Finns, placed considerable strain on both pilots and machines during the Winter War. The Navy believed that the faulty performance of Soviet pilots reflected insufficient training; however, the problems went deeper than that. Many of the aircraft in the Soviet inventory were obsolete by the time they came to the Baltic fleet. Likewise, communications and limited load capacities further restricted Soviet naval aviation's record in the war.

The Winter War demonstrated to the Soviet Navy that Baltic Fleet airfields had to be improved and become all-weather in order to meet the Navy's mission of defending the coastal frontier and supporting the Army. Likewise, the Soviets concluded that aircraft designs had to be upgraded, especially in regard to armaments and communications, to compete with other potential adversaries. Only the praxis of war would tell whether the Soviets were successful in their efforts.

SOVIET NAVAL AVIATION DURING THE GREAT PATRIOTIC WAR

During the Second World War, 1941–45, Soviet naval aviation demonstrated its importance by providing air cover for the Navy's surface forces as well as vital air support for the Red Army in the life-and-death struggle against the *Wehrmacht*. By utilizing their interior lines of communication, the Soviet command sent naval aviation units from one front to another in order to support ground operations and frontal aviation in the Baltic and as far south as the Ukraine.

When the Germans attacked the Soviet Union on 22 June 1941, the Soviet naval air arm had a total of 3,020 aircraft in its inventory. This included all aircraft in the Baltic, White, Black, and Pacific Fleets on the eve of the German invasion. But only 1,441 aircraft, including fighters and bombers, were in service for the Baltic, Northern, and Black Sea Fleets by June 21. Another 148 aircraft were in repair, storage, or were in use for training and were not available on the first day of the war. In comparison, the Soviet Air Force (VVS) numbered 5,700 aircraft of various types in the European Military Districts on the eve of the Russo-German War.

Many of those aircraft were obsolete and no match for the *Luftwaffe*. For instance, the Baltic, Northern, and Black Sea Fleets had only 73 modern fighter aircraft among them, including MiG-1s, MiG-2s, MiG-3s, YaK-1s, and LaGG-3s, on 22 June, but 361 biplane fighters. Clearly Soviet naval aviators were at a disadvantage when they challenged the Germans for command of the skies. Although many of the claims of Soviet naval aviators during the war were questionable, no one can doubt their *élan*.

Soviet naval aviation conducted offensive air operations at sea, defended naval bases, co-operated with the army in the coastal zones, and secured friendly sea lanes while disrupting the enemy's naval communications. As the Germans pushed the Soviets closer to Leningrad in the summer of 1941, the 1st Mine-Torpedo Regiment conducted a successful air strike on Berlin on 7– 8 August 1941, an event not readily recognized in the West.

In the Baltic, the naval air arm provided not only air defense for Leningrad, but also kept the ice road across Lake Ladoga open during the siege. During 1943 and 1944, Baltic Fleet pilots worked with Soviet artillery units in suppressing German artillery fire during the waning days of the German siege. In order to reduce German artillery effectiveness, additional *Shturmovik* and bomber forces were transferred from the Black Sea fleet. The Baltic fleet air units were so successful that the Soviets regained mastery of the air, which allowed the Soviets to redirect naval pilots' attention to German transports in the Gulf of Finland.

As the fortunes of war shifted and the Germans were in retreat along the length and breadth of the Eastern Front, Baltic Fleet aviation was ordered to disrupt German evacuations from east Prussian ports. Mine-torpedo aviation units, assisted by Soviet fighters, broke up German naval movements from Pillau, Danzig, and Konigsberg. By early spring 1945, Baltic Fleet pilots roamed at will and met little opposition from the *Luftwaffe*, interdicting German naval traffic as well as laying mines along the eastern coast of the Thousand Year Reich.

Before the onset of the Second World War, the principal threat for the Black Sea fleet had come not from the Rumanians, who were a minor menace for the Soviets, but from the *Luftwaffe*'s 4th Air Fleet and their 450 front-line aircraft. In addition to the 350 planes of the Rumanians, the Axis managed to deploy 750 aircraft in the southern theater against the 684 aged aircraft of the Black Sea fleet's 62nd Fighter and 63rd Bomber Brigades, the 119th Aviation Regiment, the ten aviation squadrons, two aviation flights, and training units assigned to the fleet.

As Army Group South advanced farther into the Ukraine, Soviet HQ, *Stavka* authorized Black Sea fleet aviation and the 4th Air Corps to conduct air strikes on the Rumanian oil fields at Ploesti, beginning on 2 July and continuing until 18 August 1941. When the naval base at Odessa was no longer tenable as the Germans approached the city, Black Sea fleet aviators provided air cover for the withdrawal of Soviet surface forces. But as the battle for Odessa came to an end, the battle for Sevastopol was about to begin.

According to prewar planning, Sevastopol was defended by coastal artillery, anti-aircraft artillery, and units of the Black Sea fleet naval aviation. The main weakness of the Soviet defense of the fleet anchorage was the land approach to the base, something not lost on the Germans. Naval pilots sought to check the *Wehrmacht*'s advance by flying over 2,217 sorties against the Germans. But fleet pilots could not halt the German onslaught, and on 31 October 1941, the Black Sea fleet was ordered to evacuate the base for the Caucasus, with naval air units providing cover for the withdrawal. Still the Germans could not storm the city, and naval air units continued to challenge the *Luftwaffe* over Sevastapol through 2 July 1942, when the Germans finally captured the port.

Black Sea pilots assisted the Red Army and frontal aviation in the North Caucasus, at Stalingrad, and the counteroffensive that followed in 1943. As the tide of war shifted in favor of the Soviets, Black Sea Fleet aviation attacked German naval communications in the eastern Black Sea and harassed the German withdrawal from the Crimea. Torpedo runs were the favored method for fleet aviation in the Black Sea to sever German naval communications with Rumania.

The pilots of the Black Sea fleet maintained a remarkable record during the war, as did the naval aviators in the Baltic. From the very start of hostilities, fleet aviators challenged the *Luftwaffe* over Odessa, Sevastopol, and the various fields of battle on land or at sea. When all else failed, Soviet naval aviators rammed German aircraft, with Black Sea fleet pilots accounting for over half of all ramming incidents in the war. In comparison with the other Soviet naval air formations, the Black Sea pilots were the most effective during the conflict.

As a consequence of various wartime agreements with his allies, Stalin committed the Soviet Union to the war against Japan 90 days following V-E day, utilizing the lessons learned in the war with Germany to overwhelm the Japanese in August 1945. In order to commence hostilities with Japan, *Stavka* had reinforced Soviet ground and air forces in the Far East, including the subordinate air units of the Pacific fleet. The Navy's objective in the Pacific war was to support the Red Army as the weight of Soviet armored and mechanized units breached Japanese defenses in Manchuria, as well as to provide air cover for Soviet landing operations. All told, the Soviets massed over 5,000 aircraft for this purpose.

Unlike in the war with Germany, when naval pilots began the war with mostly obsolete aircraft, the naval air units received the best the Soviets had to offer as well as what Lend-Lease could provide. Those models included Pe-2s, Il-4s, Yak-9s, La-7s, Tu-2s, P-39 Aircobras,

War, but also altered the growth of Soviet naval aviation, which stood at approximately 4,000 planes, over half being fighters. As N. S. Khrushchev consolidated his power over the party, he reversed much of Stalin's influence in the military as well. As Stalin admired major weapon systems, Khrushchev looked upon much of the navy as obsolete, especially in an age of jets and missiles. According to Khrushchev, Admiral Kuznetsov was not the man to command the navy during this period of reorganization, which could be interpreted as a return to a 'neo Young School' policy. But who was to lead the Navy?

THE AGE OF GORSHKOV

In 1955, after Admiral Kuznetsov was replaced, Khrushchev thought that Admiral S. G. Gorshkov was a man who would follow orders and would not challenge political leadership. Khrushchev believed that as a result of the military-scientific-technological revolution, the age of the battleship and traditional command of the sea was over. The Soviets believed that the aircraft-carrier also, as a consequence of the new weapons systems, had lost its status as the premier vessel of power projection. So during this early phase of Gorshkov's tenure as commander-in-chief, Soviet naval aviation relied principally upon land-based aircraft and submarines to respond to American naval preponderance.

Despite Khrushchev's views, Gorshkov maintained that naval aviation was even more important in the postwar era than it had been during the Second World War, even after the transfer of naval aviation's 1,500–2,000 fighters to the National Air Defense Units (PVO). Following the Cuban missile crisis in 1962, and the introduction by the United States of the Polaris SLBM (submarine-launched ballistic missile) system, the Soviets re-examined their position on aircraft carriers. With Khrushchev's fall and replacement by L. I. Brezhnev in 1964, a new era for the Soviet Navy and naval aviation was initiated, which involved a balanced fleet, including new and improved surface vessels, naval infantry, and an improved naval air arm.

Likewise, the Soviets during the Brezhnev era took an active role in the Vietnam conflict by sending over 985 pilots and additional aviation support personnel to North Vietnam from 1965 through 1972. These men were listed as 'volunteers', as happened in previous Soviet adventures abroad in Spain and Korea. Many of the pilots, like their American counterparts, were classified as 'advisers', but often sortied against American aircraft conducting air strikes over North Vietnam. With the collapse of South Vietnam in April 1975, Hanoi rewarded

Moscow with port and airfield rights at Cam Ranh Bay, the former American base.

After watching the ease with which the United States projected its might during the Middle East crisis in 1958, the Cuban crisis in 1962, and in Vietnam after 1964, the Kremlin looked for a means to protect the nation from the threat posed by the American Polaris submarine system. Initially, the Soviets made the decision to embark upon the 17,500-ton *Moskva*-class helicopter antisubmarine cruiser program, which the Soviets limited to only two vessels, since the *Moskva*'s limited helicopter capability made it impossible to keep pace with the new American *Poseidon*-class SSBNs (ballistic missile submarines). As the Soviets were about to embark upon the development of a sea-based air capability in 1965, the naval air arm numbered 800 land-based aircraft, 400 bombers and 400 transports and auxiliary planes, while the Air Force's strength at the time was 10,500 aircraft of all types.

Another class of aircraft carriers was required, and as in all naval construction, the Kremlin overlooked the cost, as did many in the West in analyzing Soviet military capabilities. The decision to embark upon the *Kiev*-class ASW cruisers was a logical progression in the Soviet naval scheme in the mid-1960s, followed by the decision to construct a true aircraft-carrier (CTOL) later in the 1980s. Yet the *Kiev* was not only capable of conducting anti-submarine missions, but also because of the complement of VTOL aircraft, capable of anti-antisubmarine tasks. As a result, the Yak 36/38 Forger was able to negate the P-3 Orion and other ASW aircraft in NATO's inventory.

The *Kiev* and her sisters permitted naval aviation to provide PVO (air defense) at sea and during amphibious operations, as well as supporting the Navy's traditional sea denial mission. But this was only possible when American carriers were not on station. The Soviets perceived that the most serious threat to their SSBN bastions came from attack submarines and aircraft-carriers conducting ASW missions that could overwhelm Soviet defenses. The question as to whether to build a new generation of aircraft-carriers, or to rely upon improved submarines continued to play in the pages of *Morskoi sbornik* (Naval Review) in 1979 and afterward. According to their chief proponent, Vice Admiral K. A. Stalbo, carriers afforded the Kremlin the ability to project Soviet power as well as to provide presence and suasion in an age of carrier diplomacy.

Building a carrier was only one obstacle the Soviets confronted in their quest for a true sea-based air capability. Another impediment involved the training and retraining of pilots to fly VTOL aircraft from the deck of a warship. The Soviets, in their efforts to meet that challenge, developed a training syllabus to meet their specific needs. Yet the training program only dramatized the inherent weakness of the

Soviet system which penalized individual initiative and rewarded the Party's collective objective. Still, after constructing three classes of carriers, the Soviets had to evade the restrictions placed upon them by the 1936 Montreux Convention, which prohibited the passage of aircraft-carriers through the Dardanelles. The Soviets managed that feat by classifying the *Kiev*-class carriers as aircraft-carrying cruisers and, later, the *Kuznetsov* as a heavy-aircraft-carrying cruiser.

During the ensuing debate among the naval hierarchy, one of the most vociferous critics of Stalbo was V. Chernavin, who had commanded the Northern fleet. In 1981, Chernavin was elevated to the post of the Navy's chief-of-staff, which indicated a changing of the guard. Chernavin argued that Stalbo and his allies neglected a consolidated naval doctrine and failed to appreciate the significance of naval concentration and combined naval operations. Gorshkov won the battle to build a real carrier, yet the debate highlighted Gorshkov's vulnerability.

GORBACHEV, THE COLLAPSE OF THE SOVIET STATE, AND AFTER

With the coming of M. S. Gorbachev and the internal acknowledgement of the weakness of the Soviet economy, an increase in the Navy was an extravagance the Kremlin could no longer afford. Shortly after the launching of the new 67,500-ton carrier *Kuznetsov* on 17 April 1982, Gorshkov retired. As a reward for his past service, the ex-*Baku*, a modified *Kiev*-class carrier, has been renamed the *Admiral Gorshkov* in his honor. Gorbachev, after this notable achievement, sought a replacement for his senior admiral, one who was younger and who would follow his line in redefining Soviet security needs. Chernavin was Gorbachev's logical choice.

From 1985 through 1991, the naval air arm was downsized, which supported Gorbachev's new defensive doctrine. The Navy's new mission in case of war was to interdict NATO's sea lines of communications (SLOCs) rather than to conduct blue-water fleet missions that Gorshkov had designed. As a consequence, the Kremlin reduced the Navy's combat aircraft inventory from 875 aircraft in 1985 to 750 by 1991. With decreased expenditures the Navy's logistical capability was unable to sustain fleet operations at sea as well. Western analysts reported that out-of-area fleet missions were curtailed between 1986–89 by approximately 6 per cent. US Naval Intelligence concurred by noting that the Soviets reduced funding for the Navy's operational tempo. Later, the Soviets began to close their foreign bases, particularly in Vietnam, where Soviet Pacific fleet air units were stationed.

As the Soviet economy was dissolving from inefficiency, the Navy selected Su-27 Flankers and MiG-29 Fulcrums for the *Kuznetsov* air component. Using satellite reconnaissance assets, Western intelligence agencies discovered that the Soviets were conducting demonstrations with Su-25 Frogfoots at a dummy airfield in the Crimea. Likewise, Western defense experts noticed further modifications on the MiG-29 and Su-27 models at the Paris Air Show as early as 1989. Those experts surmised that both models were to be assigned to the new carrier's air complement.

Many in the West looked upon this development with apprehension as a new round of Soviet–American rivalry, but by then, the die was cast. Gorbachev's domestic programs and determination to reduce friction between the United States and the Soviet Union made that unlikely. Gorbachev's efforts at a new openness and restructuring of the Soviet economy did not have the desired effect. Instead of revitalizing the regime, Gorbachev's economic and political reforms only hastened the demise of Soviet Communism, and with it, the dangers posed by Soviet naval aviation.

With the break-up of the Soviet Union, the destiny of the Black Sea fleet was at stake, since the Ukraine declared its intention to form its own navy and appointed Rear Admiral Boris Kozhin, the former commander of the Sevastopol Naval Base, as the Commander of the Ukrainian Navy. The *Kuznetsov*, in order to avoid seizure by the Ukrainian government, managed to escape and joined the Northern fleet in December 1991. By April 1992, Moldova, Georgia, and Azerbaijan all laid claim to naval assets in their jurisdiction. Moldova even went so far as to seize a naval aviation regiment in order to establish a national air force. Likewise, the *Minsk*, and *Leningrad* were placed in reserve in 1992. The *Novorossiysk* and the *Kiev* followed suit in 1993 and 1994. *Jane's* has reported that while those vessels were listed as being 'in reserve', it is highly unlikely that those vessels will ever go to sea again and a timely 'scrapping policy continues with the aim of having 65 per cent of operational ships less than 20 years old.'

The unfinished *Varyag*, an aircraft-carrier (CTOL), rusted at the Nikolaev shipyard while Moscow and Kiev negotiated her fate. The construction of the *Ulyanovsk*, the first true Russian CVN with a steam catapult to be designated as an 'aircraft-carrier', was ordered stopped in February 1992 and sold for scrap despite the protests from the Russian naval establishment. The Yeltsin government, in order to raise much-needed revenue, had embarked upon an ambitious foreign weapon sales program and was negotiating with the Indian government for the sale of the *Varyag*, *Gorshkov*, and the naval fighter, the YAK-141, in 1994 and 1995. As Russia slipped further into economic chaos, naval aviation was reduced proportionally from five carriers,

1,354 aircraft, and 312 helicopters in 1991, to one carrier, 396 aircraft, and 250 helicopters by 1996. Only time will tell whether the Red Phoenix will ever again challenge the West over the seven seas.

How will the Russian Navy respond to those challenges? Will the Russian economy permit the Navy to embark on another construction program comparable to that of Admiral Gorshkov's plan? If Clio, the muse of history, can serve as a guide for future Russian naval developments, I may be so bold as to assume that the Russians will never again contest the West on the same level or with the same intensity as during the Soviet era. The Communist hierarchy never understood that Soviet power was not limitless, but then again, many in the West have yet to understand that lesson as well.

The Aviation Industry, 1917–97*

JOHN T. GREENWOOD

The dissolution of the Union of Soviet Socialist Republics (USSR) in December 1991 led directly to the dismembering and subsequent and continuing restructuring of the former Soviet aviation industry. Although the industry retains much of its former framework built around specific designers, design bureaux, and production plants, very significant changes have taken place that have completely altered the nature of the former aviation industry and have threatened its very existence.

The Soviet aircraft industry differed significantly from that of other countries. Not only was it exclusively state-owned under a national-level ministry, but also for most of the Soviet era it was divided into a number of largely independent but closely interrelated sectors – the state ministerial administration; the *Tsentral'nyi aero-gidrodinamicheskyi institut* (Central State Aero-Hydrodynamic Institute, TsAGI) and the scientific and technical research institutes that were closely linked to the state testing facilities (covered later in this chapter); the experimental design bureaux (called *opytniye konstruktorskiye byura*, OKBs) (covered in a separate chapter); and the aircraft, engine, and components manufacturing plants. In many respects this was an ideal situation in that designers were freed from many pressures, notably financial, and could concentrate on design and development. As aircraft became more complex from the early 1930s on, the bureaux increasingly specialized on one general type of aircraft. Never lacking for continuing large domestic military and civilian requirements for new aircraft, the industry did not have to compete or even seek any significant export sales to sustain itself until the 1960s. The sale or bartering of military and civilian aircraft to Warsaw Pact, client, or Third World states was more a facet of the Soviet Union's

* This chapter is based on 'Patterns in the Aircraft Industry', which appeared in *Soviet Aviation and Air Power* (1977). The author wishes to thank Dr Von Hardesty of the National Air and Space Museum and Mr Aleksei Druzhilov, a graduate of the Moscow Aviation Institute and currently with the World Bank, for their very extensive and helpful comments.

global foreign and military policies during the Cold War than a result of any economic necessity to sustain the aircraft industry.

Since 1992, the aviation industry has undergone major changes as the Russian economy is restructured from centrally planned state ownership into a competitive market economy. Moreover, the country's economic situation has resulted in vastly reduced funding and priorities for the Russian armed forces, all but eliminating for now one of the aviation industry's most critical customers.

Some of the major design and production organizations are now economic enterprises of the independent successor states of the Commonwealth of Independent States (CIS). No longer can these enterprises count on the benefits of state ownership and large, guaranteed state contracts to support research, development, and production of their new aircraft designs. Now they must compete with the other members of the former Soviet aircraft industry for the few new orders for military and civilian aircraft that are forthcoming in the CIS. As a result, most of these new enterprises have had to enter the fiercely competitive world aircraft market to sell their products to support their continued existence. The change from the cozy, well-structured, and fully nourished existence of the Soviet aircraft industry under the Ministry of Aviation Industry to the real world of today's international aviation market has confronted the Russian aircraft industry with challenges much more difficult than those it ever faced throughout its long and often arduous but largely successful history.

This chapter provides an overview of that history and then briefly looks at the structure of research and testing without which the industry and design bureaux could not function. The chief designers, their design bureaux, and their aircraft are covered in Chapter 7.

I. HISTORICAL OVERVIEW

From the Civil War to the First Five-Year Plan (1917–28)

During the turbulent summer of 1917, efforts were made to bring order to the chaotic situation in the Russian aviation industry. In August, Russian flyers, technicians, engineers, and aircraft factory workers gathered in the *I Vserossiiskyi aviatsionnyi s'yezd* (1st All-Russian Aviation Congress) and established the *Vserossiiskyi sovet aviatsii* (All-Russian Aviation Council, *Aviasovyet*). The *Aviasovyet* sought to stabilize the uncertain situation in an aviation industry, which, none the less, produced 1,099 aircraft and 374 engines during the year.

Soon after the October Revolution in 1917, aviation committees were formed in the Petrograd and Moscow soviets and the *Upravleniye voyenno-vozdushnogo Flota* (Administration of the Military Air Fleet, UVOFLOT) was established to assure political control of Russian aviation affairs. The Bolshevik *Narodnyi Komissariat po voyennym i morskim delam* (People's Commissariat for Military and Naval Affairs, *Narkomvoyenmora* or NKVM) converted the UVOFLOT into the *Vserossiiskaya kollegiya po upravleniyu Raboche-krest'yanskim krasnym vozdushnym flotom* (All-Russian Board for the Administration of the Workers' and Peasants' Red Air Fleet), whose task was to reorganize Russian aviation, take control of the aircraft and engine factories, and to collect aircraft, engines, and spare parts for the defense of the fledgling Soviet regime.

The start of the Civil War provided new impetus for creating a military air force. In May 1918, the *Glavnoye Upravleniye Raboche-krest'yanskogo krasnogo voyennogo vozdushnogo flota* (Chief Administration of the Workers' and Peasants' Red Military Air Fleet, *Glavvozdukhflot* or GURKKVVF), was formed under *Narkom-voyenmora*. In June the *II Vserossiiskyi aviatsionnyi s'yezd* met in Moscow and the new *Aviasovyet* became the highest controlling organ for aviation matters. Then, on 28 June, the Soviet government issued a decree nationalizing all aircraft factories and workshops.

Before the end of 1918, two additional major central state organizations were established that would be critical to the future development of the aviation industry. On 1 August 1918, the *Komissiya po organizatsii aviapromyshlennosti* (Commission for the Organization of the Aviation Industry) was created within the section on defense industry of the Department of Metals of the *Vysshyi Sovyet Narodnogo khozyaistva* (Supreme Council of the National Economy, VSNKh). On 14 December the Metals Department and its aircraft enterprises came under the *Revolyutsionno-voyenny sovyet* (Revolutionary Military Council, or *Revvoyensovyet*) and the GURKKVVF, and the *Glavnoye pravleniye ob'yedinennykh aviatsionnykh zavodov* (Chief Administration of the Unified Aircraft Factories, *Glavkoavia*) was established. Directly under the *Revvoyensovyet* and VSNKh, *Glavko-avia* was the first official central state organization set up to direct the nascent Soviet aviation industry and to manage its overall development under the direction of the *Sovet narodnykh komissarov* (Council of People's Commissars or cabinet, *Sovnarkom*). The *glavki*, or central chief administrations, which operated under the VSNKh until it was abolished in 1932, the various people's commissariats, and later the ministries after 1946, had very broad powers to plan and manage their respective industries or sectors of the economy. The other major change was the establishment of the above-mentioned *Tsentral'nyi aerogidrodinamicheskyi institut* (Central State Aero-Hydrodynamic

Institute, TsAGI) that will be discussed in more detail in Part II of this chapter.

The establishment of administrative organs of central state control in 1918 set the pattern for the entire Soviet era. The aviation industry in the Soviet Union evolved in a significantly different way from that in the rest of the world due to the social, economic, political, and defense policies of the first Communist state. With a totally nationalized and centrally directed economy, the state owned and managed every aspect of the aviation industry after June 1918. The central government in Moscow funded, controlled, staffed, and directed all aeronautical research institutes, design bureaux, production plants, and test centers and organizations. The key body in running the entire structure and assuring the development and production of advanced aeronautical systems for the Soviet military and civilian agencies was the central state organization responsible for the aviation industry, for it alone co-ordinated the separate activities under its control. This highest state administration for aviation was restructured a number of times after 1918 until finally settling in as a major *vsesoyuznyi* (All-Union) ministry associated with the Ministry of Defense Industry and the other defense production ministries in the mid-1950s. Although its names may have changed over the years, the basic principal remained the same – strictly centralized state direction and control.

During the Civil War, the Reds scraped together about 350 aircraft, and their six small factories only produced 668 new planes and 264 new engines but rebuilt 1,574 aircraft and 1,740 engines for the Red Air Fleet's use. After the Red victory in 1921, the Soviets turned to rebuilding the economy and the defense of the country. Completely surrounded by hostile neighbors, the Bolsheviks undertook an immediate program to build a strong military air arm. The Soviets still had to depend upon foreign suppliers and purchased a wide variety of foreign aircraft, including various Fokker aircraft from Holland, Martinside fighters from England, and Junkers Ju-13 passenger planes from Germany.

Aircraft left behind by the Allies during the intervention were carefully copied and produced beginning in the early 1920s. In 1922, however, 90 per cent of all aircraft were still purchased abroad and the nascent aircraft industry devoted most of its resources to repair and reconstruction of the older foreign and Russian aircraft. Thus, very early on the Bolsheviks adopted the czarist tradition of borrowing aircraft and aviation technology from the more advanced West. This tradition of technology transfer flourished throughout the Soviet era and was critical to the continued growth and development of Soviet aviation design and production.

A number of absolutely critical infusions of Western technology contributed very significantly to the development of Soviet aviation. In the early 1920s, the purchase of Western aircraft and the Junkers connection were crucial to early developments, and in the late 1920s the introduction of Hispano-Suiza 12Y, Gnome-Rhône K14 Mistral, BMW VI, and Wright R-1820 Cyclone engines provided much needed technology and spawned entire generations of Soviet piston engines. In the late 1930s the DC-3 licensing agreement and introduction of Douglas loft-mold construction techniques, along with the transfer of German aircraft under the Soviet–German agreements, provided important new techniques and a close look at advanced Western designs. The copying of the Boeing B-29 in 1945–46 and the exploitation of captured German aeronautical designs, technology, jet and turboshaft engines, and engineers after the Second World War, combined with the purchase of 55 British jet engines in 1946–47 to jump-start Soviet aviation into the jet age. Since 1992, this tradition has openly flowered once again as numerous Russian aviation firms have established co-operative relationships with many leading Western aircraft and engine firms.

During this early period, important clandestine military collaboration began between the Germans and the Soviets. The *Reichswehr* circumvented restrictions of the Versailles peace agreements by training its pilots and tank crews in Russia, and the Soviets sent Red Army officers to Germany for schooling in modern tactics and doctrine. Agreements in 1922–23 led to the signing of a contract granting the Junkers Corporation concessions for the manufacture of engines and all-metal aircraft at the former Russian-Baltic Fili plant (later *Zavod* [Factory] No. 22) near Moscow. Here Junkers produced the all-metal Ju-20 and Ju-21 and other models from 1923 through 1927, and A. N. Tupolev (1888–1972) learned much about how Junkers built all-metal aircraft using metal airframe structures and corrugated duraluminum sheets. This plant later became Tupolev's primary design and production facility where he made his all-metal aircraft, and it remained the only plant in the Soviet Union that could produce all-metal aircraft until the late 1920s.

At Lipetsk, 300 kilometers southeast of Moscow, the Germans laid the basis for their own future air force. A steady stream of aircraft, engines, spare parts, and technicians secretly entered the Soviet Union. Fokker fighters, ammunition, and bombs came in by sea, while some aircraft were flown in. A variety of experimental aircraft arrived at Lipetsk for testing, and Soviet designers and engineers worked closely with the Germans in developing and testing aircraft and sharing plans and designs. Thus, both nations benefitted from this clandestine co-operation until it ended early in the 1930s.

During the early 1920s and the period of the New Economic Policy

(NEP), significant changes took place in the structure of the Soviet air-craft industry that established patterns for the future. In March 1921, *Glavkoavia* was reorganized for peacetime development as the *Glavkoavia Sovyeta Voyennoy promyshlennosti* (Chief Administration of Unified Aviation Factories of the Council of Defense Industry) under VSNKh. In June 1921, the new organization came under the control of the *Glavnoye upravleniye voyennoy promyshlennosti* (Chief Administration of Defense Industry, VPU or *Voyenprom*) as the *Pyatyi (Aviatsionnyi) otdel* [5th (Aviation) Department] of its *Tekhnichesko-proizvodstvennoye upravleniye* (Technical-Production Administration). The close and enduring connection among the avia-tion industry, the defense industry, and the air force was cemented in 1924 when Pavel I. Baranov (1892–1933) was appointed chief of the newly renamed *Voyenno-vozdushnye sily rabochye-krest'yanskoi krasnoi armii* (Air Forces of the Workers' and Peasants' Red Army, VVS RKKA) and also chief of *Glavkoavia*. In this one increasingly powerful person was united control of the Air Forces and the entire Soviet aviation industry.

On 28 January 1925, a radical restructuring of the aviation industry took place with the transformation of the *Aviatsionnyi otdel, VPU*, into the *Gosudarstvennyi trest aviatsionnoy promysheslennosti Glavnogo upravleniya metallopromyshlennosti* (State Aviation Industry Trust of the Chief Administration of the Metal Industry, or *Aviatrest*) under VSNKh. The new *Aviatrest* had wide responsibilities for aircraft design, development, and construction. On its formation, *Aviatrest* had 11 production enterprises – four aircraft plants, four motor plants, and three subsidiary component enterprises. It came under the 'direct jurisdiction' of the *Glavnoye upravleniye metallopromyshlennosti* (Chief Administration of Metal Industry) and was closely aligned with the Commissariat of Military and Naval Affairs because its primary products were still military aircraft. In 1926, *Aviatrest* created the *Tekhnicheskyi Sovyet* (Technical Council) with members from the Metal Industry, *Voyenprom*, VVS, TsAGI, *Nauchnyi avtomotornyi institut* (Institute for the Scientific Study of Motors, NAMI), and leading designers to coordinate aviation matters across the industry and among the users. To strengthen the defense sectors of industry, during the following year *Aviatrest* was transferred to the *Avtoaviaotdel* (Automobile and Aviation Department) of the Chief Administration of the Metal Industry.

Baranov's dual appointments assured that Commissar of War M. V. Frunze's admonition to develop the VVS and 'take measures to estab-lish a national aircraft and aeroengine industry' would be fully carried out. Indeed, it was Frunze more than anyone else who pushed for Soviet self-sufficiency in aircraft production.

Other major milestones of the mid-1920s were the creation of Tupolev's *Aviatsiya, gidroaviatsiya i opytnoye stroitel'stvo* (Aviation, Marine Aviation, and Experimental Construction Department, AGOS) at TsAGI in 1925 and the formation on 19 March 1926 of the *Tsentral'noye konstruktorskoye byuro, Gosudarstvennyi trest aviatsionnoy promysheslennosti* (Central Design Bureau, State Aviation Industry Trust, or *TsKB Aviatresta*). Within the TsKB, three departments were formed – the *Otdel sukhoputnogo samolëtostroyeniya, Opytnyi otdel-1* (Department of Landplane Construction (OSS), OPO-1) under N. N. Polikarpov (1892–1944) at Zavod No.1 and later Zavod No. 25 in Moscow; the *Otdel morskogo opytnogo samolëtostroyeniya, Opytnyi otdel-3* (Department of Marine Experimental Aircraft Construction (OMOS), OPO-3) under D. P. Grigorovich (1883–1938) in Leningrad at Zavod No. 23; *Krasnyi Lëtchik* (Red Flyer), which moved to Zavod No. 22 in Moscow in 1927; and the *Otdel opytnogo motorostroyeniya, Opytnyi otdel-2* (Department of Experimental Motor Construction (OOM), OPO-2) under A. D. Shvetsov (1892–1953) at the design bureau of Aircraft Motor Zavod No. 24 in Moscow. The creation of the TsKB was intended to concentrate and co-ordinate the work of the various sectors of experimental aircraft construction under centralized direction. Through the late 1920s and into the first years of the First Five-Year Plan of 1928, the Polikarpov, Grigorovich, and Tupolev design bureaux were the principal centers of aircraft design in the Soviet Union. It was within these centers that most of the young designers coming out of the various aviation institutes were nurtured, and new aircraft developed.

Table 6.1

AIRCRAFT AND ENGINE PRODUCTION (1918–1927/28)

Year	18	19	20	21	21/22	22/23	23/24	24/25	25/26	26/27	27/28
Aircraft	255	137	166	67	44	186	173	327	339	545	608

Total: 2,847

Year	18	19	20	21	21/22	22/23	23/24	24/25	25/26	26/27	27/28
Engines	79	77	81	15	12	50	70	157	342	285	762

Total: 1,930

Source: G.S. Byushgens (ed.), *Samolëtostroyeniye v SSSR 1917–45 gg. Kniga I* (Aircraft Construction in the USSR, 1917–45. Book I) (Moscow: Izdatel'skyi Otdel TsAGI, 1992), pp. 20, 28, 431.

The developments of the 1920s established basic organizational patterns in the aviation industry that would endure throughout the Soviet era. The relationship with the emerging military-industrial complex was solidly set. Influential design bureaux under powerful chief designers came to the forefront, but they were divorced from the production plants that built their aircraft and the scientific and research institutes that provided scientific and technical information. With still very limited capabilities, production plants specialized in one type or model of aircraft. In time, however, these features would become weaknesses.

First Five-Year Plan to the Eve of War (1928–41)

Soviet design and development efforts had gone hand-in-hand with foreign collaboration throughout the 1920s, but with the clear intention of attaining self-sufficiency in the future, just as Frunze had called for. In 1928 Stalin initiated the first of his Five-Year Plans with the primary goal of developing a modern economic structure based on heavy industry so that the Soviet Union would be fully capable of producing the modern weapons needed to defend itself. Very impressive targets were established to increase total industrial output by 250 per cent, heavy industry by 330 per cent, and to triple pig-iron production, double coal output, and quadruple electric power. The next year the original quotas were revised upward, and a decision was made to complete the First Five-Year Plan in four years. Although the entire plan was not fulfilled, its actual achievements were tremendous, but paid for at the tremendous cost of the brutal collectivization of agriculture.

Another dark side of Stalin's Five-Year Plans was the attempt to stamp out foreign influence and the older generation of 'non-Soviet' engineers. This nascent 'purge' began in 1929–30 with the Shakhty case and subsequent *Prompartiia* show trials, but it also encompassed much of the aviation industry when the *Ob'yedinyennoye gosudarstvennoye politicheskoye upravleniye* (Unified State Political Administration, OGPU, or Secret Police) arrested Grigorovich on 1 September 1928 and then Polikarpov on 25 October 1929. These arrests effectively paralyzed design work in both the TsKB's OPO-1 and OPO-3.

On 1 December 1929, in a portent of the future, the OPGU organized an internee design bureau with Polikarpov, Grigorovich, and many others from the TsKB at the Butyrka prison in Moscow under the designation TsKB-39 OGPU. In February 1930 the OGPU moved TsKB-39 to work with what was now called the *Tsentral'noye konstruktorskoye byuro* at the infamous *Zavod* No. 39 *imeni V. R. Menzhinskiy*, the former *Aviarabotnik* factory that had been renamed the V. R. Menzhinskiy plant in honor of the OGPU's chief from 1926 until his death in 1934. Among the prisoners under the direction of Polikarpov and Grigorovich were B. F. Goncharov, I. M. Kostkin, and A. V. Nadashkevich (1897–1967). They worked alongside S. A. Kocherigin (1893–1958), A. S. Yakovlev (1906–89), V. P. Yatsenko (1892–1964), and other non-prisoners who formed the TsKB. To regain his freedom, Polikarpov teamed with Grigorovich to complete the VT-12, for *vnutrennyaya tyur'ma* (internal prison), which became the successful I-5 biplane fighter.

The VVS and the aviation industry were central to Stalin's military and strategic thinking as well as his plans for industrialization, so a great deal of attention was lavished on them. In March 1930, the *Vsesoyuznoye aviatsionnoye ob'yedinyeniye Aviatresta* (All-Union Aviation Organization of the State Aviation Industry Trust, VAO) was established. On 15 April 1930, *Aviatrest* was abolished and the new VAO assumed responsibility for the entire aircraft industry directly under the Metal Industry. In June 1930, Baranov maneuvered the transfer of VAO to *Narkomvoyenmora*, and then, in February 1931, he had VAO placed directly under the VSNKh. He continued to control both the VVS and the VAO until June 1931 when the two posts were separated, with Baranov remaining as the head of the All-Union Aviation Organization (VAO) and joining the Presidium of the VSNKh, while Ya. I. Alksnis took over the VVS. In December 1931, the *Glavnoye upravleniye aviatsionnoy promyshlennosti* (Chief Administration of the Aviation Industry, *Glavaviaprom* or GUAP) was established directly under VSNKh. Upon VSNKh's demise in January 1932, during the continuing restructuring of the economy in the early Five-Year Plans, *Glavaviaprom* was transferred to the *Narodnyi komissariat tyazhëloi promyshlennosti* (People's Commissariat of Heavy Industry, *Narkomtyazhprom* or NKTP) where Baranov added duties as the special deputy in charge of aviation industry to his portfolio. Baranov's untimely death in an aircraft accident in September 1933 deprived the entire Soviet aviation industry of a forceful and forward-looking leader at a most critical time.

One of the important goals of the First Five-Year Plan was for the Soviet aircraft industry to 'rid itself finally of foreign dependence'. This goal was certainly overly ambitious and never accomplished, in the 1930s or later, and the Soviets would continue to lean heavily upon Western designs and technical information. However, a native aviation industry took more definite shape during these years and began to produce planes in quantity. At the beginning of the First Five-Year Plan 12 aircraft industry enterprises were at work, but by its end there were 31 plants, mostly new, reconstructed, or still under construction. Nine of these were for serial aircraft production and six for engines and engine components. During the first two Five-Year Plans, major new aircraft plants were begun at Voronezh (No. 18) in 1930; Leningrad (No. 169) and Novosibirsk (No. 153) in 1931; during 1932 at Gorky (now Nizhnii Novgorod) (No. 21), Tushino (Moscow) (No. 62), Khimki (Moscow) (No. 84), Kazan (No. 124), Irkutsk (No. 125), and Komsomol'sk-na-Amure (No. 126); and in 1934 Tushino (Moscow) (No. 81). Workers actually employed in the aircraft construction industry increased from 8,695 in 1928 to 24,497 in 1932/33 and to 38,729 in 1936.

In January 1933, Stalin, describing the accomplishments of the First Five-Year Plan, could justifiably declare: 'We had no aviation industry. Now we have one.' During the First Five-Year Plan new aircraft assembly, engine, and component factories were built, 56 experimental aircraft and 17 experimental motors were developed and 11 of the aircraft and 5 motors entered serial production, and annual output grew from 608 planes in 1928 to 2,509 in 1932. The Soviets had tested and adopted a series of their own designed and built planes. Polikarpov's I-3 wooden biplane fighters and R-5 reconnaissance planes and Tupolev's all-metal TB-1 bombers had entered squadron service with the VVS.

In August 1931, an effort was made to combine all experimental aircraft design and construction under the VAO's TsKB, which was transferred to TsAGI and combined with Tupolev's AGOS. S. V. Ilyushin (1894–1977), then deputy chief of the *Nauchno-issledovatel'skii institут VVS* (Scientific Research Institute of the VVS, NII VVS), was made the chief and Tupolev his deputy. This change was supposedly intended more effectively to organize and use the production base in the various aircraft factories. It is also possible that the OGPU may have been attempting to establish a greater centralized governmental control over aircraft design that was so typical of the increasing centralization that took place during the Five-Year Plans. Tupolev vigorously protested this restructuring because he believed that such a large experimental bureau was too difficult to manage effectively with many different types of aircraft under simultaneous design and construction.

In May 1932, this experiment ended with the creation of the first of two major design offices with brigades established under lead designers for specific types of aircraft and systems. The formation of the *Sektor opytnogo stroitel'stva* (Section of Experimental Construction, SOS) in TsAGI under Tupolev signaled that a new direction would be taken in organizing aircraft design. Then, in January 1933, the Chief Administration of the Aviation Industry organized the experimental aircraft construction design office, the TsKB, at the Menzhinskiy (OGPU) plant under Ilyushin to focus on fighters, ground-attack aircraft, and tactical bombers. Several design teams from TsAGI's design departments were transferred to beef up the new Central Design Bureau.

At the same time, Tupolev's newly formed SOS was combined with TsAGI's *Zavod opytnykh konstruktsii* (Experimental Construction Factory, ZOK), established in Moscow the previous January as Zavod No. 156, to form the new *Konstruktorskyi otdel opytnogo samolëtostroyeniya* (Design Department for Experimental Aircraft Construction, KOSOS TsAGI) that specialized more in designing multi-engined, long-range heavy bombers and maritime aircraft. In addition, KOSOS now became GUAP's main prototype construction factory. Tupolev's office was then established as an autonomous experimental and design organization under TsAGI's overall purview. TsAGI continued with its basic scientific research, leaving Tupolev and his associates to develop new types of multi-engined planes.

KOSOS's design brigades were reorganized under lead designers who were now charged with different categories of aircraft rather than an aircraft subsystem. For example, formerly V. M. Petlyakov (1891–1942) had responsibility for all wing design for Tupolev aircraft but now he headed Brigade No.1 that handled heavy aircraft. P. O. Sukhoi (1895–1975) took over Brigade No. 3 (Fighters and Experimental Aircraft), A. A. Arkhangel'skiy (1892–1978) had No. 5 (High-Speed Military Aircraft and Passenger Models), and V. M. Myasishchev (1902–78) had No. 6 (Experimental Aircraft).

At the TsKB, Ilyushin adopted a similar system of independent design brigades specializing in types of aircraft, aircraft systems, and components. S. A. Kocherigin headed Brigade No. 1, which developed reconnaissance and ground-attack aircraft, while Polikarpov headed Brigade No. 2, which worked on fighter aircraft. Seaplanes were the responsibility of I. V. Chetverikov (1909–87) and later G. M. Beryev (1903–79) in Brigade No. 5. Design Brigade No. 3 was under Ilyushin's direct control, and its principal efforts were concentrated on developing a high-speed, long-range bomber for the VVS. The aircraft that grew out of this effort was the twin-engined DB-3 (later DB-3F and Il-4) that served as a mainstay of the VVSs long-range bomber force throughout the Second World War. In a step that would soon produce important results, in 1934–36 the TsKB began aligning design brigades with serial production plants. For example, in 1934 Beryev and Brigade No. 5 began working with Zavod No. 31 at Taganrog in a relationship that lasted until his retirement in 1968 and actually continues today, and Sukhoi worked with Zavod No. 135 at Kharkov in 1939–40.

As the workload of the two major design bureaux expanded during the mid-1930s, a number of new OKBs were formed to expedite the development of new aircraft designs. Design teams at the major OKBs also gradually evolved into independent experimental design offices that were often spun off under individual designers such as Petlyakov,

Yakovlev, A. I. Mikoyan (1905–70) and M. I. Guryevich (1892–1976), Sukhoi, Polikarpov, and others. These new OKBs often began as an *otdel osobykh konstruktsii* (Special Design Department, OOK), a *sektsiya osobykh konstruktsii* (Special Design Section, SOK), an *opytnyi konstruktorskii otdel* (Experimental Design Section, OKO), or simply an *opytnyi otdel* (Experimental Department, OPO). After successful designs were authorized, these design offices often became full OKBs and were associated with a production plant. For example, the Mikoyan and Guryevich (MiG) OKO, which began within the Polikarpov OKB, was set up at Zavod No. 1 in Moscow in 1940 and in March 1942 moved as a full OKB to Zavod No. 155 in Moscow where MiG remains today. This governmental policy produced a rapid growth in the number of OKBs from eight in 1935, to 14 the next year, 26 in 1938, and 30 in 1939.

During the 1930s, the defense industry assumed an ever greater importance in Soviet strategic thinking and national planning. As a result of the 1936 Stalin Constitution, the new *Narodnyi komissariat oboronnoy promyshlennosti* (People's Commissariat of Defense Industry, *Narkomoboronprom* or NKOP) was established on 8 December 1936 specifically to manage the various component industries of the burgeoning military-industrial complex. As part of this ongoing restructuring, on 21 December 1936, *Glavaviaprom* was attached to the new NKOP. M. M. Kaganovich, the former director of the Rybinsk Motor Building Zavod No. 26, who had become the chief of GUAP late in 1935, continued to head the Soviet aviation industry until October 1937 when he replaced the purged M. L. Rukhimovich as head of NKOP. In February 1938, a further change in NKOP broke *Glavaviaprom* into two *glavki*, the *Pervoye glavnoye upravleniye* (*samolëtostroitel'noe*) (First Chief Administration (Aircraft Construction)) that was charged with aircraft production and another, the 18th Chief Administration, responsible for engine production.

The aviation industry achieved even greater success under the second Five-Year-Plan from 1933 to 1938, as the output again increased as the plants started during the First Five-Year Plan were completed and began serial production, and new plants were added at Rostov (No. 168) and Ulan-Ude (No. 136). The annual output of aircraft grew steadily throughout the middle and late 1930s (except for 1935), growing from the 2,509 of 1932, to 4,455 in 1934, to 6,039 in 1937, and to 10,342 in 1939 (see Table 6.2).

For a number of political and military reasons, large, multi-engined aircraft held much of the spotlight during the initial Five-Year Plans, but single-seat reconnaissance, trainer, and fighter biplanes still accounted for most of the production. They showed the technological success of Stalin's regime, had the range to conquer Russia's vast inner spaces and tie the country together, and also fulfilled the requirements of the Soviet military strategy of deep operations and combined-arms warfare that M. N. Tukhachevskiy and his colleagues were developing. To strike at the distant heart of the 'capitalist encirclement' required multi-engined aircraft with considerable range and bomb-loads, not hordes of tactical bombers and single-engined frontal fighters. VVS leaders, especially Ya. I. Alksnis, VVS commander, and V. V. Khripin, his chief of staff, also emphasized the importance of strategic bombing, and pushed for the development and production of more bombers.

Table 6.2

SERIAL AIRCRAFT PRODUCTION, 1929–39

Year	29	30	31	32	33	34	35	36	37	38	39	Total
Military	732	905	1,076	1,259	2,311	2,864	1,805	2,506	3,882	4,685	7,410	29,435
Trainers	180	203	283	942	1,381	1,100	327	968	1,947	2,695	2,675	12,701
Passenger and other aircraft	0	41	130	308	423	491	397	796	210	310	257	3,363
Total:	912	1,149	1,489	2,509	4,115	4,455	2,529	4,270	6,039	7,690	10,342	45,499

Source: G. S. Byushgens (ed.), *Samolëtostroyeniye v SSSR 1917–45 gg. Kniga I* (Aircraft Construction in the USSR, 1917–45. Book I) (Moscow: Izdatel'skyi Otdel TsAGI, 1992), pp. 431, 432–5; *Kniga II* (Book II) (Moscow: 1994), pp. 235–7.

The entire Soviet air establishment suffered heavily from Stalin's bloody purges of 1937 and 1938 which effectively decapitated the armed forces and seriously harmed the entire aviation industry. The total consequences of the purges are difficult to estimate even today. The air force lost heavily not only in the purge of its leaders but also from the loss of numerous civilian aeronautical engineers and designers who fell victim to the terror. N. M. Kharlamov (1892–1937), who was both the director of the TsAGI since he had replaced S. A. Chaplygin (1869–1942) in 1932, and the chief of scientific research and experimental construction at VAO, was purged and replaced as TsAGI chief by M. N. Shul'zenko in 1937, who, in turn, lost the post in 1940 to I. F. Petrov (1897–), then deputy chief of the NII VVS. Thousands employed throughout the aircraft industry were dismissed, and many research, testing, design organizations as well as production plants lost all or most of their key personnel or were simply closed. Moscow's *Zavod* No. 1, for example, lost its director, chief and deputy chief engineers, two design brigade chiefs, and four shop chiefs.

L. L. Kerber, a close associate of Tupolev's for many years who later wrote the important work *Tupolevskaya sharaga* ('Tupolev's

Prison Workshop) about their shared internment, estimated that the *Narodnyi komissariat vnutryennykh del'* (People's Commissariat of Internal Affairs, NKVD, the OGPU's successor) interned 450 designers, engineers, and technicians from 1934–41. As many as 300 were put to work in various *osobyye tekhnicheskiye byura pri NKVD* (Special Technical Bureaux of the NKVD, *Ostekhbyuro* or OTB), 100 died in the infamous camps of the *Glavnoye upravleniye lagerei* (Chief Administration of Correctional Labor Camps, GULAG), and 50 were executed. K. A. Kalinin (1889–1938), an established and respected designer who was responsible for the advanced swept-wing aircraft designs, was shot after four Communist Party members were killed in a crash of one of his experimental aircraft. Tupolev, accused of sabotage and espionage after a visit to the United States and Germany in 1936, was arrested in October 1937. The chiefs of four of KOSOS's design brigades soon joined him in the NKVD's care. Tupolev was eventually sentenced to a five-year imprisonment in an NKVD *sharaga* (prison workshop), designated TsKB-29 NKVD, and his ANT planes were redesignated.

A. S. Yakovlev, in his autobiography, *Tsel' zhizni*, as well as his history of Soviet aircraft construction, contended that certain serious mistakes were made before the Second World War – too few light bombers and reconnaissance aircraft were produced; in fighter design, speed and firepower took second place to maneuverability; and the Soviets became smug and boastful after the initial successes of their fighters in the Spanish Civil War. The Germans, on the other hand, had profited by their Spanish experiences and began taking urgent steps to improve their aircraft. Yakovlev's criticism was primarily aimed at Tupolev and Polikarpov, whose design bureaux specialized, respectively, in multi-engined aircraft and light, maneuverable fighters. Tupolev, however, was fulfilling directives from higher authorities and, moreover, after October 1937 he was incapable of doing much to produce new aircraft, resting, as he did, in the gentle clutches of the NKVD.

While serving his term in TsKB-29 NKVD at his own Zavod No. 156 in Moscow, Tupolev and his fellow prisoners designed the Tu-2 frontal bomber, and Petlyakov completed his Pe-2 twin-engined dive bomber that formed the backbone of the Soviet tactical bomber force during the Great Patriotic War. Many of those imprisoned only won their freedom after their new designs were accepted and entered serial production, or after Adolf Hitler invaded the Soviet Union in June 1941. Tupolev and his colleagues were released in August 1941.

As the war clouds gathered over Europe, the Soviet leadership became increasingly concerned about not only the combat capabilities of their aircraft compared with those of potential enemies but also the damage that the purges were doing to the entire aviation industry. Sobering lessons learned during the last year of air combat in Spain contributed to this growing anxiety. Polikarpov's I-15 and I-16 fighters and Tupolev's SB-2 fast bombers lost their former superiority over earlier German and Italian combat aircraft. The Messerschmitt Bf 109Es presented the most serious problem because they could outfly the I-15 and I-16 fighters while handily shooting down the SB-2 and SB-2bis fast bombers. By the late 1930s, with the experiences of the Spanish Civil War and the purging of many strategic bombing proponents, a renewed emphasis was placed on the fighters, ground-attack aircraft, and tactical bombers that could support air superiority and frontal ground-support operations.

Responding to this situation, in early 1938 the Central Committee of the Communist Party of the Soviet Union and the government jointly called a conference of aviation and air force leaders. The still-free stars of Soviet aviation attended: engine designers V. Ya. Klimov (1892–1962), A. A. Mikulin (1895–1985), A. D. Shvetsov; M. N. Shul'zhenko, then head of TsAGI; and the design bureau chiefs Yakovlev, Polikarpov, Arkhangel'skiy and Ilyushin (remember Tupolev, Petlyakov, Myasishchev, and many others were then in NKVD custody). While not arriving at any final recommendations for overcoming the sudden backwardness of Soviet aviation, this conference set the stage for similar meetings later in the year and encouraged the development of new OKBs and designs.

Attention was now heavily focused upon the rapid procurement of the latest design types, so the OKBs began pumping out new designs and modifications in great numbers. From 1938 through 1940, OKBs projected, built, and tested 115 new aircraft and 83 modified models. Of these, 31 new aircraft and 36 modifications reached state acceptance trials, but only 20 of those were approved for serial production. Fortunately for the program initiated at the 1938 meetings and the designs that were accepted, the Soviet aircraft industry received considerable attention in the Second Five-Year Plan (1932–37). The Third Five-Year Plan (1938–42) envisioned extensive use of this existing capacity, plus a significant expansion in both airframe and aero-engine production for increased output of military aircraft.

On 11 January 1939, the People's Commissariat of Defense Industry was split up and *Glavaviaprom* became a separate People's Commissariat of Aviation Industry (*Narodnoi komissariat aviatsionnoy promyshlennosti, Narkomaviaprom* or NKAP) under M. M. Kaganovich, the former Commissar of Defense Industry and head of *Glavaviaprom*, with S. I. Ilyushin as a deputy. This was an effort to improve the organization and management of aircraft production due to the great importance now placed on preparing the VVS for a possible conflict with Nazi Germany and Japan. On 17 September 1939, the *Sovnarkom*'s *Komitet oborony* (Defense Committee) directed the NKAP to construct or rebuild 18 aircraft and engine plants (nine of them to be built deep inside the USSR) in order to increase annual output to 17,000 aircraft by 1 July 1941.

Unhappy with the progress being made, Stalin removed Kaganovich in January 1940 and replaced him with A. I. Shakhurin (1904–75), a Party leader with significant aviation industry background, who directed further expansion and the conversion of seven civilian industrial plants to aircraft production. Much of this expansion was ticketed for the more secure regions of Western Siberia, the Volga region, Central Asia, and the Urals because of the two-front threat posed by Japan in the Far East and Germany in Europe. In 1939, production was underway at 17 aircraft and five engine serial plants. By 1941, 24 aircraft – 15 for fighters and light aircraft and nine for bombers and ground-attack aircraft – and seven engine assembly plants were already at work as a direct result of these actions. More important, the new and converted aircraft and engine plants in such places as Irkutsk, Kazan, Ulan-Ude, Ufa, Perm, Novosibirsk, and Komsomol'sk-na-Amure were responsible for nearly 33 per cent of the total production and almost one-half of the output of the older western factories at Moscow, Leningrad, Kazan, Kiev, Kharkov, and Taganrog.

Shakhurin's appointment to head NKAP reflected a trend that really began with Kaganovich (who had been an aircraft engine plant manager) – the selection of trained and experienced production plant managers and aeronautical engineers to run the central organization. Shakhurin surrounded himself with able and experienced deputies – P. V. Dement'yev (1907–77), his immediate deputy, had been chief engineer at Zavod No. 1; P. A. Voronin (1903–84), a former manager of Zavod No. 1, was responsible for fighters; A. I. Kuznetsov handled bomber production; A. A. Zavitayev was in charge of aircraft engine production; V. P. Kuznetsov had experimental engine development; A. S. Yakovlev was chief of prototype aircraft development; and M. V. Khrunichev (1901–61) was responsible for plant buildings and raw material supplies. While not always appreciated by Stalin, all of these experienced and technically competent decision-makers played critical roles during the war and in the postwar era.

Paralleling the rapid enlargement of the aircraft industry was the movement toward greater decentralization in aircraft design and development. Tupolev's KOSOS, which had effectively ceased to function with his arrest and that of many of his colleagues in 1937–38, and Ilyushin's loosely structured TsKB were the main design organizations until the late 1930s. Realizing the increasing inadequacy of the design base, the Soviet government had encouraged the leading designers in each bureau to form their own teams. Thus, a number of young and talented but unproven designers answered the Central Committee's call for new fighters when it was issued at the 1939 Kremlin conference. But nothing could hide the fact that little positive happened in aircraft development in the years immediately following the imprisonment of Tupolev and so many other leading designers and technicians.

A. S. Yakovlev, A. I. Mikoyan and M. I. Guryevich, and S. A. Lavochkin (1900–60), assisted by M. I. Gudkov and V. P. Gorbunov, submitted new single-engine fighter designs which became the Yak-1, MiG-1 and LaGG-1. All three were soon approved for prototype development and flight testing prior to state acceptance trials that would determine their suitability for full-scale production and introduction into the VVS. While the Yak-1 and LaGG-1 were air-superiority fighters destined for tactical use, the I-61 of Mikoyan and Guryevich was primarily intended as a high-speed, high-altitude interceptor for air defense.

Although these new Soviet aircraft appeared crude by contemporary Western standards, they were carefully tailored to Soviet needs. These aircraft were economical and easy to build in large numbers, and easy to maintain. All Soviet combat aircraft were designed and built to operate in the primitive field operating conditions and severe winter weather found in many parts of the Soviet Union. They were designed to use limited resources efficiently and, thus, relied heavily on wood and composite materials. Owing to persistent metal shortages, airframes were generally built of wood and steel tubing covered with fabric or *shpon*, bonded birch strips. This made the aircraft too heavy for the available engines and severely degraded their performance. The LaGG-1 was built with highly flammable *delta-drevesina* (delta wood), plywood sheathing, and bakelite ply, and burned so easily that the pilots sarcastically joked that LaGG in Russian stood for *Lakirovannyi Garantirovannyi Grob* (varnished guaranteed coffin). The Yak-9 and La-7 fighters of 1943 were the first to have metal wing spars, and only in 1944 did the Yak-9U, the first fighter with an all-metal airframe, enter production. Lavochkin's first all-metal airframe fighter, the La-9, did not reach operational units until 1946. Thus, few Soviet aircraft of the war period had all-metal airframes or stressed-skin construction.

Compared with Western models, these aircraft were definitely underpowered, largely as the result of lagging Soviet engine development. For this reason, the performance of the aircraft was generally inferior to that of the German Bf 109E/F or British Spitfire. Indeed, this deficiency holds true for most Soviet aircraft of the time because the Soviet aircraft engine industry continued to be heavily dependent upon improvements to older Western designs and had yet to develop sufficiently powerful new engines. The chief engine designers, Shvetsov, Mikulin, and Klimov, worked to correct this shortcoming, and a series of improved and more powerful engines emerged from their efforts during the war years. However, the deficiencies in engine design and production limited the output of the entire wartime aircraft industry and, indeed, have continued to bedevil the industry up to the present day.

During 1940–41, the Soviet aircraft industry began building the modern planes that had been designed and approved for series production in 1939–40. Tooling up and establishing production lines took precious time, while the aircraft suffered the inevitable problems associated with the hasty commitment of prototypes to production before thorough testing detected all defects. Considering the critical two-front threat facing the VVS in 1940 and the shoddy Soviet performance in the Russo-Finnish War (1939–40), however, the decision to move quickly into full-scale production was the only one possible.

Between 1 January 1939, and 22 June 1941, the total production of the newly introduced Yak-1, MiG-1 and 3, LaGG-3, Pe-2, and Il-2 ground-attack aircraft represented only 2,839 of the approximately 22,000 military aircraft built. Production of these models numbered only 186 in 1940 but jumped to 2,653 in January–June 1941 as more new plants came fully on line and technical problems with the new aircraft were resolved. Obviously, then, the vast majority of the aircraft turned out were such increasingly obsolescent types as the I-153 and I-16 fighters and SB bombers, whose production tapered off or was halted only in 1941 when the new types came on line. Of the 16,288 military aircraft then in or entering serial production that were completed from 1939 through 1940, 13,195, or 81 per cent, were already obsolescent or quickly becoming so – 1,304 I-15bis models, 3,373 I-153s, 4,545 I-16s, and 3,973 SBs (see Table 6.3 below) – while Ilyushin's DB-3/DB-3F/Il-4 bombers accounted for another 2,065 aircraft or 12.7 per cent. On the positive side, the Soviet aircraft industry had clearly been gearing for war during the last years of peace. From 10,342 aircraft built in 1939, output dropped to 10,137 in 1940 as the older models were phased out, new models were introduced on the production lines, and new plants and plant expansions were completed. Aircraft production reached 5,858 in January–June 1941 and 15,735 aircraft and 28,707 engines for all of 1941, despite the severe disruptions during the war's opening months.

Table 6.3

SERIAL PRODUCTION OF OLDER MODEL COMBAT AIRCRAFT, 1939–41

Type:	1939	1940	1941	Total
I-15bis	1,304			1,304
I-153	1,011	2,362	64	3,437
I-16	1,835	2,710	356	4,901
SB	1,778	2,195	237	4,210
Total	5,928	7,267	657	13,852

Source: G. S. Byushgens (ed.), *Samolëtostroyeniye v SSSR 1917–45 gg. Kniga I* (Aircraft Construction in the USSR, 1917–45. Book I) (Moscow: Izdatel'skyi Otdel TsAGI, 1992), pp. 431, 432–5.

The Great Patriotic War (1941–45)

The heavy attrition of combat aircraft in June and July 1941, along with the beginning of plant evacuations, spurred severe rationalization of Soviet aircraft production in the following months. The VVS decided to concentrate almost exclusively on the fighter and attack air arms, and production plans followed suit. Unnecessary duplication of types was eliminated and production was cut to a few types with designs frozen to facilitate large-scale output.

The anticipated growth of aircraft and engine production for July–December 1941 declined sharply as 85 per cent of the existing aircraft assembly and engine plants located in western Russia were evacuated and existing industrial space was converted to the building of new models. In a monumental effort, assembly plants and highly skilled workers were evacuated from Leningrad to Novosibirsk (No. 23) and Kazan (No. 169); from Kiev to Novosibirsk (No. 43); from Voronezh to Kuibyshev (now Samara) (No. 18); from Moscow to Tashkent (No. 84), Irkutsk (No. 39), Kazan (No. 82), Omsk (No. 156); Kharkov to Perm (No. 135); and from Taganrog to Tbilisi in Georgia (No. 31). Although output was maintained at respectable levels through October, in November production dipped to 627 aircraft, just 27 per cent of the September level of 2,329. This precipitous drop denied the VVS sufficient planes to recoup its continuing heavy losses.

The combination of front-line demands and limited production capacity led to the discontinuation of a number of aircraft and engine models, freeing valuable plant space for a few essential aircraft until the evacuated plants were back in operation and new ones had been built. Engine production was restricted to four basic air-cooled and three liquid-cooled power plants. In their improved 1,600–1,850 hp variants, Shvetsov's ASh-82 (M-82), Klimov's VK-107 (M-107), and Mikulin's AM-38 engines were used in most Soviet combat aircraft, including all Yak and Lavochkin fighters as well as the Pe-2 and Il-2 series.

Superfluous aircraft models were quickly dropped. Sukhoi's excellent Su-6 attack aircraft was discontinued; available plant capacity could not support both it and Ilyushin's Il-2 which was already in production and operational. The Pe-8 four-motored bomber was completely dropped by 1944 in favor of more Pe-2 light bombers. The only entirely new aircraft to be produced during the war was Tupolev's Tu-2 tactical bomber. Except for 22 aircraft built in 1942, MiG-3 production ended in November 1941 when its Mikulin AM-35A 1,350-hp engine was terminated to allow expanded production of the 1,600-hp AM-38 engine for the Il-2.

Aircraft output greatly benefitted from a concentration on single-engine fighters and attack planes and from the rationalization program that limited production to a few basic models of engines and combat aircraft to take full advantage of standardization and mass production. This approach, which characterized most Soviet wartime armament programs and would continue into the postwar years, simplified the persistent Soviet spare parts problem and facilitated maintenance and training. Once the production lines were operating at full capacity, ample aircraft became available to re-equip the existing air regiments and to outfit new units added in the expansion of the VVS.

By early 1942, most aircraft and engine factories were relocated and series production had resumed, but the plants were still short of over 215,000 workers, half of them skilled. Production expanded greatly during the year, and total output in 1942 reached 38,002 engines and 25,436 aircraft, of which 21,577 were combat aircraft. New and improved aircraft, the La-5, Yak-7B, Yak-9, and Il-2m3, entered production and combat that year.

The most significant improvements in aircraft came in the VVS fighter force. Older models, such as the LaGG-3 and Yak-1, gave way to re-engined, upgunned, and aerodynamically refined versions of the same types. Re-engined with improved versions of the Shvetsov air-cooled radial ASh-82 engine, the La-5, La-5FN, and La-7 replaced the LaGG-3. The basic Yak spawned an infinite variety of ever-improving offspring, culminating in the Yak-3 of 1943 and the Yak-9U in 1944. These aircraft had the greater range and firepower required to support the offensive operations of the Red Army. The Yak-3, Yak-9U, and La-7 were equal to, if not better than, the standard German Bf 109G-6 and FW 19OA-8 fighters in speed and maneuverability, especially below 10,000 feet, but generally tended to remain inferior in firepower.

The continuing development of the Soviet aircraft industry provided the means for the VVS's seizure of air supremacy in 1943 and its
retention during 1944–45. Safe from German interference, Soviet aircraft enterprises profited from factory expansions and refinement of
mass-production techniques. The average monthly output increased
from 2,100 aircraft in 1942 to more than 3,350 in 1944 and to 3,480 for
January–June 1945.

The large yearly increases in aircraft production allowed the VVS
not only to make up for its heavy combat losses but also to add new air
units as well as more aircraft to the existing regiments. The growth
in numbers and in front-line combat strength was the result of the
industry's concentration on building combat aircraft. In 1944, when
40,241 aircraft were produced, 17,895 (44.4 per cent) were fighters,
11,110 (27.6 per cent) were attack planes, 4,200 (10.4 per cent) were
frontal and long-range bombers, and the remaining 7,036 (17.4 per
cent) were transports and trainers. Thus, 33,205 aircraft, or 82.5 per
cent, of total output were combat aircraft. Significantly, the combat
aircraft built after 1942 were predominantly of good design and
quality, easily equal or even superior to German models.

The operational requirements of the air war in the east placed overriding importance on the production of fighters, ground-attack aircraft, and frontal bombers. Thus, out of a total production of 142,775
aircraft from January 1941 through December 1945, 118,041 all told
were divided as follows: 60,657 (42.5 per cent) were fighters, 38,719
(27.1 per cent) ground-attack aircraft, and 18,665 (13.1 per cent)
multi-engined bombers.

Table 6.4

SOVIET WARTIME AIRCRAFT AND ENGINE PRODUCTION,
JANUARY 1941–DECEMBER 1945

Year	1941	1942	1943	1944	1945	Total
Fighter	7,081	9,918	14,627	17,895	11,136	60,657
Bomber	3,754	3,534	4,057	4,200	3,120	18,665
Ground-attack	1,542	8,229	11,193	11,110	6,645	38,719·
Transport	257	469	1,241	1,543	1,231	4,741
Trainer	3,101	3,286	3,766	5,493	4,347	19,993
Total aircraft	15,735	25,436	34,884	40,241	26,479	142,775
Total engines	28,707	38,002	48,825	52,776	40,565	208,875

Source: G. S. Byushgens (ed.), *Samolëtostroyeniye v SSSR 1917–45 gg. Kniga II* (Aircraft
Construction in the USSR, 1917–45. Book II) (Moscow: Izdatel'skyi Otdel TsAGI, 1994), pp.
215, 222, 226, 233, 234.

Yakovlev's fighter aircraft were by far the most numerous in the
VVS fighter arm. From 1941 through 1945, a total of 35,085 Yak

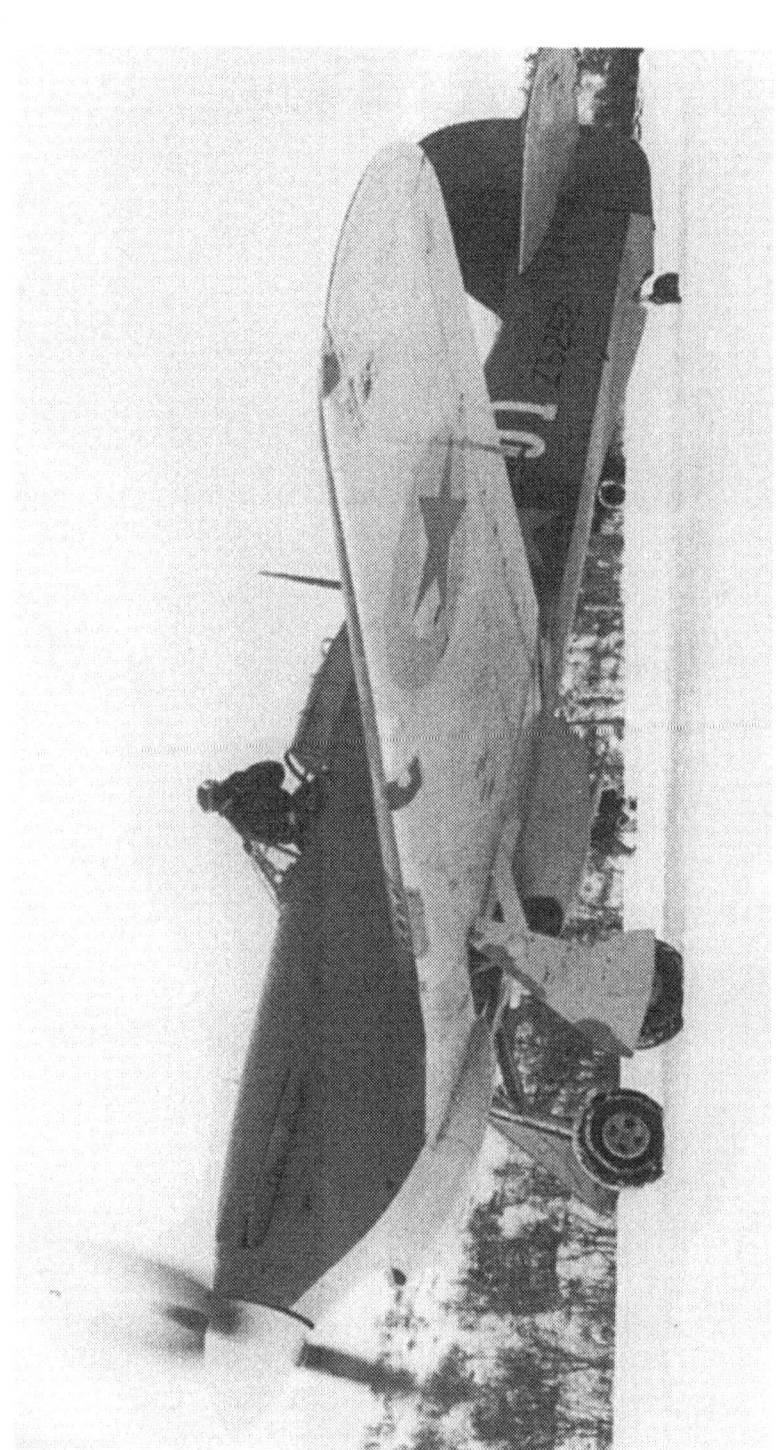

24 The British equipped two Soviet Air Force squadrons with Hawker Hurricane fighters at the start of the Second World War. This Hurricane with new red-star insignia flew from a Soviet air base near Murmansk. Courtesy of the National Air and Space Museum.

25 Soviet Air Force aces gather in front of a P-39 Aircobra in the North Caucasus: (left to right) K. Vishevskiy; A. I. Pokryshkin; N. Livitskiv; D. B. Glinka; I. Drushov; I. I. Babak; and G. A. Rechkalov. Courtesy of the National Air and Space Museum.

26 Soviet ace Alexander Pokryshkin (59 victories) flies a P-39 from an autobahn near Berlin in 1945. Courtesy of the National Air and Space Museum.

27 Lend-Lease Douglas C-47s and the Soviet-built licensed copies of the DC-3 were the backbone of Soviet air transport capability during the Great Patriotic War. This aircraft is equipped with a dorsal turret, with a 7.62 mm machine gun. Courtesy of the National Air and Space Museum.

28 A. S.
Yakovlev and
S. I. Ilyushin
seated below a
picture of Stalin
in 1945.
Courtesy of the
National Air and
Space Museum.

29 The MiG-9 became the Soviet Air Force's first operational jet fighter in the post-1945 period. Courtesy of the National Air and Space Museum.

30 A North Korean defector flew this MiG-15 jet fighter to an American airfield during the Korean War. This fortuitous event allowed the US Air Force to test the capabilities of the Soviet-designed fighter against the F-86 Sabre. Courtesy of the National Air and Space Museum.

31 The Il-28 jet bomber saw extensive service in the Soviet Air Force in the post-1945 period. Courtesy of the National Air and Space Museum.

32 The Be-6 flying boat, with its gull-wing design, mirrored the Martin Mariner influence. The Soviet Navy required flying boats to perform numerous roles. Courtesy of the National Air and Space Museum.

33 The Yak-36 (Forger) flew as a VTOL aircraft with the Soviet Navy. The design allowed Soviet aircraft carriers to project air power at remote spots beyond the territorial waters of the Soviet Union. Courtesy of the National Air and Space Museum.

34 The An-12, a four-engine air transport, possessed the capability for large-scale air drops and airborne operations. Courtesy of the National Air and Space Museum.

35 The Soviet *Kiev*-class helicopter-carrier in the Mediterranean Sea, July 1976. Courtesy of the US Navy.

36 Soviet MA-25 Hormone helicopter alongside a US Navy SH-30 SeaKing in the 1970s. Courtesy of the US Navy.

37 Over the North Pacific in 1971 a US Navy F-4 Phantom escorts a Soviet Tu-20 Bear long-range turboprop reconnaissance bomber. Courtesy of the US Navy.

38 A Backfire bomber with a standoff weapon over the Baltic escorted by a Swedish Air Force Viggen fighter. Courtesy of the Royal Swedish Air Force.

39 An Aeroflot Il-18 at Frankfurt Airport in the late 1950s. Courtesy of Frankfurt Airport.

40 The Il-62 went into Aeroflot long-range services in 1967. Courtesy of Aeroflot.

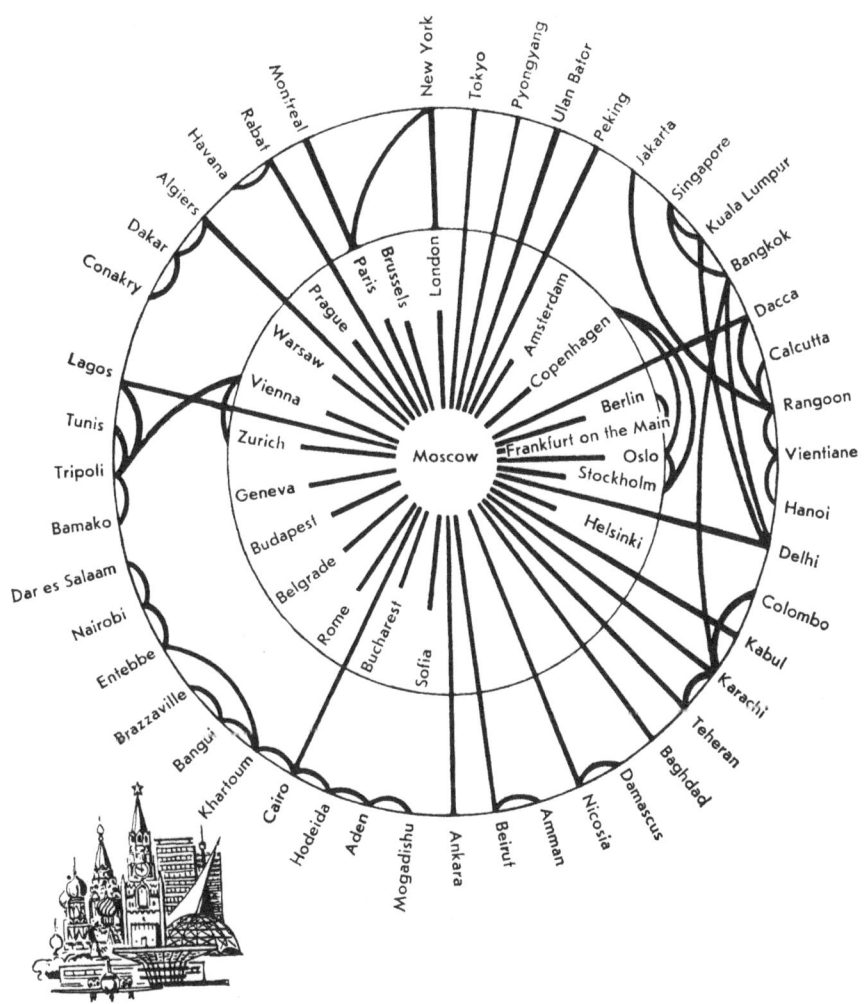

41 Aeroflot services in the mid-1970s. Courtesy of Aeroflot.

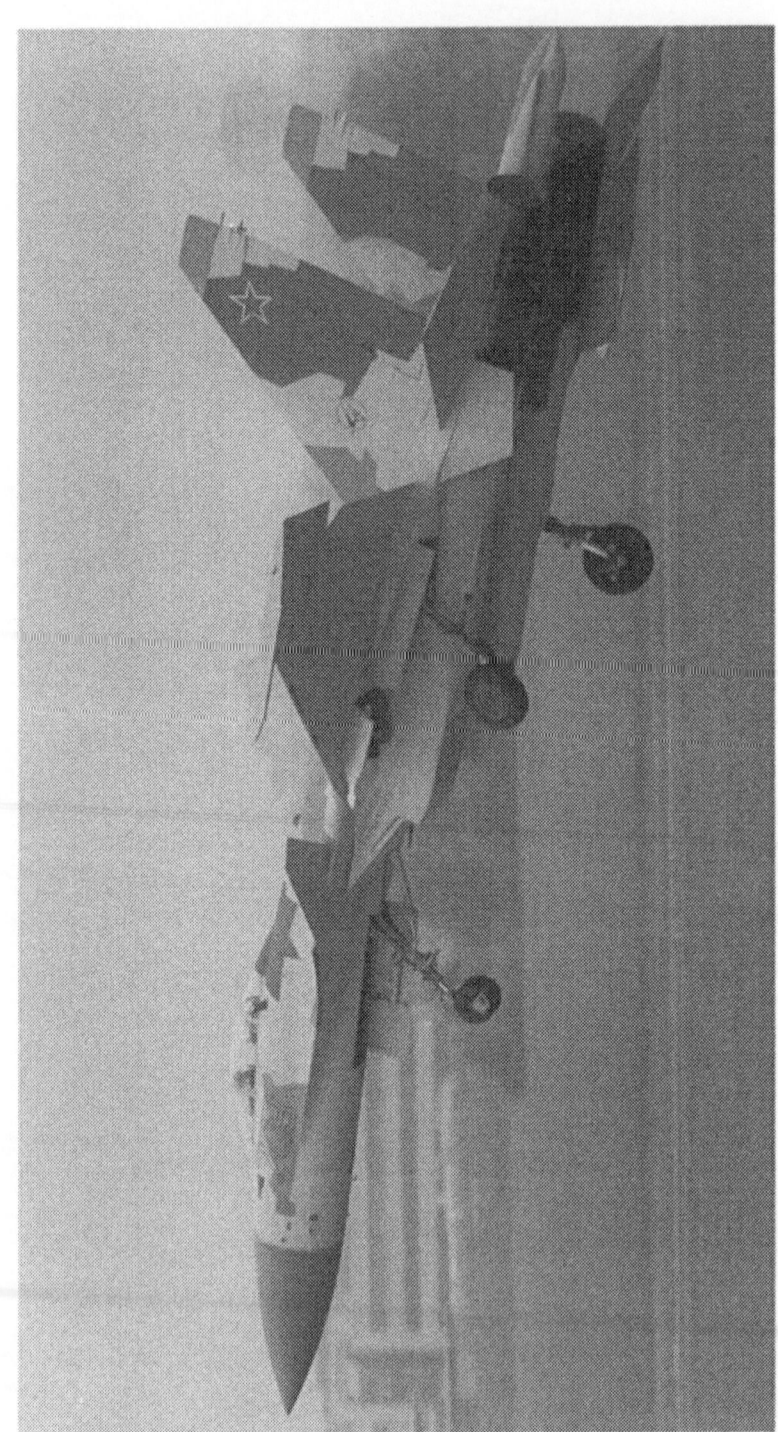

42 Shown at the Berlin Air Show in 1996, the Sukhoi-35 represented the latest Russian fighter development and one now for export. Chris Sorenson, photo courtesy of ILA '96 Berlin.

fighters were built, accounting for 57.8 per cent of all fighters and 24.5 per cent of all aircraft produced, while 22,436 Lavochkin fighters were completed (36.9 per cent of all fighters and 15.7 per cent of all aircraft). Some credit for this should probably go to Yakovlev's position throughout the war as a deputy director of the *Narkomaviaprom* where he was able to influence state production decisions. Il-2s and Il-10s accounted for all ground-attack production, and the 11,201 Pe-2s made up the vast majority of the frontal bombers until the Tu-2 appeared in numbers in 1944–45. The Il-2 accounted for 36,163 and the Il-4 for 4,528 aircraft. The yearly production figures for selected key aircraft during the war years can be found in Table 6.5.

Table 6.5

PRODUCTION OF SELECTED AIRCRAFT MODELS, JANUARY 1941–DECEMBER 1945

Year	1941	1942	1943	1944	1945	Total
Yak-1	1,332	3,476	2,720	1,128	–	8,656
Yak-3	–	–	–	2,180	2,380	4,560
Yak-7	207	2,431	3,296	465	–	6,399
Yak-9	–	59	2,493	7,831	5,087	15,470
LaGG-3	2,463	2,771	1,065	229	–	6,528
La-5/5FN	–	1,129	5,048	3,826	–	10,003
La-7	–	–	–	2,236	3,669	5,905
Il-2	1,542	8,229	11,193	11,110	4,089	36,163
Il-10	–	–	–	–	2,556	2,556
Il-4/DB-3F/	911	858	1,568	706	485	4,528
DB-3	–	–	–	–	–	–
Pe-2	1,671	2,524	2,428	2,944	1,634	11,201

Source: G. S. Byushgens (ed.), *Samolëtostroyeniye v SSSR 1917–45 gg. Kniga II* (Aircraft Construction in the USSR, 1917–45. Book II) (Moscow: Izdatel'skyi Otdel TsAGI, 1994), pp. 235–6.

In general, Soviet sources discount the wartime impact of Lend-Lease. However, deliveries from England and America were significant to the air war. Of 14,798 American aircraft allocated to the Soviet Union, 14,018 were delivered, while the British eventually supplied 2,952 Hurricanes, 143 Spitfires, and about 30 other aircraft. The 17,000 Lend-Lease aircraft (18,303 according to a recent excellent Russian study by G. S. Byushgens) did not represent a large proportion (11.9 per cent) of the 142,775 produced in the USSR during the Second World War. What is often overlooked, however, is that the program contributed $500 million in machine tools and factories (including an aluminum rolling mill) in addition to 2.25 million tons of steel, 400,000 tons of copper, and 250,000 tons of aluminum (the latter equaling two years' Soviet production at 1945 rates). The raw materials and equipment undoubtedly facilitated the changeover from wooden or composite aircraft to metal aircraft late in the war.

During the war, the plant relocations and new construction had significantly altered the geographical distribution of the aviation industry which was now much more widely dispersed throughout the country. Aircraft production had increased by 30 times in eastern Siberia, in the Urals by eleven times, and by fivefold in the Nizhnyi Novgorod (then Gorky)–Kazan–Samara (then Kuibyshev)–Saratov area along the Volga River. New aircraft and engine plant complexes were now located in Tashkent, Ufa, Omsk, and Tbilisi, while the new prewar plants in Irkutsk, Novosibirsk, Ulan-Ude, Perm, and Komsomolsk-na-Amure had greatly expanded. All of these plants would remain major production centers of the Soviet aircraft industry during the postwar years.

The Soviet aviation production plants produced 142,775 aircraft (117,501 combat) from January 1941 through December 1945. Despite its major wartime achievements, the Soviet aircraft industry still had major shortcomings. Of the 66,720 aircraft that the Soviet aviation industry produced in 1944–45, 84.9 per cent were single-seaters and all of them were piston-engined. Owing to the early loss of major aluminum production plants and the wartime scarcities, the airframes were mostly made of wood, laminated plywood, and composites, such as bakelite, so that the assembly plant workforce gained relatively little experience in building modern all-metal, stressed-skin aircraft. The concentration on single-engined aircraft since 1939 meant experience in building multi-engined aircraft had faded. Piston-engine technology was a serious area of weakness, and jet engine research had made small but important progress during the war. Thus, despite its very obvious wartime successes and vigorous development, when the war ended the Soviet aviation industry still lagged behind the West.

The Cold War Years (1945–91)

Well before the fighting was over, the Soviet Union urgently set about studying captured and interned aircraft and the technology of both friend and foe as it tried to redress the balance in its favor. In 1944 three American B-29 bombers landed in the Soviet Far East, forced down by combat damage or lack of fuel, and the crews were interned. Stalin assigned Tupolev the job of dismantling and copying the B-29s from top to bottom as the highest national priority, second only to the development of an atomic bomb. The Soviet version of the B-29, designated Tu-4, appeared at the Tushino air show in 1947. This entire effort was as critical to the overall postwar development of instrumentation, piston engines, airframe construction, and airborne armament as the exploitation of captured German technology. For Tupolev it was absolutely essential to the successful development of his impressive postwar family of strategic bombers.

In the Soviet governmental reorganization of 1946, the NKAP became the *Ministerstvo aviatsionnoi promyshlennosti* (Ministry of Aviation Industry, MAP) under M. V. Khrunichev, one of Shakhurin's former deputies. MAP largely led an independent existence within the Soviet military-industrial complex. It designed and produced military and civilian aircraft under powerful professional administrators, such as V. P. Dement'yev (1953–77), V.A. Kazakov (1977–81), I. S. Silayev (1981–85) and A. S. Systsov (1985–91), until the collapse of the Soviet Union in 1991. As during the war, MAP was responsible for the overall national aviation industry and directed the Soviet push into the jet age.

The war years also winnowed the design bureaux and left only several major OKBs for each type of aircraft. Fighter aircraft were largely the domain of Yakovlev, MiG, Lavochkin, and Sukhoi. Tupolev and Ilyushin specialized in multi-engined bombers and transports, and had already begun to branch out into civilian airliners. Ilyushin handled ground-attack aircraft, and the new Antonov OKB was directed into transports. Helicopters would later become the responsibility of the Kamov and Mil' OKBs. With a few notable exceptions, such as the demise of the Lavochkin OKB and brief existence of Myasishchev's OKB in the 1950s and early 1960s, this pattern remained consistent until 1992.

Even before the war ended, the Soviets had begun a high-priority examination of German jet aircraft to overcome their own low-level wartime efforts. Although the Soviet Union was the only major power which had not produced a jet fighter by the end of the Second World War, it had experimental engines and aircraft under development. Early in 1944, MiG, Yakovlev, Lavochkin, and Sukhoi were ordered to develop jet aircraft, but the lack of a true turbojet engine severely handicapped their efforts. A hybrid engine composed of a VK-107A compressor drive shaft, and special booster nozzle, was available for use, and around this engine MiG designed its I-250 which first flew on 3 March 1945. But this technology was outdated before it was ever used in a production aircraft. The immediate answer rested with advanced German technology. In 1945, Mikoyan headed a Soviet team that went to eastern Germany to study wartime jet aircraft designs. The Me-262, the world's first operational jet fighter, especially interested them.

The Soviets were naturally eager to exploit this German technology.

In 1945 MiG and Sukhoi went to work on twin-engined designs (which became the MiG-9 and Su-9, respectively), and the Yakovlev and Lavochkin OKBs were to develop single-engine models (the Yak-15 and La-150). Mikoyan proposed an innovative arrangement with both BMW 003 jet engines mounted side by side within the fuselage for his MiG-9 while Sukhoi's Su-9 strongly resembled the Me-262 with a Junkers Jumo 004A engine mounted under each wing. For his Yak-15, Yakovlev merely modified a Yak-3 airframe to take a Jumo 004B engine in a pod-and-boom configuration like the MiG-9 and La-150.

Just before his removal in December 1945, Shakhurin raised the question of copying the Me-262 at a Kremlin conference of the VVS and NKAP which ended up rejecting the concept of producing Germany's advanced aircraft. Yakovlev argued that the Me-262 was just too complex for the Soviet aviation industry and demanded too much of pilots and maintenance crews. He was also concerned that such outright copying would harm the Soviet designers who were then just beginning to work on jet aircraft. Although the Soviet leaders resisted the temptation to build an exact copy of the Me-262, they nevertheless copied its Junkers Jumo 004 axial flow engine as well as the captured BMW 003 jet engine.

Asher Lee has pointed out that the Soviets found in defeated Germany perhaps the largest and most advanced pool of scientific and engineering skills that any victorious nation has ever acquired in modern war. Nearly 80 per cent of Germany's aircraft production centers fell into Soviet hands, including the Junkers plant at Dessau, the Siebel plant at Halle, and the Heinkel plants at Oranienburg and Rostock-Warnemünde. They were quickly stripped, and their equipment sent to the Soviet Union. In addition, the Soviets carted off hundreds of aviation and rocket specialists to the USSR when they found they could not entice them to work for the Soviet Union.

The prototype MiG-9 (I-300) and Yak-15 both had their first flights on 24 April 1946, and with them the Soviet aviation industry officially entered the jet age. Both models were ordered into state acceptance trials and then approved for mass production by October 1946. The La-150 first flew five months after the MiG and Yak models which soon entered serial production, so Lavochkin's first jet fighter was abandoned in April 1947. Sukhoi's Su-9 flew late and suffered the same fate as the La-150.

With their basic jet fighter foundation set, the Soviets turned to British engines for their next group of jet aircraft. Mikoyan had studied jet technology in England, and in 1946–47 the postwar British Labour Government allowed the Soviets to purchase 30 Derwent V and 25 Rolls-Royce Nene II centrifugal-flow jet engines which were far superior to and more powerful than the German axial-flow engines. German engines used for the early Yak-15 and MiG-9 jet fighters were abandoned, and the British engines and their Soviet variants were installed in the MiG-15, La-15, and Yak-23 single-engine fighters and a twin-engine tactical bomber, the Il-28. The design offices of Klimov, Mikulin, and A. M. Lyulka (1908–84) worked on the development of new Soviet jet engines based on the Western models.

Jet propulsion brought Soviet designers many headaches in addition to the problem of where to locate the engines. In the MiG-15 (1948) and the Yak-19 (1949) the designers replaced the previous pod-and-boom approach with the standard straight-through jet fuselage design. Undercarriages were redesigned to fit into fuselages instead of inside the new slim, swept wings; the delta wing was investigated; and tailplanes were redesigned.

The MiG-15 was produced in large numbers and proved to be a superior fighter. Its design included swept wings, tricycle landing gear, and an ejection seat. It was armed with one 37 mm and two 23 mm cannon. Other equally impressive fighters of the Mikoyan and Guryevich family – the MiG-17, 19, 21, 23, 25, 29, and 31 – would follow the MiG-15 in the years to come. Yakovlev and Sukhoi also designed and produced a number of jet fighters, including Yakovlev's all-weather Yak-25, his tactical general purpose Yak-28, and Sukhoi's Su-7, Su-9, and Su-11.

In addition to these aircraft, Tupolev built a family of impressive jet aircraft: his twin-engine bombers, the Tu-16, seen in numbers in 1954, and the Tu-22, first seen at Tushino in 1961 along with the large Tu-28 fighter-interceptor. His Tu-95 swept-wing, turboshaft-engined bomber startled the West in 1955, but it was an efficient and effective aircraft – and remained in production in its refined Tu-142 model until 1988.

Bomber technology also helped to create a fleet of civil aircraft. In the mid-1930s the Soviet Union built the Douglas DC-3 under license as the Li-2, named after B. P. Lisunov, the chief engineer who organized its production. In late 1943, converting the Yer-2 bomber into a transport and passenger plane was considered and rejected. Realizing that a suitable civilian passenger and cargo plane would be needed after the war, Ilyushin began working on his twin-engine Il-12 which appeared in 1947. It was followed by the twin-engined Il-14 in 1950 and the four-engine Il-18 in 1957. As Western nations began introducing the first jet airliners into commercial service, the Soviet Union felt itself seriously lagging behind. Therefore, the government turned again to its bomber design bureaus with orders to produce a jet passenger plane, and the successful Tupolev and Ilyushin airliners discussed in more detail in Chapter 7 were the results.

Soviet designers always emphasized such essentials as clean lines, singleness of purpose, maintainability, and the necessity of volume production. As a result, they have produced aircraft which are highly suited to both their operational climate and their air crews and which are, in the long run, much less wasteful than comparable British, American, or French designs. While it might be argued that the Russian system of designing competitive prototypes was wasteful, such a system ensured success through permitting comparison at independent testing establishments of real, not paper, machines.

The 'Golden Age' of Soviet aviation was in the 1950s and 1960s when national strategic priorities provided continuing requirements for new military and civilian aircraft and almost unlimited funds to support research, development, and production. Inefficient, loud, gas-guzzling engines were tolerated because they were powerful and gave Soviet fighters a better thrust-to-weight ratio compared with Western jet fighters. Aeroflot was never concerned about the costs of aviation fuels because the state covered those costs under the planned economy. The leading aircraft designers were national and even international celebrities, and the engineers and technicians were an élite in an industry that enjoyed among the highest national priorities during the Cold War.

Beginning in the 1970s and continuing through the 1980s, the economy was less able to afford multiple, competitive design procurements. Fewer prototypes were ordered, and those were much more complicated and complex than ever before in an age of sophisticated electronics and exotic metal compounds that required careful shaping and extruding. The Soviet aviation industry could design and build an aircraft, such as the Mach 3 MiG-25, but it was built of welded steel construction – completely unlike anything that would have been built in the West. Vacuum tubes rather than microchips often inhabited even the most advanced aircraft. Powerful but inefficient and uneconomical engines on civilian aircraft often could not meet international noise and pollution standards, and reduced the attractiveness of the less expensive Soviet aircraft to potential buyers.

By the late 1970s, centralized control over the aviation industry had been lost. The industry's continued progress now depended on bureaucratic decisions in the central ministries which were made under heavy lobbying pressures to maintain development and production of specific aircraft and projects. The erosion of central authority lessened the MAP's ability to integrate the work of the design bureaux and production plants. The powerful design bureaux then stepped in to take over this function by leveraging their own vertically integrated, informal networks of suppliers and production plants. In an effort to improve management and to consolidate planning and supply within the Soviet economy, production plants were grouped into *proizvodstvenniye ob'yedinyeniya* (production associations, or POs) under an association director beginning in the mid-1970s. Some of these new POs were formed in MAP, with aircraft plants grouped into an *aviatsionnoye proizvodstvennoye ob'yedinyeniye* (aircraft production association) and engine plants into a *motorostroitel'noye proizvodstvennoye ob'yedinyeniye* (engine building production association). By the late 1980s, the major aircraft and engine production plants were consolidated into POs, some with their own associated design bureaux, such as those with their main offices at Kazan (Kazanskoye APO, KAPO), Irkutsk (Irkutskoye APO, IAPO), Moscow (Moskovskoye APO, MAPO), and so on.

Despite the concerted efforts at change in Mikhail Gorbachev's *glasnost* and *perestroika* policies, the inherent weaknesses and contradictions of the Soviet system finally could take the internal and external burdens no more. In the early 1990s the Soviet Union collapsed, and with it went this centrally controlled and managed aviation industry.

A New World: The Post-Soviet Era (1992–)

In the years following the dissolution of the Soviet Union in 1991, Russia's economic collapse crippled the aviation industry which struggled to right itself with a series of privatization initiatives and little overall governmental guidance. Hundreds of formerly state-owned research institutes, OKBs, production associations, and production plants received few if any new airplane orders from a Russian Federation Air Force that now found itself without funds and an Aeroflot that disintegrated into a mix of new national and regional airlines ('mini-flots') with the creation of the Commonwealth of Independent States (CIS).

In 1992 the Russian Government reduced its defense orders by 70 per cent, and Tupolev's military work alone dropped by 85 per cent. The aviation industry was divided up among the new countries of the CIS, with Russia and the Ukraine receiving by far the lion's share. Russia inherited nearly 85 per cent of the former Soviet aviation industry, with over 300 research institutes, design bureaux, production associations, and series production plants and three million workers. By 1994–95 total production had dropped 60–70 per cent compared with the levels of the mid-1980s.

New joint-stock companies, such as *Tupolev*, *Aviastar*, and *Rostvertol* (Mil' helicopters), were formed to venture into the unknown, many of them directly derived from the existing POs that had been formed from the 1970s on. By 1995, over 70 per cent of the aviation industry had become joint-stock companies, and 21 per cent remained state-owned, including those forbidden to privatize. In 1996, ten major design bureaux and 34 aircraft production plants remained in the CIS.

In Russia after 1992, a number of different governmental agencies supervised the dismantling of the Soviet Ministry of Aviation Industry and its subordinate *glavki* and their enterprises. Most prominent among them were the *Roskomoboronprom* (Russian Federation Committee on Defense Industry until May 1996, when it was redesignated the Ministry of Defense Industry), the Ministry of Defense, the State Committee of Industrial Policy, and the Department of Aviation Industry. At first part of the State Committee of Machine Building and later attached to the Ministry of Defense, the Department inherited many of the old Ministry of Aviation Industry's duties. Frequent changes of personnel and structure and a lack of co-ordination prevented the development of any coherent or long-term policies for the industry, so *Roskomoboronprom* and the Ministry of Defense assumed the leading roles until 1997.

In January 1997, the Russian Ministry of Economics assumed responsibility for the aviation industry from the Ministry of Defense and thus ended over 70 years of the aviation industry's alignment under some form of control by the military-industrial complex. This was a clear sign to the aviation industry that it no longer had its old security blanket of large orders from the military-industrial complex. Both Sukhoi and MiG had already turned to the international market to seek sales of their advanced Su-27/30/35/37 and MiG-29/35 models to replace their former VVS and Warsaw Pact customers that would sustain the development of advanced, fifth-generation aircraft. At home, Sukhoi and MiG were already fighting it out for what would likely be limited support for the design, development, and acquisition of future advanced combat aircraft for the VVS.

Ilyushin and Tupolev, the primary designers of large civilian passenger airliners and transport aircraft, have tried to go their own ways and establish vertical organizations formed around their design bureaux and the production plants that have built their aircraft in the past. Tupolev is aligned with the *Kazanskoye aviatsionnoye proizvodstvennoye ob'yedinyeniye imeni S. P. Gorbunova* (S. P. Gorbunov Kazan Aircraft Production Association, KAPO), builder of the Tu-4, Tu-16, Tu-104, and Tu-160, and the Il-62 and Il-62M. KAPO began series production of the Tu-204-200 passenger airliner in March 1992 and in the future will handle the Tu-330. For its Tu-334 series, Tupolev will use plants in Kiev and Taganrog. Ilyushin works with the Tashkent Aviation Production Association in Uzbekistan which produces advanced variants of the Il-76 transport and the Il-114 and Il-114T regional passenger airliners. The new Il-96T transport, which first flew in May 1997, is produced at the *Voronezhskoye aviatstionnoye proizvodstvennoye ob'yedinyeniye* (Voronezh Aircraft Production Association, VASO) plant in Voronezh. Both Ilyushin and Tupolev have rejected any discussion of mergers and have stated their view that the design bureaux and production companies are better off working as self-contained enterprises that keep their own traditions and expertise.

Airlines in Russia and other CIS states face significant problems with their aging fleets of Russian-built aircraft. Some estimates are that 50 per cent of the 8,000 Russian-built aircraft now in service will have to be retired by 2000. In 1996 the civil fleet had completed 72 per cent of their active lives and some had even reached 80 per cent. Thus, many of these airlines have turned to Western suppliers to lease or purchase new aircraft while Ilyushin and Tupolev have struggled to get new generations of commercial airliners into production and to re-engine older aircraft, such as the Il-76s, Il-86s, and Tu-154s, with quieter and more fuel-efficient engines. Sales of new and cheaper Russian-built airliners and transports to a potentially large domestic and international market could help revive the entire Russian aviation industry in the years to come.

The Ministry of Economics is now (in 1998) pushing for the creation of several Western-style, integrated, market-oriented companies, similar to Boeing/McDonnell Douglas and Lockheed Martin, that will combine research, design, manufacturing, and marketing. A start in this direction was made with the regrouping of the military aircraft sector into two large alliances of research, design, and production enterprises.

The new stated-owned conglomerate enterprise *Moskovskoye aviatsionnoye promyshlennoye ob'yedinyeniye-MiG* (Moscow Aircraft Production Organization-MiG, MAPO-MiG) was first formed in 1995 by combining MAPO with the MiG OKB. Then the *MAPO Voyenno-promyshlennyi kompleks* (Military-Industrial Complex, MAPO VPK) was established by Presidential decree in January 1996 and opened in May as a 'unity state enterprise' incorporating civilian and military aircraft manufacturers. It was intended to develop, manufacture, and market advanced Russian aircraft and weapon systems internationally and thus preserve and consolidate the national defense potential. MAPO VPK combined 12 separate aviation enterprises, including MAPO-MiG (jet fighters, etc.), Kamov (helicopters), Klimov (engines), Chernyshev Machine Building Enterprise (engine production), avionics, testing facilities, an overhaul plant, and the *Aviabank* Incorporated Commercial Bank.

The new chairman of MAPO VPK was Aleksandr Bezrukov

(b. 1964), who graduated from the Moscow Aviation Institute and was first deputy director general of MAPO-MiG. The general designer is Mikhail Valdenberg, also a graduate of MAI, who was the deputy chief designer for the MiG-21U, MiG-23, and MiG-27 before becoming the chief designer of the MiG-29 in 1982. Vladimir Kuzmin (b. 1937), another MAI graduate and general director of MAPO-MiG since 1991, was appointed the chairman of the board of MAPO VPK.

The second of these groupings, built around the Sukhoi and Beryev design bureaux and the three production plants at Komsomol'sk-na-Amure, Novosibirsk, and Irkutsk, was designated the *Aviatsionnyi voyenno-promyshlennyi kompleks Sukhoi* (Aviation Military-Industrial Complex Sukhoi, AVPK Sukhoi). Like MAPO VPK, AVPK Sukhoi was a state enterprise with the fully state-owned Novosibirsk and Komsomol'sk plants and partly privatized Irkutsk plant, and includes the Perm engine enterprise as well as component manufacturers.

AVPK Sukhoi's new general director, Aleksei I. Federov, former director of the Irkutsk Aviation Industrial Association, is engaged in a running battle with the chief designer of Sukhoi OKB, M. P. Simonov (b. 1929), who is fighting the vertical integration that AVPK Sukhoi and MAPO VPK represent. Simonov believes that only through the supremacy of the design bureau and general designer can the Russian aviation industry survive and prosper. Federov views things quite differently, and contends that the old system was destroyed but has not been replaced with a successful free-market system built on private investment. Thus, he contends that government leadership is required to push through radical restructuring that will produce the vertical integration of design and production. 'Up until recently in Russia,' Federov has pointed out,

> the relationship between design bureax and production plants differed radically from the rest of the world in their separation, even segregation, from each other. The absence of united aircraft design and manufacturing companies hampers the competitive capabilities of our industry on the international, and now also on the domestic, market. The creation of AVPK Sukhoi is the second effort, after the establishment of VPK MAPO, to establish an integrated aerospace company in the internationally accepted meaning of the word.

It will be interesting to see whether the old 'Soviet' view held by Simonov and supported by nearly 80 years of experience or the new 'Russian' view of Federov wins out. Which way will the aviation industry turn in this new world?

II. TSAGI AND THE RESEARCH AND
TESTING INSTITUTES

The role that the *Tsentral'nyi aerogidrodinamicheskyi institut* (Central State Aero-Hydrodynamic Institute, TsAGI) and the galaxy of other aviation research institutes has played in the history of Soviet and Russian aviation since 1918 demands special attention. Without some knowledge of the complex interaction of TsAGI and these institutes with the designers and production plants, much of the unique character of the Soviet aviation industry cannot be understood.

Nikolai Yegorovich Zhukovskiy (1847–1921), the man whom Vladimir Il'ich Lenin called the 'father of Russian aviation', set up one of the first wind tunnels in Europe at Moscow University in 1902. In 1904 he founded the Institute of Aerodynamics in the village of Kuchino and conducted extensive research on lifting capacity, wing contours, propellers, hydraulics, hydrodynamics, astronomy, and mathematics. During the First World War he worked on bombing techniques, artillery ballistics, and navigation, while continuing his work on theoretical mechanics.

Zhukovskiy opted to remain in Russia after the revolution. In March 1918 he founded the *Letuchaya laboratoriya* (Flight Laboratory) at the Moscow Higher Technical School. Zhukovskiy and some of his best students, including Andrei Nikolaevich Tupolev and I. A. Rubinskiy, then stressed the need to develop a scientific center upon which to base the development of Soviet aviation. They proposed the establishment of a single center for aerodynamics and hydrodynamics that could study, design, build, and test aircraft, motors, and other high-speed vehicles.

On 1 December 1918 Lenin approved the merger of the Flight Laboratory and personnel from the Red Air Fleet's design bureau to establish TsAGI which Zhukovskiy then headed until his death in 1921. TsAGI soon became the principal center for the study of aircraft aerodynamics, design, engineering, and construction and the training of engineers, designers, technicians, and managers for the aviation industry. Among those who helped Zhukovskiy found TsAGI and were to play central roles in the development of the Soviet aircraft industry were A. N. Tupolev, A. A. Arkhangel'skiy, A. I. Putilov (1893–1979), N. S. Nekrasov (1883–1957), and A. M. Cheremukhin (1895–1958). Zhukvoskiy also established a course of instruction for military pilots which became the basis for the *Institut inzhenerov Krasnogo vozdushnogo flota* (Institute of Engineers of the Red Air Fleet, II KVF) in 1920. Following his death in 1921, the institute was renamed the N. Ye. Zhukovskiy Military Aviation Engineering Academy and became the primary VVS center of aeronautical education and training for military pilots and engineers, which it still is to this day.

TsAGI has remained the heart of the Soviet and Russian aviation industry and over the years has explored virtually every aspect of aeronautical and later also astronautical science. From it have come many of the research scientists, aircraft designers, aeronautical engineers, production managers, and aviation industry ministers who were so instrumental in the development of the Soviet aircraft industry. During the 1930s, the first self-contained and isolated *naukograd* (science or defense city) of Zhukovskiy was built southeast of Moscow to accommodate all of TsAGI's laboratories, large wind tunnels, test facilities, airfield, and residential areas to support the expanding staff. From the research of TsAGI and other research institutes came the constant flow of scientific and technical reports, articles, and data that gave the design bureaux the latest information for the design of airframes, wings, structures, etc. TsAGI eventually grew to over 9,000 personnel by the early 1990s, but the current economic difficulties of Russia and its aviation industry have confronted TsAGI with some very hard times and difficult choices as it cuts its workforce and restructures to preserve itself.

In addition to TsAGI, the new Soviet Government of the early 1920s soon developed other special research institutes to investigate all aspects of aircraft development. In 1922, the *Nauchnyi avtomotornyi institut* (Institute for the Scientific Study of Motors, NAMI) was established to study and develop automobile and aircraft engines. NAMI's aircraft engine department along with TsAGI's engine department were responsible for engine development. An important step in Soviet engine development was the combining of the NAMI and TsAGI engine departments in September 1930 into *otdel aviamotorov* (Department of Aircraft Engines, OAM) in TsAGI under B. S. Stechkin (1891–1969). In December 1930 OAM TsAGI was united with the experimental engine department of Zavod No. 24 in Moscow to form the *Institut aviatsionnogo motorostroyeniya* (Institute of Aircraft Engine Construction, IAM) under I. E. Mar'yamov with two departments – gasoline engines under V. R. Klimov and diesel engines under A. D. Charomskiy (1899–1982). In August 1931 IAM was renamed the *Tsentral'nyi nauchno-issledovatel'skii institut aviamotorstroyeniya* (Central Scientific Research Institute of Aircraft Engines, TsIAM), now the *Tsentral'nyi institut aviamotorstroyeniya*, which eventually assumed all responsibility for research and development of aircraft engines.

Another important institute was created in June 1932, when TsAGI's experimental aircraft materials department was split off to form the *Vsesoyuznyi nauchno-issledovatel'skii institut aviatsionnykh materialov* (All-Union Scientific Research Institute of Aviation Materials, VIAM) to deal with materials related research and experimentation. Like TsAGI and TsIAM, VIAM came under the overall control of Baranov's VAO.

Over the years, numerous other institutes were created as new technologies were developed – for instance, the *Reaktivnyi nauchno-issledovatel'skii institut* (Jet Engine Scientific Research Institute, RNII). During the war, the relocation of TsAGI from Moscow to Kazan and Novosibirsk in Siberia led to the establishment of the *Gosudarstvennyi soyuznyi Sibirskii nauchno-issledovatel'skii institut aviatsii* (State Siberian Scientific Research Institute of Aviation, GosSibNIA) which became a sister research institute and even took over some of TsAGI's missions. In 1961, the importance of developing fabricating aluminum, magnesium, and titanium alloys for aircraft led to the establishment of the *Vsesoyuznyi institut lyogkykh splavov* (All-Union Institute of Light Metals, VILS) at Moscow. During the postwar period, the *Nauchno-issledovatel'skii institut aviatsionnoi tekhnologii i organizatsii proizvodstva* (Scientific Research Institute of Aviation Technology and the Organization of Production, NIAT) was established. To look after the planning and construction of the extensive network of research, testing, and production facilities, the *Gosudarstvennyi institut po prochtirovaniyu zavodov aviatsionnoi promyshlennosti* (State Institute for Design of Aviation Industry Factories, *Giproaviaprom*) was set up in the 1930s. The *Gosudarstvennyi nauchno-issledovatel'skii institut aviatsionnykh sistem* (State Scientific Research Institute for Aviation Systems, GosNIIAS) was established in 1946 from departments of TsAGI and LII.

Aviation educational institutes, such as the large *Moskovskii aviatsionnyi institut* (Moscow Aviation Institute, MAI) and *Kharkovskii aviatsionnyi institut* (Kharkov Aviation Institute, KhAI), also conducted structural and wind tunnel research for the design bureaux. Over the years these institutes and others have trained thousands of aeronautical engineers and technicians for the aviation industry.

The Soviet central government and the customer commissariats (after 1946 ministries) of Defense and the Civil Air Fleet (civil aviation) also created research and testing institutes and organizations that focused on their own special requirements and programs. The VVS early on realized the importance of flight testing aircraft designed for it to assure their conformity to the stated requirements and suitability for service. In 1920 the *Nauchno-opytnyi aerodrom KVF* (Scientific Test Airfield of the KVF, NOA KVF) was established at the Central Airfield at Khodynka, Moscow, under the control of the *Nauchno-tekhnicheskyi komiteta upravleniya KVF* (Scientific and Technical Committee of the Chief Administration of the KVF, NTK KVF and later NTK VVS after 1925). In 1926, NOA VVS was redesignated the *Naucho-issledovatel'skii institut VVS* (Scientific Research Institute of the VVS, NII VVS). Also under the NTK's control were the *Nauchno-ispytatel'nyi institut aviatsionnykh priborov VVS* (Scientific Test Institute for Aviation Instruments, NIIAP VVS), and the *Nauchno-ispytatel'nyi institut aviatsionnykh vooruzheniya VVS* (Scientific Test Institute for Aviation Armaments, NIIAV, VVS), which conducted their own research as well as state acceptance trials for these critical aircraft subsystems and components.

In 1930 the Civil Air Fleet added its own flight-test institute, the *Gosudarstvennyi nauchno-issledovatel'skii institut, Grazhdanskii vozdushnyi flot* (State Scientific Research Institute, Civil Air Fleet, GosNII GVF) to test new aircraft developed for civil aviation. In 1954, GosNII GVF became the GosNII *Grazhdanskaya aviatsiya* (GosNII GA) when the term *Grazhdanskaya aviatsiya* (Civil Aviation) replaced Civil Air Fleet. In March 1941, *Narkomaviaprom* created the *Lyotno-issledovatel'skii institut Gosudarstvennogo komiteta aviatsionnoi promyshlennosti* (Flight Research Institute of the State Committee on Aviation Industry, LII GKAP) from several TsAGI elements near TsAGI at Zhukovskiy to handle aerodynamic flight research. This institute is now known as the Gromov LII in honor of the former NII VVS chief test pilot, record-setting world flyer, and air commander of the 1920s, 1930s, and 1940s, M. M. Gromov (1899–1985).

The NTK and NII VVS established requirements for future military aircraft that were based on the current and projected military strategy and doctrine of the Red Army. These were then submitted to the central aviation industry organs for review and forwarding to OKBs. When new design proposals were received, the engineers of NII and NTK reviewed them for scientific and technical adequacy. NII VVS's responsibilities included conducting the flight testing and state acceptance trials for all military and civil aircraft at the huge Zhukovskiy complex and at the *Nauchno-ispytatel'nyi aerodrom VVS* (Scientific Test Airfield of the VVS) at Chkalov LII southeast of Moscow.

TsAGI and the research institutes operated under the overall control of the Ministry of Aviation Industry. Specialization of research and the careful centralization of the resources under MAP management were key components of the entire system. MAP authorized, co-ordinated, and funded research projects; oversaw standardization programs; organized the flow of information among the component *glavki* (chief administrations) that actually ran the separate subministerial parts of MAP; and put together the Five-Year Plans.

The structure of this system of aviation research institutes remains largely intact today, but central funding has largely evaporated as Russia struggles to reorganize its economic and governmental systems. The key institutes, such as TsAGI, TsIAM, Gromov LII, GosNIIAS, and others, are now called *Gosudarstvenniye nauchniye tsentry Rossiiskoy Federatsii* (State Scientific Research Centers of the Russian Federation). Many of the institutes have downsized and others have sought Western or commercial partners for their ongoing research and experimentation work.

To maintain the scientific and technical infrastructure of the military-industrial complex for the future, the Russian Government and State Committee for Defense Industry in 1996 decided to give up any procurement of new military aircraft for some years to come so that funds could be channeled into sustaining the scientific and technical base and developing new aircraft. It is still uncertain what will happen to this vital sector of the Russian aviation industry. What is critical is that after several years of drift the Russian Government finally determined that, without a fully functioning scientific-technical base and new generations of scientists and engineers, the future would be bleak for the entire aviation industry.

SUMMARY

This excursion through the history of the Soviet and Russian aviation industry is but a short introduction to a complex and important story that remains to be told in its entirety. Russian aircraft designers and manufacturers have taken scientific, technological, and organizational approaches to the development of their aircraft that are very different from those adopted in the West. Western aeronautical and military experts were repeatedly surprised by the effectiveness of Soviet aircraft that to their eyes appeared crude and unsophisticated. A careful reading of this chapter might well convince one that as soon as the present 'Time of Troubles' is past, a new and vibrant Russian aviation industry will once again assume its rightful and hard-earned place.

7

The Designers: Their Design Bureaux and Aircraft*

JOHN T. GREENWOOD

Starting from almost nothing, with few experienced designers and limited production facilities, the nascent Soviet aviation industry slowly coalesced around a handful of key designers whose numerous contributions laid the foundations of Soviet military and civilian air power. In addition to the purely technical aspects of aircraft design and construction, Soviet aircraft designers also faced challenges that were unknown to their Western colleagues.

The foremost of these was the political environment in which they worked, for politics always came before design and production considerations for the Communist Party. Party leaders, especially I. V. Stalin but also N. S. Khrushchev and Lavrenti Beria (head of the NKVD from 1937–53), dabbled in aircraft development and often interfered in design selection and priorities, production schedules, and the design bureaux to push their own ideas, favorites, and personal priorities. Moreover, Stalin and his cohorts in the 1930s and 1940s often pursued irrational purges that may have sustained some perceived political purposes but seriously harmed the aviation industry.

All aircraft designers must consider the basic aerodynamic requirements for flyable air vehicles in developing their creations. After that, many other factors shape their individual designs, such as the designer's personal education, training, and experience; the availability of materials, power plants, instruments, and other components; the skills and training of the assembly line and maintenance workforces; the quality and size of the production base; the intended operational mission requirements; the climatic operating environment; the

* This chapter is based on 'Patterns in the Aircraft Industry', which appeared in *Soviet Aviation and Air Power* (1977). The author wishes to thank Dr Von Hardesty of the National Air and Space Museum and Mr Aleksei Druzhilov, a graduate of the Moscow Aviation Institute and currently with the World Bank, for their very extensive and helpful comments.

social, political, and economic system; and so on. Over many years, Soviet aircraft designers evolved their own distinctive, indigenous design philosophy that took all of these factors into consideration and that differed significantly from that of their Western colleagues. Aleksandr S. Yakovlev, a leading Soviet aircraft designer for many years, summarized this basic Soviet design philosophy in his auto-biographical *Tsel' zhizni* (The Aim of a Lifetime) simply as: (1) maximum simplicity for ease of production and reliable operation; (2) evolutionary development to minimize risk and potential impact on production; (3) minimum requirements for field maintenance to operate without regard to climatic conditions and the availability of developed airfields and field support. Hence, Soviet aircraft were simple, rugged, and reliable as well as easy and inexpensive to manufacture, maintain, and modify.

The most prominent and persistent design characteristics of Soviet aircraft are evolutionary design, component commonality, design heredity, adaptability, and standardization. Proven designs and components were often reused in new designs and model improvements to simplify production and assure reliability. A quick look at Sukhoi's Su-7/17/20/22 and Su-9/11/15 series as well as the now almost infinite number of Su-27 variants demonstrates the wisdom of evolutionary design, design heredity, and component commonality – no great technological leaps are risked, production is not seriously interrupted for retooling and retraining, and service requirements are met.

The Russians are masters of improvisation, and successful aircraft designs built for one purpose were frequently adapted for important new uses. Tupolev used this approach throughout his career, and the conversion of the Tu-16 and Tu-95 bombers into the very successful Tu-104 and Tu-114 civilian airliners stand out as excellent examples of adaptation of proven designs for alternate uses.

Standardization of major components and subsystems played an important role in all Soviet aircraft design after 1940. Prior to that, designers in the various OKBs used their own rules for designs, component and subsystem selection, construction, and testing. Large OKBs like Tupolev and Polikarpov with many highly skilled personnel may not have needed regulations, but the sudden appearance of a number of OKBs in the late 1930s and the rush to get new aircraft to production resulted in a number of crashes of test models. This prompted the preparation of the initial edition of *Rukovodstvo dlya konstruktorov* (Designers' Handbook, RDK) in 1940 and its publication in 1941 to standardize the design, construction, and testing of new aircraft.

Leading aeronautical scientists, engineers, and test pilots contributed to this new design manual that covered everything from aerodynamics to engines, strength of materials, landing gear, and arma-ment. The RDK funneled the latest scientific, research information, and data on available technology and equipment to the designers and engineers to assure the development of the best possible designs using standard data and components. At the same time it helped to assure standardization and uniformity because the *Narkomaviaprom* and later the Ministry of Aviation Industry executed it as policy at the national level across all OKBs and production plants. Design, production, maintenance, training, and the production and supply of spare parts were all greatly simplified because the components and subsystems were centrally developed and produced for distribution and use throughout the aviation industry and by the customers.

Men such as Tupolev, Petlyakov, Myasishchev, and Sukhoi, all heavily influenced by Zhukovskiy and TsAGI, combined with Polikarpov, D. P. Grigorovich, K. A. Kalinin, and others to form the older, first generation of designers whose educational origins predated the Bolshevik seizure of power. A younger group of almost purely Soviet-bred designers formed around Ilyushin, Yakovlev, Mikoyan and Guryevich, Antonov, and Lavochkin. Ilyushin, Yakovlev, and Mikoyan also represent a different tradition in aeronautical engineer-ing as they were trained at the Zhukovskiy Military Air Engineering Academy and formed the distinctive VVS engineering influence among the designers. Because their ideas, design bureaux, and aircraft have shaped Soviet and now Russian aviation, the key designers and design bureaux will be surveyed.

N. N. POLIKARPOV (1892–1944)

Nikolai Nikolayevich Polikarpov was the son of a priest from Orel. In 1916 he graduated from the Petrograd Polytechnical Institute and began his engineering work with the air navigation department of Igor Sikorsky's Russo-Baltic plant in Petrograd. He remained in Russia after the Revolution and worked for the Chief Administration of the Red Air Fleet from March to August 1918 before becoming the tech-nical director at the former Duks Factory in Moscow, which became *Gosudarstvennyi aviationnyi zavod* (State Aircraft Factory, or GAZ) No. 1 in the early 1920s.

Through 1925, Polikarpov worked as the chief designer at GAZ No. 1. There he copied the Avro 504K, put the SPAD VII into produc-tion, and produced Russian versions of the DeHavilland DH4, 9, and 9A. His earlier R-1, really a variant of the DH9, was the first Soviet aircraft to be mass produced, with over 2,700 being built. In 1923 he also designed the I-1 (IL-400), which was not only the first Russian-designed and -built fighter but also the first monoplane fighter. Remaining at GAZ No. 1, in 1926 he joined the newly established *Tsentral'noye konstruktorskoye byuro, Gosudarstvennyi trest aviatsionnoi promyshlennosti* (Central Design Bureau, State Aviation Industry Trust, or *TsKB Aviatresta*) as the chief designer of its newly organized Department of Landplane Construction (OSS).

Polikarpov's work as OSS director complemented his continued extensive design activity. Like Tupolev, in his direction of design work Polikarpov was autocratic and tried to control his staff much as Stalin controlled the *Politburo*. Such a hierarchical structure severely limited creative work to small subareas of design and frustrated both new and old engineers. Polikarpov's design philosophy, however, differed markedly from Tupolev's in that he always strove to design an aircraft equal to if not better than those then serving in the West. He constantly pushed this approach which strained the bounds of current technology, and he was repeatedly frustrated with the inability of the Soviet aircraft industry to build his aircraft.

An excellent multi-purpose Polikarpov biplane aircraft, the simple and ubiquitous *Uchebnyi* (trainer) 2 (U-2), was produced in large numbers from 1929 on and over 30,000 were built before its production finally ceased. Redesignated the Po-2 (Polikarpov-2) after the designer's death in 1944, the U-2 was the notorious puddle-jumping, 'stealthy' night bombing 'Washing Machine' and 'Bed Check Charlie' that harassed first German troops throughout the war in the East, and later United Nations' forces during the Korean War.

Polikarpov's situation changed dramatically in the late 1920s when the OGPU arrested first Grigorovich and many of his associates in September 1928, and then in October 1929 Polikarpov, who was accused of sabotaging ongoing work after a number of his prototypes crashed. Interned in the OPGU's special design bureau, the TsKB-39 *Menzhinskiy*, Polikarpov created the successful I-5 biplane fighter. Capable of speeds up to 286 kph (180 mph), the I-5 entered series production in 1931 and 803 were built through 1934. Widely used for many years, the I-5 even saw service in the Second World War. The success of the I-5 obtained Polikarpov's release from captivity but not his escape from the influence of the OGPU as he remained associated with the TsKB as chief of design Brigade No.1. He become chief designer at Zavod No. 36 and his OKB was transferred to Zavod No. 21 in Gorky in 1935.

Now under the overall direction of Ilyushin's TsKB, Polikarpov became the leading designer of Soviet fighter aircraft. In 1933–34, Polikarpov designed the biplane I-15 fighter, with a maximum speed of 360 kph, which he later upgraded to the I-15bis (1938) and I-153 (1939) fighters. Finally came the I-16, the first monoplane fighter with a retractable landing gear and controllable pitch propeller to be accepted by the Soviet air force. The I-15 and I-16 entered service at about the same time and saw extensive action in the Spanish Civil War and later against the Japanese at Khalkhin Gol in 1939.

The I-15's maneuverability was outstanding, and several modifications improved its speed and upgraded its armament. The I-153, for example, built in 1938, attained speeds up to 443 kph (278 mph) because of its streamlining and retractable landing gear (it was the only Soviet biplane ever so equipped). At Khalkhin Gol in 1939 the I-153, mounting four machine guns and rockets, proved to be an outstanding fighter.

Polikarpov's I-16 also was continually improved over the years. Eventually the plane was fitted with a movable canopy, and, for the first time, armor plate was installed behind the pilot. It was continually upgraded with ever more powerful air-cooled radial engines in each succeeding model until the I-16 had a 930–1,100 hp M-63 engine tipped with a controllable-pitch propeller that gave it a maximum top speed of 462 kph. After several years the various modifications, including adding more armor plate, became a problem because the 30 per cent increase in take-off weight and wing loading hampered the plane's maneuverability. Despite these problems, Polikarpov's fighters of the mid-1930s were well ahead of those of rest of the world.

Only 384 I-15s were produced in 1934–36, but the other models were produced in large numbers – 2,408 I-15bis in 1938 and 1939, 3,437 I-153s from 1939 through 1941, and 9,450 I-16s from 1934 through 1941. The 15,679 Polikarpov fighters were the Soviet front-line fighter force from the mid-1930s through the beginning of the Great Patriotic War.

Polikarpov's later designs, specifically the I-180 (1939) and I-185 (1941), were largely unsuccessful and suffered many crashes which seriously harmed his reputation. He fell out of favor with Stalin due to a disagreement over fighter designs in 1939, and this led the *Narkomaviaprom* to detach a number of his key personnel, such as A. I. Mikoyan and M. I. Guryevich, to form a new design team in 1939. Polikarpov's OKB moved to Novosibirsk in 1941 during the relocation of the aviation industry, and he returned to Moscow in 1944 where he died on 30 July. His OKB was disbanded after his death, with Sukhoi taking over much of his experimental work.

A. N. TUPOLEV (1888–1972)

The most successful and influential of Zhukovskiy's students, Andrei Nikolayevich Tupolev, came from a middle-class provincial family.

While attending the Moscow Higher Technical College, he developed an interest in aviation. While still a student working with Zhukovskiy, Tupolev helped design the first Russian wind tunnels and also designed and built training gliders. Graduating from the college in 1918, he became a co-founder of the TsAGI with Zhukovskiy and held key leadership positions there from 1918–35. He headed TsAGI's design departments, AGOS and then KOSOS, until 1936 when his design bureau was separated from TsAGI and established at Zavod No. 156 in Moscow as a major design bureau and prototype construction factory for the Chief Administration of Aviation Industry.

Of TsAGI's founders, Tupolev had the longest and most distinguished career and the greatest influence on Soviet aviation. On 1 January 1919 the *Aviatsionnyi otdel* (Aviation Section) under Tupolev was established to study aircraft design and construction. His early studies convinced him that cantilever monoplanes with metal structures of aluminum alloy rather than wooden biplanes would be the best designs. He closely followed foreign aeronautical developments and knew of the Junkers' work on cantilever all-metal monoplanes with skins of corrugated aluminum alloy.

Following Zhukvoskiy's death in March 1921, he became deputy to S. A. Chaplygin, the new director, and began working on the introduction of light-alloy metals in aircraft construction. He assisted in the development of Kol'chug aluminum, similar to Duraluminum, that was lightweight and strong and could be rolled in sheets and sections. In September 1922, TsAGI appointed Tupolev as the chairman of a new commission which was charged with the design and construction of an all-metal aircraft for the Red Air Forces. The commission included TsAGI's aviation materials testing division and Tupolev's own design section. Within the new organization he set up an experimental base to test aircraft because he very much believed, contrary to others at TsAGI, that a design team required test aircraft to develop fully new concepts and designs. Critical to Tupolev's efforts was the establishment of the Junkers factory at Fili in 1923 that was to design and build all-metal aircraft using Duralumin and to train a Russian workforce.

In 1922, the first Tupolev aircraft, the light sports plane ANT-1, took to the skies and launched an amazing career that spanned the next half century. Tupolev's team then designed the first all-metal aircraft built in Russia, the ANT-2, which flew in 1924. In 1925, Tupolev designed and built the first all-metal combat aircraft, the two-seat *razvedchik* (reconnaissance) R-3 (ANT-3). That November he produced the *tyazhyolyi bombardirovshchik* (heavy bomber) TB-1 (ANT-4) twin-engined, monoplane, which firmly established Tupolev as a leading designer of large, multi-engined aircraft.

As Tupolev's design work increased in the mid-1920s, his design team became the *Aviatsiya, gidroaviatsiya i opytnoye stroitel'stvo* (Aviation, Marine Aviation, and Experimental Construction Department, AGOS) of TsAGI in September 1925. He split AGOS into groups, later designated brigades, under lead designers that were responsible for specific design tasks. For example, A. A. Arkhangel'skiy and A. I. Putilov designed fuselages while V. M. Petlyakov handled wings and N. S. Nekrasov had tail units. Each project had a chief designer who was responsible for integrating all facets of design. Thus, Pavel O. Sukhoi was the lead designer of the *istrebitel'* (fighter) I-4 (ANT-5) that appeared in 1927.

In 1929 a ten-passenger, three-engine Tupolev ANT-9 with a complement of Moscow correspondents flew to Berlin, Paris, Rome, London, and Warsaw, a distance of 9,000 kilometers, at an average speed of 180 kph. After this flight the plane was put into serial production for the emerging civil air fleet.

Tupolev's TB-1 (ANT-4) bomber was one of the largest planes built in the 1920s and was in production from 1926–32. An all-metal monoplane powered by two water-cooled engines in the wings, the TB-1 could carry a payload of three-and-a-half tons. With a ton of bombs, the plane had a range of 1,350 kilometers (850 miles) and a speed of about 200 kph. The TB-1 was the first plane to be equipped with boosters to reduce takeoff distance. Based on their experience with the TB-1, Tupolev, Petlyakov, and Arkhangel'skiy determined that all-metal cantilever monoplanes much larger than the TB-1 could successfully be built. The next design was a four-engined aircraft with a wingspan of 130 feet (40 meters) and weighing 35,000–40,000 pounds. That aircraft, the TB-3 (ANT-6) bomber, was built and first flew in 1930, was in series production from 1932 to 1937 and could carry five tons of bombs; reducing the load to two tons gave a range of almost 2,500 kilometers. With a crew of eight, the TB-3 defended itself with eight machine guns. The TB-3 also set a design pattern for all large aircraft by placing the engines across the wing's leading edge. Soviet factories built 216 TB-1 and 818 TB-3 bombers.

In 1933, Tupolev became head of KOSOS TsAGI, which replaced the former AGOS TsAGI and had three aircraft construction workshops in a large assembly hall and five separate OKBs. As Polikarpov did with the TsKB, Tupolev ran a large, tightly controlled design enterprise. He believed that a number of small design bureaux only dispersed rather than concentrated the limited talent available. Only large organizations like AGOS and KOSOS could gather the many skills and numerous engineers and technicians needed to design the complex multi-engined aircraft of the day. Tupolev's aim was not to compete with the Western designers but rather to design aircraft that the industry could produce and that fulfilled the requirements of the customer, be it the VVS or Aeroflot.

Tupolev was the pre-eminent Soviet designer of the 1920s and 1930s, specializing in large, multi-engined aircraft. His design bureau was in the midst of a run of major design successes that included the TB-3 (ANT-6), I-4 (ANT-5), PS-9 (ANT-9) transport and air-liner, R-6 (ANT-7), SB (ANT-40), the ANT-20 *Maxim Gorky*, and the RD (ANT-25). His aircraft set records for size (ANT-20) and long-distance flights (ANT-25). Under his tutelage a number of important designers developed, including Petlyakov, Myasishchev, Arkhangel'skiy, and Sukhoi, who were later to head their own inde-pendent design bureaux.

Based on the success of his ANT-4, ANT-6, and ANT-16 designs and the belief that aircraft with wingspans up to 650 feet (200 meters) were feasible, Tupolev designed and built the ANT-20 *Maxim Gorky*, a giant eight-engined plane with a wingspan of 206 feet (63 meters). Described by Soviet historians as an agitation and propaganda aircraft intended to popularize Stalin's regime, the ANT-20 made its first flight in June 1934 and could carry 80 passengers at a speed of 280 kph. Weighing 40 tons, it had a radio station, printing press, photo labora-tory, telephone switchboard, telegraph office, and motion-picture pro-jectors. In 1935, it crashed after a mid-air collision with another plane, killing 35 people.

The first Soviet ground-support bomber to be put in series pro-duction was Tupolev's ANT-40 *skorostnoi bombardirovshchik* (SB) high-speed bomber. Designed by Arkhangel'skiy's brigade, the SB-2 twin-engined bomber of 1935 was state of the art, highly maneuver-able, and faster than most fighters it confronted early in its service with the Republican forces during the Spanish Civil War (1936–39). Manu-facturing began in 1934 and ended in 1941, and eventually 6,831 SBs came off the Soviet assembly lines in Moscow and Irkutsk, and Czech-built versions brought the production total to 6,992. Its two 860-horsepower engines gave the SB a speed of 420 kph, a range of 1,000 kilometers, and a bomb load of 500 kilograms (1,100 lbs). Unlike earlier Tupolev aircraft, which had corrugated skins, the SB was made completely of smooth Duralumin. Production of all other Tupolev air-craft up to the ANT-40 totaled only 2,084 aircraft, so the SB-2 definitely was a milestone for Tupolev and KOSOS. When the Great Patriotic War began in June 1941, SBs accounted for 94 per cent of the VVS's operational tactical bomber strength.

Tupolev's ANT-25, also known as the *Rekord dal'nosti* (Record Distance) or RD, was produced in 1933. This all-metal, single-engine monoplane with a very large wing span gained fame in the summer of 1937 when V. P. Chkalov, G. F. Baidukov, and A. V. Belyakov flew it for 63 hours nonstop from Moscow to Portland, Oregon, in the United States, a distance of 8,504 kilometers (5,673 miles). Initially designed as the *dalnyi bombardirovshchik* (long-range bomber) DB-1, its low speed (200 kph) ruled out such employment. However, mostly under P. O. Sukhoi's design brigade, the basic RD design was continually revised in the mid-1930s and included the DB-1 (ANT-36) (*RD-Voyennyi variant*, RD-Military Variant), and the DB-2 (ANT-37) twin-engined bomber which never reached production.

In January 1936, Tupolev was appointed the chief designer of the *Glavaviaprom* under *Narkomtyazhprom* and in July KOSOS-ZOK, whose staff had grown to 4,391, was separated from TsAGI and became Zavod No. 156 of the NKTP. At this time he was also appointed as chief engineer of *Narkomtyazhprom* and made responsible for production plant construction. Arkhangel'skiy's brigade then became a separate OKB and moved to Zavod No. 29.

The Soviet Union in 1937 was in the midst of the Great Purges, and on 21 October the NKVD arrested Tupolev and a number of his colleagues. So many designers were imprisoned that work at KOSOS effectively ended and that in much of the aviation industry virtually came to a halt. Thus, I. V. Stalin and Nikolai Yezhov, then head of the NKVD and himself soon to become a victim, organized the Special Technical Bureau of the NKVD that included special design bureaux and research facilities in the various prisons and concentration camps under control of the GULAG.

In late 1938, Tupolev and a number of his close associates were assigned to the NKVD's Central Design Bureau No. 29 (TsKB-29) that was set up at the old KOSOS offices on Radio Street and at Zavod No. 156 in Moscow. While in what came to be called *Tupolevskaya sharaga* (Tupolev's Special Prison Workshop), Tupolev worked on Project 103 (ANT-58) that later became the excellent Tu-2 twin-engined tactical bomber; Petlyakov designed Project 100, the VI-100, that became the Pe-2 frontal bomber and won Petlyakov his release in 1940; Myasishchev was on Project 102 for a new long-range, high-altitude bomber until his release in 1941; and Tomashevskiy worked on Project 110 for a new fighter.

Under this severe handicap of imprisonment from 1937–41, Tupolev rebuilt his design team and designed the excellent twin-engine tactical bomber, the Tu-2, as a replacement for his SB series. It appeared in 1941, too late to compete seriously with the Pe-2, which was already in full production. The Tu-2 was also handicapped by the need to convert to ASh-82 air-cooled radial engines, because the AM-37 water-cooled, high-altitude engines it was designed to use had been canceled so that more AM-35 engines could be built for the Il-2s. The re-engining cut the Tu-2's top speed from a remarkable 643 kilometers per hour to 560–580 – a loss of 80 kph and a significant tactical disadvantage compared to the German fighters. Hence, the new Tu-2 did not see major commitment to combat until 1944 and was the only new aircraft design to be built during the war. The 1,216 Tu-2s built from 1942 to 1945 represented only 10.6 per cent of the 11,426 Pe-2s of all varieties produced before and during the war, but it remained in production until 1950 and over 3,000 were eventually built. The Tu-2 was the standard tactical bomber in the postwar VVS inventory until the advent of the Il-28 jet bomber in the early 1950s.

With the success of the Tu-2, Tupolev regained his freedom in August 1941 as the design team moved to Zavod No. 166 at Omsk. Tupolev's OKB was re-established and returned to Zavod No. 156 in Moscow in 1943.

Tupolev's postwar career included the complete copying of the Boeing B-29 to produce the Tu-4, the first truly strategic bomber in the Soviet inventory. Even before the Tu-4 flew, Tupolev had converted the Tu-4 into the 66-passenger Tu-70 by moving the wings from a mid- to the low-wing position and increasing its length. The Tu-70 was not adopted for production because there was then neither an Aeroflot requirement for such a civilian airliner nor many Soviet airports it could use. The creation of the Tu-70 from the Tu-4 established Tupolev's pattern of converting his bomber designs into civilian passenger and transport aircraft with only minor modifications.

Tupolev's second-generation, swept-wing jet bomber, the twin-engined Tu-16, first appeared at the May Day flyby in 1954 and was also easily adapted for passenger use. The Tu-16 was soon modified into the civilian Tu-104 which first flew in June 1955. Again, the civilian version differs only slightly from its military twin – the wing was moved from the mid- to low-wing position for civilian service and the military model had a different, slimmer fuselage. Because the Tu-16 preceded the Tu-104, its design and components from wing-form and tail unit to undercarriage and engines were largely carried over to the civilian clone.

Tupolev continued his dual development with the Tu-95 and Tu-114. The swept-wing Tu-95 was truly an exceptional aircraft. It was designed in response to the Ministry of Defense and Stalin's requirement for an intercontinental strategic heavy bomber capable of atomic strike missions against the United States. Tupolev's design was a four-engined, propeller-driven, mid-wing aircraft with a gross weight of 342,000 pounds, a speed of 500 mph at 41,000 feet, and an estimated operational range of 7,800 miles with a 25,000-pound payload. The Kuznetsov NK-12M turboshaft engine (12,000 shp) was originally developed by Junkers engineers under Dr Ferdinand Brandner at Kuibyshev (now Samara) from 1950–54 and later modified by N. D. Kuznetsov. The eight-blade, contra-rotating propellers ran at 750 rpm with a supersonic tip speed. The Tu-95 first appeared in May 1955 and entered operational service with the Strategic Forces in 1957 and later with the Naval Air Forces. The Tu-95's range and endurance made it an excellent maritime reconnaissance aircraft that often tracked Western naval forces in the Atlantic, Pacific, and Mediterranean from the early 1960s on. From 1954–69, 172 production model Tu-95s were built, of which 53 were strategic reconnaissance versions. All Tu-95s are now out of service or scrapped as a result of the Strategic Arms Limitation Talks (SALT) agreements.

A variant of the Tu-95 resulted from a Soviet Navy requirement for a long-range anti-submarine and maritime patrol aircraft. Tupolev appointed Nikolai Bazenkov, chief designer of the Tu-95 and Tu-114, as the chief designer for the project, which was designated the Tu-142. Bazenkov stripped the Tu-95 of its strategic equipment, redesigned the wing to increase the span so it could carry more fuel, and removed all defensive armament. The Tu-142 first flew in July 1968 and entered production that year and continued until 1988. It is believed that 225 Tu-142s were built, and all but eight, which went to the Indian Navy in the 1980s, were assigned to Long-Range Naval Aviation units based in the Northern and Pacific areas and also in Cuba and Vietnam until 1990. The Tu-142 has a combat load of 25,000 pounds, including missiles, depth charges, and torpedoes; a maximum range of 7,779 miles; and a maximum speed of over 550 mph.

New strategic bombers were also coming from Tupolev's drawing boards during the 1960s and 1970s. His Tu-22, Tu-22M, and Tu-160 supersonic bombers replaced older Soviet bombers.

Soviet desires to expand Aeroflot's services beyond the Soviet frontiers to enhance its influence in less developed areas of the world required a long-range airliner. In 1952 Tupolev put Nikolai Bazenkov to work building such an airliner based on his Tu-95 design. By taking the wings, engines, tailplane, and undercarriage from the bomber and adding a new double-deck fuselage designed for passengers, Bazenkov built a prototype in just 18 months. Tupolev's second airliner, the Tu-114, could carry 220 passengers in a single cabin tourist arrangement on routes up to 6,000 miles. The Tu-114's normal Aeroflot configuration carried 168 passengers and was powered by four NK-12M (14,000 shp) or MV (14,795 shp) turboprop engines which gave it a normal cruising speed of 480 mph at 29,500 feet and a maximum speed of 541 mph. The Tu-114 entered serial production at Factory No. 18 at Kuibyshev (now Samara) in 1958 and a total of 32 aircraft were built before production ended in 1964. The Tu-114 was the world's largest passenger aircraft until the Boeing 747 appeared and, except for Howard Hughes's *Spruce Goose*, was the largest aircraft to enter service in the world until the Antonov An-22 flew in 1965. Aeroflot's introduction of the Il-62 in March 1967 resulted in the eventual phase out of the Tu-114s, most of which were withdrawn in 1975–76, although a few continued to operate until 1980.

Soviet planners in the late 1950s realized the need to replace the piston-engined airliners then serving the short- and medium-length routes with a jetliner. Impressed with the performance of the Tu-104, the planners selected Tupolev to develop the new medium-range jet. Tupolev assigned this task to Dmitrii Markov who used the Tu-104 as the basis for the new, smaller Tu-124. Completing its first flight in March 1960, the Tu-124 was powered by two Solovëv D20P 11,905-pound thrust engines mounted in the wing roots. Although it resembled a scaled-down Tu-104, the aircraft was essentially completely new. Initial deliveries were made to Aeroflot in 1962, and 163 of the 111 passenger models were produced at Kharkov through 1966. The Tu-124 remained in Aeroflot service until 1979.

The former Soviet design, development, and production process for civil aircraft is very clearly demonstrated in the case of the Tu-154. The *Ministerstvo grazhdanskoi aviatsii* (Ministry of Civil Aviation, MGA) wanted a replacement for the three different types of medium-range aircraft then in service – the Tu-104, Il-18, and An-10. The Ministry developed its requirements – 150–160 passengers and 2,000–2,175 miles range – and sent them to the Ministry of Aviation Industry which asked the Tupolev, Ilyushin, and Antonov design bureaux to submit proposals. Tupolev's totally new design, the tri-jet Tu-154, was accepted and work began in 1965 under Sergei Yeger.

The Tu-154 added a number of new features for a Soviet aircraft, among them high-efficiency wing systems with slats, triple-slotted flaps, and spoilers; control boosters on all surfaces with multiple redundancy; multiple redundancy for the electrical, hydraulic, and control systems; and a six-wheel undercarriage to reduce stress on the runways. The first flight took place in 1968 and Aeroflot-initiated passenger service began in 1972. Production of the new aircraft began at Kuibyshev (Samara) as Tu-114 production ended and continues today. By the end of 1975, the Aeroflot's 110 Tu-154s carried over 10 million passengers. Through 1994, 602 Tu-154A/B production models had been built along with 307 Tu-154Ms that had the more fuel-efficient Solovëv D30KU-154 engines replacing the old NK-8s.

During a state visit to France in 1960, Nikita S. Khrushchev flew in a Sud Caravelle, the first jetliner with tail-mounted engines, which was significantly quieter than the Tu-104 with its engines mounted in the wing roots and next to the passengers. On his return to Moscow,

Khrushchev directed Tupolev to design an aircraft similar to the Caravelle. Markov again received the job, and used the Tu-124 as the basis for the new aircraft, originally designated the Tu-124A and later the Tu-134 in 1963. The new aircraft first flew in 1963 but suffered from a number of technical problems. It finally entered service in 1967 with Aeroflot and was soon sold to a number of eastern European and client states. Production continued from 1963 to 1984 and 852 Tu-134s were built. The Tu-134 continues in service today with 608 still operating and 410 of those in airline service. By current operating standards, the Tu-134's fuel economy is poor for a basic 72-passenger aircraft, but for many the airlines of the CIS there are few choices available for replacements.

In 1990, Aeroflot's 540 Tu-154s teamed with its 450 Tu-134s to carry over 75 per cent of the airline's 137.5 million passengers. Since then Aeroflot has broken up into numerous airlines, and only 435 Tu-154s remained in service with Russian airlines at the close of 1996. Many Tupolev airliners which are rapidly approaching the end of their design lives continue to fly, and as many as 270 will have to be withdrawn from service within the next five years. In addition, other Tu-154s have been withdrawn already due to the steep drop in Russian passenger traffic since 1991. Those Tupolev airliners still in service continue to carry a large amount of the Russian and CIS passenger traffic.

One of Tupolev's most ambitious postwar undertakings was the design of a supersonic transport (SST). In 1963 he assigned his son, Aleksei, himself an accomplished designer, the task of designing the new aircraft. A great many problems confronted the design team, and the Soviet government made sure that Tupolev had the support required. The first prototype was completed in the summer of 1968 and the world's first SST flew on 31 December of that year. The Tu-144 broke the sound barrier for the first time on 5 June 1969 and exceeded Mach 2 on 26 May 1970. However, all was not well with the world's first SST. Fuel consumption was much higher than previously anticipated. While added fuel costs were not a significant factor for Soviet planners, the range dropped from 4,039 miles to just 2,175 miles so that refueling stops were necessary to reach distant cities in eastern Siberia and Central Asia.

The Soviet competitor to the Anglo-French Concorde, the Tu-144 was also cursed with bad luck when the second production aircraft crashed before 300,000 horrified spectators at the Paris Air Show in June 1973. The accident was not apparently a result of an aircraft failure, but the negative publicity it generated hurt the Tu-144. Aeroflot grew less enthusiastic about the Tu-144 after the Paris crash. The restrictions on its use meant that Tu-144 saw only limited service in

1977–78 before being withdrawn after the first production model Tu-144D with more fuel-efficient engines crashed after an in-flight fire in May 1978. Of the 16 Tu-144s built, ten remain, and four of them are in museums.

The Tu-144's story does not end there, however, for in 1993 Tupolev entered a contract with Boeing, McDonnell Douglas, Rockwell Collins, NASA, and others to make a Tu-144 into a research vehicle for future supersonic transports, including a proposed Tu-244. Flights tests of the refurbished Tu-144LL (*letayushchaya laboratoriya*, flying laboratory) began in November 1996, and the first supersonic flight (Mach 1.42) took place during the sixth test flight on 21 May 1997.

In the 1980s, the Ministry of Civil Aviation sought replacements for the Tu-134s and Tu-154s then serving as mainstays in Aeroflot. In 1983 the Tupolev bureau proposed a new Tu-204 as the second generation replacement for the Tu-154. Lev Lanovskiy, chief designer, intended to give the Tu-204 the latest technology in engines, electronics, and systems. The new aircraft was completed in the summer of 1988 and first flew in January 1989. Continuing technical problems, new initiatives to replace the Solovëv PS-90A high bypass turbofan engines with Pratt and Whitney PW2240 or Rolls Royce RB211-535 engines, and the unraveling of the Soviet Union and its aviation industry after 1991 severely hampered the entire Tu-204 program. The emergence of new aviation companies, such as the Tupolev Joint Stock Company; Aeroflot's smaller successor, ARIA (Aeroflot-Russian International Airlines); *Aviastar*, the former Ul'yanovsk aviation production factory; and various joint ventures such as BRAVIA (British–Russian Aviation Corporation) that equipped one Tu-204 prototype with the RB211 engines as the Tu-204-220, only added more obstacles to overcome.

All of these difficulties dramatically slowed development of the Tu-204. Expected to enter service in 1990, the first Tu-204 was not received until 1995 and entered service with Vnukovo Airlines and ARIA in 1996. The Tu-204 can carry 214 passengers and has a range of 2,800 miles.

Another new Tupolev airliner, Tu-334, was designed in response to a Ministry of Civil Aviation request in 1985 for a replacement for the aging fleet of Tu-134s. It, too, suffered from the severe dislocations in the aviation industry after 1992. Rollout of the prototype took place in August 1995, two years behind schedule, and funding shortages prevented flight testing until 1996.

After Tupolev's death in 1972, his son Aleksei Andreyevich, became the general director of the Tupolev design bureau, which was renamed the *Aviatsionnyi nauchno-tekhnicheskii kompleks imeni ANT* (Aviation

Scientific and Technical Complex, named after A. N. Tupolev, ANTK ANT) in 1989. By 1972, the Tupolev OKB had grown from 3,397 workers in 1945 to 15,240, and over 4,000 of these worked at the *Zhukovskaya lëtnaya i dovodochnaya baza* (Zhukovskiy Flight Test and Development Base, ZLiDB) that was established in 1949–51 for flight testing Tupolev-designed aircraft. With a talented group of designers around him, Aleksei Tupolev continued to develop new aircraft. With the breakup of the Soviet Union, a new Tupolev Joint Stock Company was organized, with the new Russian Government holding a strategic interest for a specific period of time and the remainder of the stock being sold on the market to raise capital. The new company was designated a 'shareholders society of the open type' and for the first time began negotiating contracts with the production plants that built its aircraft. In 1992 the stockholders of the new company appointed Vladimir Klimov as the general designer to take it through the turbulent times, and Aleksei Tupolev became the chief designer. In mid-1997, Klimov himself gave way to Sergei Shevchuk.

In June 1995, Russian President Boris Yeltsin signed a decree announcing the formation of the Russian Aviation Consortium that was composed of Tupolev, the *Aviastar* Production Factory at Ul'yanovsk, *Aviadvigatel* (designers and builders of the Perm PS90A engines), ARIA, and Universal Scientific Products Center. The government provided the RAC with financial concessions and promised government assistance to help solve the problems that were delaying the completion of the much needed new Tu-204, Tu-214, and Tu-334 airliners.

During his long career, Tupolev is credited with over 100 aircraft designs, many of which established significant Soviet and international records and pioneered major technological advances. He nurtured new generations of designers who now form a major part of his continuing legacy to Russian aviation. Without question, Andrei Nikolayevich Tupolev is rightly the dean of Soviet and Russian aircraft designers.

S. V. ILYUSHIN (1894–1977)

Sergei Vladimirovich Ilyushin, born to a peasant family from the Vologda province, was 20 years old when he saw his first plane at the St Petersburg Aerodrome. In 1914 he was called into the army, and in 1916 he became involved with military aviation – first as a workman in a hangar, then as an engine mechanic. In 1917 he passed his pilot's examinations. When the Red Army was created, he enlisted, and in 1921 he was sent to the TsAGI, where he designed and built training gliders. Ilyushin graduated from the Zhukovskiy Military Aviation

Engineering Academy in 1926, and until 1931 was assigned to the *Nauchno-issledovatel'skii institut VVS* (Scientific Research Institute of the VVS, NII VVS), which established aircraft design requirements for the VVS and reviewed all designs submitted by the design bureaux. From 1931 until his retirement he was fully engaged in aircraft design.

Ilyushin assumed direction of the loosely structured Central Design Bureau (TsKB) in 1933 and co-ordinated the work of other OKBs, such as those of Polikarpov and Yakovlev, through the remainder of the 1930s. His DB-3 twin-engine bomber first appeared in 1935 through the personal intercession of Stalin, and entered production and service in 1937. Continually modified to improve its range, bomb load, and overall performance, the DB-3, redesignated the DB-3F in an improved version and later the Il-4, went into production in 1937, and 6,883 were built through 1945. The Il-4 formed the backbone of the VVS and naval air force (VVS/VMF) medium-bomber units throughout the war.

Ilyushin served briefly in 1939–40 as a deputy chief of the *Narkom-aviaprom* before resigning to return to design work. His greatest claim to fame must remain the Il-2 Shturmovik ground-attack aircraft of the war years. Work was already underway on new model ground-attack and bomber aircraft when the Kremlin conference met in 1939. Ilyushin and Sukhoi worked on suitable ground-attack aircraft in an effort to satisfy the VVS's requirement for a new tactical armored attack aircraft. Ilyushin's TsKB-55 was completed first and flew, albeit somewhat indifferently, in 1939. Military officials were not pleased with the sluggish performance of the two-seat prototype, and it was re-engined with Mikulin's more powerful AM-38 engine (1,600 hp) and redesigned into a single-seat aircraft without defensive protection. This new version, the TsKB-57 (Il-2), began flight testing in 1940 and successfully completed state trials in March 1941. Later known as the Shturmovik or *Ilyushka*, the heavily armed and armored Il-2 was already in limited production and slated to replace the I model fighters that had been pressed into temporary service and the Su-2s then in the ground-attack units.

The most important addition to the VVS in 1941 was Ilyushin's single-seat Il-2 Shturmovik. Although it was a rugged aircraft with good armament of two 7.62 mm Shkas machine guns, two 20 mm SHVAK cannon, and excellent armor protection, the early version of the Il-2 was sluggish because of its maximum takeoff weight of over 11,000 pounds. Only 249 Il-2s had been completed by the time the war started. Production rose slowly to 1,293 for July–December 1941 but fell far short of meeting the insatiable demands of the front for attack aircraft.

The VVS ground-attack regiments were equipped with the finest aircraft of its type in any air force – the Il-2m3 Shturmovik. With the attainment of air superiority, and with such fighters as the La-5/5FN and Yak-3 providing cover, the Il-2s operated in ever-larger swarms with great effect against German infantry and armored units. For more effective antitank operations, in May 1943 the GKO ordered Il-2s equipped with two wing-mounted NS37 37 mm cannon. This new Il-2m3M (modified) version made its combat debut in the Kursk cauldron, fighting with telling effect. The Shturmoviks contributed substantially to the Soviet air and ground effort and remained in series production from 1941 through 1944; total output of the Il-2s reached 36,163, representing 25.3 per cent of all aircraft built during the war.

His postwar achievements were largely in the area of transports and airliners but also included the Il-28 twin-engined medium jet bomber that appeared in the early 1950s, and saw extensive service in the VVS and other Soviet Bloc air forces. The twin-engined Il-12 and Il-14 airliners formed the backbone for Aeroflot's growth in the early 1950s, along with the Il-18 turbine-engine airliner of the late 1950s. Ilyushin entered the jet airliner era with the 186-passenger transcontinental Il-62, powered by four rear-mounted NK-8 engines, which entered service with Aeroflot in 1967, and followed that with the four-engine widebody Il-86 in the late 1970s. The Il-76 military and civilian heavy-lift transport appeared in the mid-1970s.

Upon Ilyushin's retirement, Genrikh Vasilyevich Novozhilov, chief designer of the Il-86, assumed the duties as general director. Today Ilyushin's attention is largely focused on programs to re-engine its Il-76s and Il-86s and to get its new Il-96T and M aircraft into production. While the Il-76 program looks successful arranged with CFM International, the financing for the Il-86 conversion has fallen through. Much hope today is placed in the Il-96M, a 300-passenger airliner, and its transport version, the Il-96T, which are Il-86 derivatives with new and better PW 2337 engines and greater range. The Il-96T first flew in May 1997, and the first production aircraft of the 3T models for ARIA was scheduled for delivery later in the year, with initial deliveries of the 17 Il-96Ms slated for 1998.

A. S. YAKOVLEV (1906–89)

Much hope was placed in a young generation of Soviet engineers and designers. One of these was Aleksandr Sergeyevich Yakovlev, author of the highly readable but not always accurate memoirs *Rasskazy aviakonstruktora* (Tales of an Aeronautical Engineer, 1957) and *Tsel' zhizni* (The Aim of a Lifetime, 1969). As a youth, Yakovlev demonstrated an interest in mechanical objects. After completing secondary school at the age of 17, he decided to be an aircraft designer. His first step was to obtain a job as an ordinary worker in the carpentry shop at the Zhukovskiy Academy while actively designing and flying gliders. Then, to get even closer to the planes, he took a job as a mechanic's helper in the academy's flight-training unit and eventually worked his way up to engine mechanic. This brought him into contact with Ilyushin, who was then finishing his studies at the Academy, and the two established a career-long relationship. In 1926, Yakovlev built his own two-seater, named the AIR-1 in honor of A. I. Rykov, then president of Aviakhim, but the always politically correct Yakovlev quickly dropped the AIR designation when Rykov was purged and executed in the late 1930s. The AIR-1 turned out to be successful and gained Yakovlev admission to the Zhukovskiy Academy, thus opening a long and rewarding career, despite some temporary reverses in the mid-1930s.

During the press for industrialization and new aircraft in the 1930s, Yakovlev joined the TsKB and worked on Polikarpov's I-5 fighter but continued to work on his own light aircraft. His own record-setting AIR-7 suffered an aileron failure and crashed, nearly costing him his design career. However, by shrewd politicking with Communist and aviation leaders, Yakovlev was allowed to re-establish his own small design bureau in a bed factory near Moscow. 'No one ever thought at the time', Yakovlev later noted, 'that this little shop was destined to turn into a leading aircraft plant ...'. He designed a series of *Uchebno-Trenirovochnyi* (advanced training, UT) aircraft, the UT-1 (AIR-14), and the UT-2 (AIR-10/20), that appeared in 1935.

Yakovlev's design bureau (OKB-115) was among a number of new OKBs that responded to the VVS requirement for new fighter designs during the late 1930s. His I-26 design of 1939, redesignated the Yak-1 in 1940, completed its state tests early and went into series production. Yakovlev's career especially benefitted at this point from an extended trip to Germany in 1939 with Polikarpov to procure copies of the record-setting Heinkel He 100 fighter and the ongoing purge of many of his contemporaries. In 1940, Yakovlev was appointed deputy chief of the *Narkomaviaprom* for research and development, a position he held until 1956, when he became the ministry's chief designer.

The production of the Yak-1 was the smallest of the three fighters accepted in 1940 due in part to plant evacuations. Only 1,332 of these planes were built in 1941 compared to 3,100 MiG-3s and 2,463 LaGG-3s. The Yak-1 was the best of the three and the forerunner of the exceptional series of Yakovlev fighters that formed the backbone of the VVS's wartime fighter force. Refinements introduced in the Yak-

7B and the Yak-7DI long-range fighter became the basis for the initial Yak-9s, which were produced in the fall of 1942 and first saw action with Guards units at Stalingrad in December of that year. This aircraft was the progenitor of an entire family of multipurpose Yak-9 fighters built after 1942. The basic Yak-9 had a ceiling of 33,000 feet and a maximum speed of nearly 370 mph at 10,000 feet, was armed with one 20 mm SHVAK cannon and a 12.7 mm UBS machine gun, and had a range of approximately 600 miles. The fast, lightweight Yak-3 of 1943 was the best pure fighter of all the Soviet wartime designs, and was excellent in low-level fighter-to-fighter combat.

Yakovlev's postwar designs continued his tradition of building light aircraft and trainers, but also included important first generation jet fighters, and even a largely unsuccessful venture into helicopters with the Yak-100 and Yak-24 in the 1950s. The Yak-15 jet fighter was basically a Yak-3 airframe fitted with a German Jumo 004B, and first flew on 24 April 1946, three hours after Mikoyan's MiG-9. The Yak-17 and Yak-23 variants followed the Yak-15 in the late 1940s and early 1950s. The Yak-25 of 1955 was the first Soviet twin-engined, multiseater, all-weather fighter to enter service and was followed in the early 1960s by the improved Yak-28.

In the 1960s Yakovlev began development of the Yak-36, vertical takeoff and landing (VTOL) powered by twin vectored-thrust turbofan engines. Similar to the Hawker Harrier, the experimental VTOL led to the Yak-38 single-seat air defense and strike fighter that served on the small carriers of the Kiev class, the *Kiev* and *Minsk*. The Yak-41, which appeared in 1989, was the world's first supersonic VTOL. Another follow-on VTOL, the more advanced Yak-141 naval fighter has recently appeared.

The Yakovlev OKB has continued to design civilian feederliners, such as the short-range tri-jet Yak-40 that appeared in 1968 and medium-range Yak-42 that entered service ten years later. More recent aircraft are the Yak-77 business-class jet that appeared in 1993, and the Yak-130 trainer prototype that was built with Aermacchi to contest for the Russian Air Force advanced trainer, and first flew in April 1996.

Yakovlev's extensive writings provided an insight into the world of the Soviet aircraft designer, but must also be read with caution. He was much more politically attuned and aligned with the Communist Party and its leadership than most of his colleagues. Yakovlev's role in Tupolev's imprisonment and criticism of Tupolev's leadership of the aircraft industry during the 1930s did not endear him to a large segment of the Soviet aviation industry and caused a bitter and continued hostility between the two men and their followers. Yakovlev died in August 1989.

A. I. MIKOYAN (1905–70) AND M. I. GURYEVICH (1892–1976)

Artem Ivanovich Mikoyan grew up in a remote Transcaucasian village, and had the good fortune to be the younger brother of Anastas Ivanovich Mikoyan, the powerful Bolshevik leader and close associate of Stalin. After working as a mechanic in the 1920s, Mikoyan entered the Zhukovskiy Military Air Engineering Academy in 1931 and graduated in 1937. He was then assigned as VVS representative at GAZ No. 1 in Moscow, which then housed Polikarpov's OKB and was producing the I-153. In March 1939 Polikarpov asked Mikoyan to assist in redesigning the I-153 production line, which brought him to the attention of P. A. Voronin, manager of GAZ No. 1, and P. V. Dement'yev, the chief engineer. After a difficult meeting with the GAZ No. 1 team and Polikarpov during the summer of 1939, Stalin personally ordered the organization of an *opytnyi konstruktorskyi otdel* (Experimental Design Department, OKO) within the Polikarpov OKB to develop a new monoplane fighter. Voronin and Dement'yev recommended Mikoyan for the demanding new assignment.

Before he would accept the mission, Mikoyan insisted that his close associate, Mikhail Iosifovich Guryevich (MiG aircraft bear their initials, the 'i' standing for the Russian word 'and') join the OKO as his deputy. Guryevich had studied physics and mathematics for two years at Kharkov University. During the revolution he emigrated to France, but returned in 1921 to study at the Kharkov Institute of Technology, graduating in 1925. As did other Soviet aviation personalities, he spent several years working on various types of gliders. In 1929 he joined the aircraft industry and worked for P. E. Richard, a French engineer who briefly worked for the Soviet government, and then S. A. Kocherigin. In 1937, Guryevich went to the US to negotiate a license to build the Douglas DC-3 and learned the Douglas loft-mold manufacturing technique which was new to Soviet aircraft production. After helping to set up the new PS-84 (Li-2) production line, he joined Polikarpov in late 1938. Guryevich's experience in aircraft manufacturing and his technical knowledge were critical to the new design team.

Mikoyan and Guryevich went to work on their I-200, and in March 1940, Mikoyan was appointed chief constructor at GAZ No. 1 and Guryevich as deputy chief constructor. The I-200 first flew on 5 April 1940, but, as with the two other new fighters, all was not well with the new aircraft (designated the MiG-1 in December 1940). Consequently, after a series of tests and changes, the new MiG-3 with improved range and stability emerged and entered production after the last of the 100 MiG-1s was completed in December 1940. MiG-3 production itself was terminated early in 1942 to satisfy the more urgent demand for Shturmoviks. Although Mikoyan's fighter was an excellent high-altitude interceptor, such an aircraft was little needed in a largely tactical air war featuring few German strategic bombing attacks. As the LaGG-3 and Yak-1 became available in numbers, the MiGs were progressively withdrawn from service at the front, where they had been used because of the lack of other more suitable fighter aircraft, and were transferred to air defense units, where they remained until 1943–44.

In March 1942, the Mikoyan OKO was reorganized with Zavod No. 155 in Moscow, with Mikoyan as manager and Guryevich as chief constructor of the new OKB. Wartime requirements meant that no MiG aircraft saw service except for the MiG-1 and MiG-3 interceptors.

That situation changed dramatically after an extensive postwar tour of occupied German aircraft factories and an examination of their jet designs. With this information to draw on, Mikoyan and Guryevich began to produce quality jet fighters, and never ceased. Designed in 1945–46 to use the German BMW 003 jet engine, the MiG-9 was the first of these fighters and the first Soviet jet aircraft to fly on 24 April 1946. The MiG-9 entered full production after state acceptance trials later in 1946 and was the predecessor of a distinguished line of MiG jet fighters that has included a vast array of aircraft, in the MiG-15, MiG-17, MiG-19, MiG-21, MiG-23, MiG-25, MiG-29, and MiG-31 series.

Mikoyan became general designer of the OKB in 1956 and held that post until his stroke of May 1969 disabled him. He died on 9 December 1970. Guryevich was chief constructor from 1956 to 1964, when he retired. He died in 1976. The Mikoyan and Guryevich OKB was indeed a genuine team in which both men contributed their strengths to the good of the whole OKB. The success of their designs and the OKB they built show that most clearly.

Rostislav Apollosovich Belyakov (b. 1919) became the general director of the MiG OKB in 1971. A 1941 graduate of the Moscow Aviation Institute, he had worked as a design engineer on a number of projects and had shepherded the MiG-29 and MiG-31 aircraft through their development phases and into operational service in 1983. With the dissolution of the Soviet Union, MiG did not go the way of Tupolev or Ilyushin but has remained the centerpiece of the new state-owned MAPO VPK, which combines the A. I. Mikoyan Aviation Research and Production Complex (ANPK MiG) with the oldest Russian aircraft manufacturing enterprise, the MAPO. In May 1995, MAPO-MiG was established by governmental decree as a 'state unitary enterprise of the military-industrial complex' and incorporated 12 companies of the aviation industry and a commercial bank. With a steadily declining domestic market for its advanced combat aircraft, MiG has increasingly turned to export sales of its thoroughly modernized MiG-29 while it continues the development of a new MiG-35 intermediate size fighter, the *Mikoyan Mnogo funktionalniy frontovoi istrebitel'* (Multi-Functional Frontal Fighter, MFI) (Project 1-42) for a large, multi-role fighter, and the Mikoyan LFI lightweight frontal fighter series of fighters. The new MiGs are intended as much to compete in the international market for fighter sales as against its archrival, Sukhoi's new Su-37 and Su-54 fighters.

O. K. ANTONOV (1906–84)

One of the younger Soviet designers who had attained some success before the Second World War was Oleg Konstantinovich Antonov, who was born in the village of Troitsa in Moscow province. He completed his training at the Kalinin Polytechnical Institute in Leningrad in 1930. Already active in gliders and sport flying, he was immediately appointed chief engineer at the Moscow Glider Factory and rose to be its chief designer before it was closed in 1938. After failing to enter the Zhukovskiy Air Academy that year, he joined Yakovlev's Sportsplane Factory and was assigned to prepare the German Fieseler Fi 156 Storch for production under license in Kaunus (Lithuania) in 1940–41. The German invasion quickly terminated that plant's work and Antonov was reassigned to work on the A-7 assault glider until he rejoined Yakovlev in 1943 as first deputy designer in charge of Yak-3 development and production at Novosibirsk.

At the end of the war he was directed to create his own OKB at Novosibirsk, which subsequently moved to Kiev, and to produce a utility airplane with a 730-hp engine. The resulting An-2 biplane won him a Stalin Prize in 1952. Since then, Antonov has concentrated on large military and civilian transports, most of them, until very recently, high-wing, rear-loading, and turbine-powered. Among Antonov's most successful designs are the twin-engined An-8, the four-engined An-10 and its military counterpart the An-12, the An-22 (which held the title of the world's largest aircraft until the Lockheed C-5A appeared in 1968), and the An-24 and An-26 medium transports. In the 1970s and 1980s, Antonov designed the two largest aircraft in the world – the four-engine An-124 (maximum takeoff weight of 890,000 pounds) and the gigantic (maximum gross takeoff weight of 1.32 million pounds) six-engined jet An-225 that appeared in 1988. Petr Vasilyevich Balabuyev (b. 1931) replaced Antonov as the general designer in 1984. A graduate of the Kharkov Aviation Institute in 1954, Balabuyev has worked on all of the recent Antonov designs and headed the An-225 team.

The Antonov design bureau, now located in the Ukraine, recently designed and built the An-70 tactical transport to replace the An-12. Another high-wing aircraft with a tail loading ramp, the An-70 is powered by four Motor-Sich/Progress D-27 propfan engines with contra-rotating propellers.

P. O. SUKHOI (1895–1975)

Pavel Osipovich Sukhoi studied under Zhukovskiy and then joined the Red Army in 1917. In the 1920s he joined Tupolev's design team at TsAGI where he designed a single-seat fighter (I-4) and worked on the ANT-4 bomber. Sukhoi's team at KOSOS designed the ANT-25, ANT-37 (DB-2) bomber, and the ANT-31 (I-14) – the first Soviet all-metal low-wing monoplane fighter with retractable undercarriage and an enclosed cockpit. In 1936 he was put in charge of the factory design department at the new Komsomol'sk-na-Amure plant and began designing the ANT-51 (Su-2) ground-attack aircraft which entered serial production in 1940. After Tupolev's imprisonment in 1937, Sukhoi established his own OKB in September 1939 at Zavod No. 135 in Kharkov, and designed the Su-6 ground-attack aircraft. Due to the decision to concentrate on producing the Il-2 as the standard ground-attack aircraft, only 877 Su-2s were built in 1940–42 before the plants were evacuated and production ended, while the follow-on Su-6 was never considered for mass production. Sukhoi's wartime work was largely experimental, but after the war he set to work in earnest on jet bomber and fighter designs.

Like Mikoyan and Guryevich, Lavochkin, and Yakovlev, Sukhoi was commissioned in 1945 to build jet fighters. Sukhoi based his Su-9 design on the proven German Me-262 twin-jet fighter using Junkers Jumo 004A engines, but the aircraft flew later than the Yak-15 and MiG-9, never entered serial production, and won him Stalin's scorn for simply copying an old German design rather than pioneering a new Soviet approach. His Su-15 twin-engined design used Rolls-Royce Nene engines buried in the fuselage and a highly sweptback wing, but a prototype disintegrated during test flights and development was discontinued. The work on another fighter, the supersonic Su-17, and a four-engined jet bomber, the Su-10, were halted when Sukhoi lost favor with Stalin and his OKB was closed in 1949. Sukhoi transferred to the Tupolev OKB as Deputy Chief Designer.

After Stalin's death, Sukhoi's OKB was revived and its new supersonic Su-7 and Su-9 series of fighters appeared in the 1950s and 1960s. Both aircraft series were extensively modified and upgraded during the 1960s and 1970s and saw considerable service in the VVS and other

Soviet Bloc air forces. They also very clearly show the Soviet tendency toward careful evolutionary design, strong design inheritance, and component commonality. Like his mentor Tupolev, Sukhoi heavily stressed design heredity and component commonality, so that the Su-7B series of ground-attack aircraft was improved into the Su-17 variable geometry aircraft, and then the Su-20 and 22 export versions by incremental changes and not by completely new redesigns. The Su-9 delta-wing fighter was improved to the Su-11 and Su-15 PVO fighter-interceptor series. The superb Su-24 multi-role aircraft entered service in 1974 and the Su-25 close support aircraft in 1981–82.

The new Su-34 began replacing the Su-24 in the mid-1990s, and the thoroughly redesigned Su-39 shturmovik is now replacing the older model Su-25s. Sukhoi's Su-27 tried and tested long-range air superiority fighter is an exceptional and versatile aircraft that comes in a wide variety of models, including a carrier version, now designated the Su-33, and the virtually all new Su-35 multi-purpose fighter derivative. The new multi-role, all-weather Su-37, which has new engines with thrust-vectoring, state-of-the-art electronics, a forward-swept wing, and much more is also now being produced by the equally new VPK Sukhoi to compete with the Mikoyan MFI large multi-role fighter (Project 1-42). Also under development is the Su-54, a lightweight fighter, comparable with the US Joint Strike Fighter, which will be the major competitor to the Mikoyan LFI.

S. A. LAVOCHKIN (1900–60)

Semen Alekseyevich Lavochkin attended the Moscow Higher Technical School while working at TsAGI in the late 1920s on the ANT-3 and ANT-4 aircraft. After working with Richard at OPO-4 in the early 1930s, he joined the *Upravleniye spetsrabot NKTP* (Chief Administration of Special Work, *Narkomtyazhprom*) in 1935 and worked under Kurchevskiy and with Lyushin on the Lavochkin Lyushin (LL) fighter. In 1936, he became a senior engineer in *Glavaviaprom*, and in 1938 he teamed up with V. P. Gorbunov and M. I. Gudkov to respond to the government requirement for new fighter aircraft.

A new OKB was set up and designed the I-22 (LaGG-1) – a single-seat fighter of all-wood construction. The Lavochkin–Gudkov–Gorbunov effort, the I-22 (later LaGG-1), first flew on 30 March 1940. An aircraft built of resin-impregnated birch and plywood sheathing, the LaGG-1 was considered of such great importance that it was rushed into series production on the basis of the initial flight test report. Problems with the original LaGG-1 required such significant reworking, however (including a new M-105PF engine of 1,210 horsepower and changes on the Taganrog production line), that the aircraft was redesignated the LaGG-3. The LaGG-1 and modified LaGG-3 became standard VVS fighters.

Although not hit as hard as that of other planes, LaGG-3 production was seriously dislocated during the late summer of 1941 as the German forces neared Zavod No. 31 in Taganrog, prompting the relocation of equipment and personnel to five factories in Tbilisi, Georgia. Despite this interruption, LaGG-3 output reached 2,141 in July–December 1941 and totaled 2,463 for the year.

These relocation difficulties were as significant as the design problems affecting the aircraft itself. The decision to emphasize mass output locked defects into the LaGG-3 that proved difficult to remedy. In this case, as in others, pressures caused by evacuation and pressing demands of the air war combined to exaggerate technical deficiencies already in the aircraft. Front-line units equipped with Lavochkin fighters encountered so much trouble that the designer, an engineering team, and test pilots had to be sent to investigate the problems and devise solutions. The earlier efforts in 1940–41 to improve the original LaGG-1 design had failed to correct its weak undercarriage and inadequate armament. The aircraft lacked maneuverability and was unwieldy in the air because its engine lacked the power to compensate for its weight of more than three tons. The LaGG-3, although comparable with the Yak-1 in overall performance (307 mph at sea level and 354 mph at 13,000 feet), was less maneuverable than Yakovlev's aircraft. Despite these problems, 2,771 LaGG-3s were built in 1942 as Lavochkin's new fighters entered production. LaGG-3 production only ended at Tbilisi in 1944 after another 1,294 were built, bringing the total production of this model to 6,528.

In an effort to solve basic problems and improve performance, Lavochkin and Gorbunov substituted Shvetsov's 1,570-hp ASh-82 air-cooled radial engine (and later the 1,600-hp ASh-82A) for the Klimov VK-105PF (1,210 hp) engine, and made other changes that resulted in the LaG-5. Redesignated the La-5 after Gorbunov left the team, the new fighter entered production in the summer of 1942, and 1,107 La-5s and 22 La-5FNs with a more powerful ASh-82FN engine were completed during that year. The La-5 went into action in August 1942 with the 287th Fighter Air Division in the defensive fighting at Stalingrad. Its maximum speed of 388 mph at 10,000 feet and its range of 390 miles quickly made the new fighter a mainstay of the VVS fighter force. The introduction of the Shvetsov radial engine lessened the aircraft's vulnerability, especially to engine fires, and other improvements made the La-5 easy to control and maneuver.

The Lavochkin La-5, La-5FN, and La-7 fighters were excellent wartime aircraft, and 10,003 La-5/5FNs and 5,905 La-7s were produced through 1945. The postwar La-9 and La-11 (a long-range version of the La-9) were the last piston-engined Soviet fighters, and the first of Lavochkin's all-metal, stressed-skin aircraft.

Lavochkin's OKB designed a number of prototype jet fighters from 1946 through 1960 – the La-150, La-152, La-154, La-156, La-160, La-168, La-174 (La-15), La-190, La-200, and La-250, but none of them were ever accepted for serial production. Despite this lack of success, the Lavochkin OKB established a number of important firsts in their designs that significantly aided Soviet aircraft design – the La-152 (1947) was the first aircraft with a laminar flow wing and the La-160 (1947) was the first Soviet fighter to sport a sweptback wing and achieve near supersonic speeds (Mach 0.92). Upon Lavochkin's death in 1960, his design bureau was simply disbanded.

V. M. PETLYAKOV (1891–1942)

Vladimir Mikhailovich Petlyakov was a student of Zhukovskiy at the Moscow Higher Technical School before the First World War. After graduating in 1920, he went to work at TsAGI and eventually joined Tupolev's AGOS. In 1925–26, he headed AGOS's 1st Design Brigade that specialized in wing design. When KOSOS was formed, he headed Brigade No. 1 (KB-1) that was responsible for heavy aircraft and specifically the ANT-20 *Maxim Gorky*. In 1936, when KOSOS became independent of TsAGI, Petlyakov took over the Experimental Aircraft Factory (ZOK) with responsibility for designing the ANT-42 (TB-7 and later the Pe-8) four-engined heavy bomber which turned out to be an excellent aircraft. Petlyakov was arrested as part of the roundup of Tupolev and his associates in 1937, and later incarcerated with him in TsKB- 29.

Petlyakov's design bureau, assigned identification number 100, was to design a twin-engined, long-range *vysotnyi istrebitel'* (high-altitude interceptor), the VI-100, for use against high-altitude bombers. The initial design was completed in 1938–39 and approved in 1939, and the prototype was built and test flown by the end of that year. The aircraft appeared in the May Day Parade of 1940 and received much acclaim due to its speed. Because the German bomber threat was considered to be minimal, the VI-100 gave way to a redesign which became the VB-100 short-range bomber, which was quickly accepted and placed into production in 1940 as the Pe-2 frontal bomber.

The Pe-2 light bomber was a well-constructed twin-engined aircraft, superior in speed, armament, and flight characteristics to other

Soviet bombers. Pe-2 output suffered from the dislocations of the last half of 1941, and only 1,405 were built and turned over to VVS units during that period, bringing production for the year to 1,671.

By June 1942, Pe-2 units were suffering such heavy losses from German fighters that the aircraft were field-modified by front-line engineering teams. Their efforts to improve the plane's defensive armor and firepower by replacing the 7.62 mm guns with dorsal and ventral 12.7 mm Beresin UBT machine guns, had the effect of increasing its weight appreciably while reducing its performance. As a result, the more powerful 1,210-hp VK-105RF engine was substituted for the 1,100-hp M-105R. The Pe-2, equipped with the Klimov engine, began coming off the production lines in February 1943. Its maximum speed of 360 mph at 16,000 feet represented an improvement over the previous 335 mph maximum, and its cruise speed was nearly 300 mph, or an increase of 35 mph. Range was improved by nearly 200 miles, from 900 to 1,100 miles. An extremely versatile aircraft that served as a dive or horizontal bomber as well as a fighter or fighter-bomber, the Pe-2 became the mainstay of the tactical bomber force, at times accounting for as much as 75 per cent of its front-line strength.

The success of the Pe-2 secured Petlyakov's release in 1940 and the re-establishment of his OKB with himself as the chief designer at the relocated factory at Kazan in 1941. In January 1942 Petlyakov was killed in an air crash on his way from Kazan to Moscow. A. M. Izakson assumed control of Petlyakov's OKB initially, but was soon replaced by A. I. Putilov and then V. M. Myasishchev in 1943. The OKB was closed in 1946 only to be reopened in 1949 on Stalin's direct order and placed under Myasishchev's direction.

V. M. MYASISHCHEV (1902–78)

Another pupil of Zhukovskiy and colleague of Tupolev, Vladimir Mikhailovich Myasishchev began working for TsAGI's AGOS in the 1920s on the ANT-6 (TB-3) bomber. With the creation of KOSOS TsAGI, he was placed in charge of Design Brigade No. 6 working on experimental aircraft. His efforts included long-range heavy aircraft, such as the ANT-42 (TB-7), on which he was responsible for the fuselage and tail design. He then spent some time in the United States working with the Douglas Company as part of the DC-3 license agreement. He was arrested along with Tupolev and his closest associates and eventually assigned to TSKB, where he headed design team No. 102 that worked on an experimental high-altitude, long-range heavy bomber, the *dalnyi vysotnyi bombardirovshchik* (DVB-102).

Myasishchev was released in 1941 and assumed responsibility for Petlyakov's designs upon his death, and then became chief designer of the Kazan plant and managed the Pe-2's production for the rest of the war. He lost his position when the OKB was closed in 1946 but went to work with Tupolev on duplicating the B-29, and then became a lecturer at the Moscow Aviation Institute.

In 1949, after approving the Tu-16 for production, Stalin asked Tupolev to develop an intercontinental jet bomber capable of round-trip missions to the United States by adding two engines to the new bomber. When the design bureau answered '*Nyet*', Stalin called in Tupolev who explained that 'with existing engines it was impossible to do this, since fuel consumption was too great'. Stalin was greatly displeased, and within days Myasishchev, who had agreed to undertake this difficult design task, was directed to establish a new OKB at Zavod No. 23 in Moscow. The new OKB soon received over 1,500 designers and technicians from the Sukhoi, Ilyushin, and Tupolev OKBs.

The four-engined, swept-wing Mya-4 that emerged in the early 1950s fell far short of the original design requirement of 9,940 miles and had a range of only 5,590 miles. None the less, it formed the backbone of the Soviet strategic bomber force for many years along with the Tu-95. Myasishchev also designed the delta-wing, supersonic M-50/52 that never went into production, and began work on the M-18 supersonic bomber that was transferred to Tupolev and became the Tu-160. The coming of the missile age and Nikita Khrushchev ended the Myasishchev OKB's strategic bomber designing days. Khrushchev believed that the funds and energies devoted to Myasishchev's unsuccessful strategic bombers were better spent on intercontinental ballistic missiles (ICBMs) and in 1960 ordered that the bureau be closed and converted to ballistic missile and space work, which it still does today. Myasishchev himself went off to head TsAGI until 1967, when he became general designer of the *Eksperimental'nogo mashinostroitel'nogo zavoda* (Experimental Machine-Building Factory), which designed and built the supersecret M-17 and VM-T high-altitude reconnaissance and research aircraft. Similar to the US U-2, these aircraft have only recently appeared in the West.

SUMMARY

The Soviet and Russian aircraft designers and their design bureaux have overcome many challenges in the past, and today face daunting new challenges that are testing them as never before. They have played leading roles in the complete restructuring of the former Soviet aviation industry from largely unifunctional *glavki* to vertically structured aviation 'companies' that include design, development, production, repair, testing, and financing of aircraft, engines, and aircraft sub-systems and components. Only time will tell how well they will adapt to their new domestic and international environments, and what role they will play in moving the Russian aviation industry into the twenty-first century.

8

The Defense of Russian Aerospace[*]

DENNIS J. MARSHALL-HASDELL

THE HERITAGE

The responsibility for command of the air over Soviet territory traditionally lay in the hands of the Soviet Air Defense Forces (SADF) – known as *Voyska PVO Strany* from 1941 to 1982 and thereafter as *Voyska PVO* – usually under the command of an artilleryman, and comprising an assortment of surface-to-air defenses (SAD), radar facilities and interceptor aircraft. Until early 1941 this had been very much an *ad hoc* affair and development of air defense (AD) was severely hampered by Stalin's purges and the erratic nature and output of the Soviet military-industrial complex. Moreover, the limitations of air power, and by consequence the invulnerability of the Soviet homeland to anything other than overland attack, further reduced the need to develop sophisticated air defenses. However, Operation Barbarossa ensured that defense of Soviet air space took on a new and significant meaning. In June 1941, the PVO (Air Defense Corps) was created to provide AD for Moscow. Six months later a new force – *PVO Strany* (National Defense of the Homeland) – brought together interceptors, anti-aircraft artillery (AAA), air observation units, radar warning facilities and communications networks under a single banner.

THE POSTWAR BUILDUP

After the war, the Kremlin realized its vulnerability to long-range air attack and the menacing appearance of a combined air and nuclear threat from US long-range bombers. Despite the concept of 'national' AD during the war there was still no comprehensive radar early warning system or satisfactory control and co-ordination between air,

* The views expressed are those of the author and not necessarily those of the UK Ministry of Defence.

surface-to-air and ground forces involved in AD operations. To redress these shortcomings, from early 1946 the development of aircraft, early warning detection and tracking radars, surface-to-air missiles (SAM), anti-aircraft artillery (AAA) and communication systems capable of destroying high-flying, fast-jet bombers went on at a great pace. Already the Mikoyan–Gurevich, Sukhoi and Yakovlev fighter design bureaux were well established, ensuring the development of a whole range of tactical fighters and strategic, long-range interceptors. An increasing number of early-warning radars, AAA, ground-control sites and airfield facilities were becoming integrated and linked into the AD network, which was extended to the western edge of eastern Europe thus giving a new dimension to Soviet security requirements.

Another important development was a simplification of command and control (C^2) by the reorganization of fronts into AD districts during 1946. Eight years later SADF was given formal status as an independent service of the Armed Forces following the appointment of a commander-in-chief, coming under the direct control of the General Staff. Its responsibilities were the AD of industrial areas, military installations, deployed troops, and other vital interests in the USSR and non-Soviet Warsaw Pact (NSWP) countries.

By the 1950s, a second generation of fighter aircraft (MiG-21 and Su-9) had appeared in the skies along with the first Soviet SAMs (surface-to-air missiles). While most Western observers suggested that the operational effectiveness of Soviet aircrew and controllers was limited and initiative was severely restricted, the overall efficiency of the developing SADF was probably better than the West speculated. The loss of Gary Powers' U-2 reconnaissance aircraft, on 1 May 1960, not only gave an indicator of Soviet AD capability but also proved the operational capability of the SA-2. Although we are now aware of the many difficulties encountered by SADF during this episode – about 14 SAMs were fired at Powers' aircraft and a MiG-19 was shot down by at least one of them – there seems little doubt that the Soviets had made great progress in the development of their AD system. Notwithstanding the rhetoric, exaggeration and blustering so evident in the early days of the Khrushchev era, the infrastructure, organization and equipment being put into place would enable the Soviets to spread their AD umbrella across the length and breadth of their empire. Even with the emergence of détente, the Kremlin remained determined to close the technology gap and continued to develop new generations of fighter aircraft, SAMs and AAA to work alongside the expanding radar warning and tracking facilities.

EVOLUTION AND DEVELOPMENT

A common trend throughout the postwar era was a continual process of trying out new techniques and modifications to improve air defenses. Research and development (R&D) into new systems – such as the anti-ballistic missile ABM-1 (Galosh) paraded in 1964 – demonstrated a hunger for countering the threat of air attack on the Motherland. Because fiscal management was not a concern and the Soviets never threw anything away, a vast inventory of equipment was accumulated and integrated into the AD system. The military indulged in huge development programs using the endless supply of resources allocated over three decades. But that was a military bonanza that would not last: by the late 1980s the influence of *perestroika* and democratization placed the military in a completely different fiscal environment in which funds dried up and the rot set in.

Throughout this period, it can be argued, there was a willingness to adapt the AD system to reflect changes in NATO policy and to respond to changes in the character of the aerospace threat as a result of technological and operational advances. There may be grounds for a 'chicken-and-egg' counter to this hypothesis because NATO's introduction of standoff weapons and low-altitude bombing penetration tactics was arguably in response to its analysis of Soviet AD capabilities. On the other hand, there is evidence suggesting that when NATO moved away from a nuclear 'trip-wire' doctrine to the doctrine of 'flexible response', Soviet military planners concentrated R & D on tactical systems operating on the combined-arms conventional battlefield. New SAM and AAA systems were introduced during the 1960s and 1970s resulting in a complex web of AD coverage forcing NATO strike attack aircrew to train at ever lower altitudes.

This complex combination of radar-laid AAA, SAMs, and tactical and strategic fighter interceptors required a comprehensive C^2 system to ensure co-ordination with artillery and mortar fire, strike aircraft and helicopters. Little was known at the time – to some extent it remains a mystery, perhaps even to Russian AD commanders – about the co-ordination or relationship of all these diverse assets. Certainly, most Soviet exercises involved detailed planning and practice of airspace management with particular attention given to the co-ordination of assets in the highly maneuverable land/air battle. The principles of Soviet tactical AD were designed to ensure that any NATO air attack would be met with a network of flexible, mobile SAD units which would not necessarily cover the whole front but be able to react quickly to new lines of attack. The low-flying enemy would be outmaneuvered by what Soviet AD commanders proudly called 'roaming AD units backed up by faultlessly organized logistic support'. Optimism remained high and even concerns about an effective look-down/ shoot-down capability for interceptor aircraft were eventually resolved – initially with the introduction of the MiG-23 and then the MiG-25. Meanwhile, strategic AD systems retained a time-honored position of importance reflecting the postwar Soviet fear of nuclear attack. It should be noted, however, that the static nature of strategic AD assets – interceptors required long, hard-surfaced runways and SAMs lacked mobility – meant survivability was severely reduced.

STRUCTURE AND ORGANIZATION

Aerospace defenses relied on centralized C^2 from Moscow, from where responsibility for AD was assumed. In order to perform a wide variety of operational tasks SADF controlled all non-ground force surface-to-air weapon systems, including the anti-ballistic missile (ABM) defenses, space, ground- and air-based early warning radar systems, fighter interception, the national warning and monitoring organization, and all civil and military air traffic control. The military elements were structured around the four main groups of Air Defense Command: Radar Troops responsible for early warning and control systems covering the entire aerospace; Anti-Aircraft Artillery Troops providing a variety of anti-aircraft, self-propelled gun systems as part of mobile AD capability of ground troops; Anti-Aircraft Missile Troops with a wide range of surface-to-air missiles (SAMs), including responsibility for ABMs; and fighter interceptors deployed along the entire length of the old Soviet borders, including the NSWP (Warsaw Pact) countries, with particular emphasis on the Leningrad/Moscow area and Baku. For obvious geostrategic reasons, the Soviet anti-ballistic missile warning or SPRN (Early Warning of Missile Attack) and the integrated conventional threat detection, tracking and reporting facilities, were all located primarily on the periphery of the USSR.

There were many other elements of this very complicated system including those less well known such as the *Protivokozmicheskaya Oborona* (PKO) holding responsibility for outer space or cosmic defense. Moreover, Fighter Aviation of the Air Force *Istrebitel'naya Aviatsiya* (IA-VVS) also included interceptor units of naval AD from each of the four (Black, Baltic, White and Pacific) fleets. To add further to the confusion, interceptor aircraft worked in close co-operation with Frontal Aviation (for example in the escort role) and on occasion with Army Aviation in co-ordination with Ground Forces providing top-cover for close air support (CAS) operations. Finally, there have been a number of occasions when the level of effort between conventional tactical and strategic AD missions have resulted in changes to SADF organization. For example, in 1986 a reorganization seemed to indicate a reversal of measures taken in the late 1970s. Quite understandably the scale of effort, resources and equipment required to put into effect the SADFs' terms of reference was colossal.

The strategic AD system comprised a mixture of launch-detection satellites, over-the-horizon radars, phased-array radars, command and control sites, communication networks, tracking stations and missile sites. The launch-detection satellite network provided 30 minutes' warning and location of the general area of any ICBM (intercontinental ballistic missile) launch. Two over-the-horizon radars had similar warning parameters but with less geographical fixing capability, but working together these systems provided a reliable warning system. A further ballistic missile warning layer was provided by about 11 large *Hen House* (3,000-mile range) detection and tracking radars again located in peripheral regions. These radars were capable of distinguishing the magnitude of an attack and to provide target-tracking data to the ABM interception units. *Dog House* installations (1,500-mile range) provided cover in the middle of the chain, and *Try Add* close-range, target-tracking radars completed a complex network of over 7,000 surveillance radars of various types located at about 1,200 sites.

Considerable effort was expended on R&D under a veil of secrecy commonplace on both sides of the Iron Curtain. The result was an output of AD weaponry that met the needs of military doctrine and strained to keep pace with the West: no 'ifs' no 'buts', just the outright pursuit and projection of military power. Even with social, political and economic collapse looming during the 1980s, the development and operational deployment of new phased-array radar systems was given high priority. The price tag on a comprehensive AD system for the Motherland and empire may have been ignored during the Soviet era, but the Russians will, no doubt, find the premiums too high for this dubious legacy.

PROBLEMS OF EVALUATION

The system was, of course, never put to the test and so the Russian legacy retains a degree of the mythological, propagandized efficiency that epitomized the Soviet era. In reality, the combined effects of suppression of enemy defenses (SEAD), the formalism and stereotyping in Soviet training, reliance on poorly trained conscripts and the lack of familiarity in combined-arms operations shown by many Ground Force commanders suggests that the system may indeed have been too complicated to work efficiently. Additionally, fighter aircrew were rarely able to train outside the strictures of close control when conducting intercepts; although the reluctance to allow free-ranging fighter operations can be partly explained by the complexity of the whole air space management problem in a fluctuating, highly mobile combat environment. Bound by a traditional reluctance to allow any form of initiative to be displayed by aircrew, controllers or commanders – 'Initiative is Punishable' was a terrible Soviet phrase – claims of a high degree of combat effectiveness among ground- and air-crews should be treated with skepticism.

To overcome these shortcomings the Soviets fashioned a training and evaluation system embracing the concept of 'Socialist Competition', which relied on falsification, formalism, fudging and fabrication to produce the archetypal 'Pilot First Class'. Because missions were conducted according to meticulous planning and close control, especially if ECM (electronic countermeasures) conditions, poor weather or night flying were likely to be encountered, many aircrew and GCI (ground controlled interceptor) controllers were able to maintain a 'first class' level of combat efficiency. Even under these conditions, not all Soviet operators were able to maintain or even achieve the 'first class' standard. Furthermore, the advent of ever more complex hardware and the increasing challenges of long-range interception, low-flying targets and rapid developments in ECM placed greater strains on personnel at all levels. Therefore, the effectiveness of AD operations involving the third generation of fighters and equipment (MiG-25, MiG-23, Su-21, etc.), which were on regimental strengths by 1974, remained a matter of concern to SADF commanders, albeit while retaining a public image proclaiming efficiency and high combat readiness.

In retrospect, it is easy to dismiss Soviet claims that their AD system was highly effective; after all we have the 'Rust affair' (1987) and the KAL 007 tragedy (1983) to demonstrate how things could go drastically wrong. The West was always quick to deride the ability of SADF to perform its strategic defense mission, but this does not account for the considerable respect shown by NATO fighter aircrew towards Soviet AD systems. However, as aerospace warfare became increasingly technologically determined – the introduction of the B-2 bomber, cruise missiles, stealth, SEAD (Suppression of Enemy Air Defense), etc. – many Western analysts believed the problems facing the Soviet air defenses were insurmountable. Of course, it could equally be argued that these Western advanced systems were in response to the effective development of Soviet counters to the previous generation of weapons. This technologically determined race to gain advantage or maintain parity in the aerospace theater is further evidenced by the arrival of the MiG-29 (Fulcrum), MiG-31 (Fox-hound), Su-27 (Flanker), AA-9, AA-10, SA-10 and SA-12, which were considered to be the fourth generation of Soviet AD equipment.

It is the ultimate dilemma in a polarized defense environment based on deterrence that each side will convince itself of the validity and effectiveness of its systems, procedures, weapons and facilities. When the debate began to be concentrated on Soviet operational deficiencies during the latter half of the Gorbachev period, it is, perhaps, more a reflection of increased 'reality' among the reformists in the Kremlin – a reaction against a background of political, social and economic restructuring – than a definitive adjustment of a doctrinal commitment to strategic aerospace defense. Whether they were positive or negative, the fact remains that Western analyses of Soviet AD were founded on conjecture, limited access to conclusive intelligence material and without the benefit of hindsight. The latter does, of course, have an important role to play when making an analysis of the Russian legacy, because knowledge in hindsight indicates that the comprehensive AD network would surely have failed to work effectively.

OPERATIONAL CAPABILITIES

For an assessment of general operational capabilities of Soviet aviation units, the Afghan War clearly demonstrated that their ability to conduct air/ground operations was severely hindered by unsatisfactory tactical command and control procedures resulting in a lack of co-ordination between air and ground forces. Admittedly, the Afghan War theater was totally unlike the conditions anticipated during full-scale conventional war in Europe and, moreover, it did not involve many AD force units; but the principles of ground/air co-ordination still applied and the complexity of airspace management remained an enduring problem for Soviet military commanders. The Gulf War was another combat testing ground for Soviet AD procedures, training and equipment: the failure of the Iraq Air Defense forces to cope with a sophisticated air threat suggested that the Soviet model, upon which the Iraqi system was based, could be called into question.

While it can be argued that, taken in isolation, many aspects of Soviet AD were tried and tested in combat environments and shown to be effective – the SA-6 (Gainful) was shown to be a formidable mobile SAM during the Yom Kippur War – it is when the whole integrated AD system comes under the microscope that questions of infallibility arise. At the height of the second Cold War, in the mid-1980s, Tony Mason argued that: 'while the comprehensive network of Soviet SADs would undoubtedly create difficulties for an attacker, it is not necessarily the impregnable barrier which its sheer size would suggest'. It has already been pointed out, as long as the Soviet leadership was willing to allocate the resources that the military demanded, SADF were able to maintain credibility among friends and foes alike. We now know that the whole system was cumbersome and experienced serious problems with command, control and communications; and yet its very existence as a demonstrably comprehensive system gave it an edge portraying it as an effective part of the Soviet war machine. Western analysis until the late 1980s continued to make the assumption that the system would have worked under war conditions, albeit with uncertain efficacy. But like many such assumptions, made by both NATO and the Warsaw Pact, it was the very nature of the uncertainty that provoked the concern and added to the deterrent value rating of any particular system. With its covers off and its resources distributed across the old empire, the Soviet AD system is a miserable legacy for the Russians to receive at a time when many in the Russian Armed Forces feel at their most vulnerable militarily, economically, politically and socially.

FROM STABILITY TO CHAOS: THE COLLAPSE OF THE USSR

Even before the disintegration of the Soviet Union, evidence can be found of serious disruption to the AD system caused by a number of factors. Not least were the efforts of regional pressure groups, one of which managed to halt the construction of a large, phased-array radar facility near Mukachevo in Ukraine. Intended to fill a gap in the coverage of the western sector, this facility would have completed the chain of nine major reporting facilities allowed under the ABM Treaty. The demise of this military complex – since 1991 it has been systematically stripped of equipment and the integrated infrastructure broken down into its component parts – graphically illustrates the effects of post-Soviet dysfunctional syndrome. Asset stripping is rife across the whole of the FSU (Former Soviet Union). Pestryalovo is a good example of the multiple uses to which former AD facilities are being put.

Both during and after the Mukachevo débâcle the General Staff were angered and frustrated – the Russians have inherited the feeling – by the influence of pressure groups, many of which clearly held fervent anti-military views. Under the influences of *glasnost, perestroika* and democratization the military had reluctantly to acknowledge it was no longer possible simply to ignore public opinion. In consequence, by the mid-1990s, amid a climate of strained relations with the Ukraine, the Russians are in no position to argue strongly for the restoration of this particular facility.

DISINTEGRATION OF THE AIR DEFENSE SYSTEM

Following the breakup of the USSR, there has been a dramatic disintegration of the unified AD system. This has disrupted, and in some areas completely decimated, many of the integrated facilities vital to its satisfactory operation; not least, the command, control and communications networks, radar tracking stations and air operations centres. Across the Former Soviet Union (FSU) many of the facilities that still remain intact are either being transferred to civilian use or allowed to rot from lack of maintenance. The problems are not restricted to equipment and facilities as many training establishments are also now on 'foreign' soil. Ukraine took control of SADF's Marshal Vasilievskiy Military Academy in 1991, leaving the Russians with no alternative but to found a new academy in Smolensk. It must also be noted that many SADF personnel were recruited from across the USSR: as a result many pilots, engineers, radar operators and the like have returned to their roots. On their return, many of these experienced, core personnel have transferred their skills to the armed forces of the newly independent former republics. In Russia, however, many have been demobilized as part of the general force reductions or have found themselves victims of the hopelessly disorganized Russian Armed Forces.

When considering the difficulties caused by the breakup of the USSR and the subsequent dispersal of assets across the FSU, we need to address not only general issues such as early warning facilities, radar reporting centers, SAM sites, etc., but also problems facing the Russian military as they experience the new phenomena of public pressure and politically inspired programs to civilianize many important military facilities. In response to these problems the Russians have been at pains to initiate talks to create a unified AD system for the Commonwealth of Independent States (CIS). In their search for co-operative partners in the AD field the Russians have encountered many obstacles; not least the provision of funding, training and resources for such a system. Attention has been given to reaching agreements with neighbors on the strategically vulnerable western borders. Here the importance of Ukraine cannot be overemphasized given its strong negotiating position and the significant influence that Russian–Ukrainian relations have on stability in the whole region.

The major problems facing the Russians in their attempts to prepare satisfactory AD are the loss of vital facilities; ranging from major elements of the extensive early warning system for ballistic missile and conventional AD threats, to individual mobile SAM systems. Following division of the Union, radar and space tracking facilities in Latvia, Ukraine, Kazakhstan and Azerbaijan were 'lost' to the Russians. Russian personnel operating these facilities soon found they were having to work under extremely difficult circumstances. Of the nine major SPRN sites in the FSU less than half are now on Russian soil, providing limited cover of the far north (the stations at Murmansk and Pechora) and south east (the station at Irkutsk). The fourth station, located at Krasnoyarsk, covers the eastern sector but because it violates terms of the ABM Treaty it has been out of commission for a number of years.

The disruption of the old AD system, which relied almost entirely on integration and co-operation between all the facilities, has meant that in the mid-1990s it is virtually impossible for Russian AD commanders to be provided with a comprehensive Recognized Air Picture (RAP). While the best way to plug the gaps in radar coverage would be to construct new facilities on Russian soil, this is highly unlikely on the grounds of cost alone. Additionally, the very nature of the role and purpose of these strategic facilities required them to be in optimal geographic locations. Thus the dilemma cannot be easily resolved, either by relocation of existing facilities or by replacement with modern radars and equipment constructed at alternative sites in Russia.

THE QUEST FOR PARTNERSHIP

The only feasible option seems to be for the Russians to find agreement with their neighbors over the joint use of facilities for the mutual benefit of all. While this has certain value to the Russians, others may not see any advantage in sharing their inherited military assets. Russian military preoccupation with the collapse of the integrated AD system has directed the debate towards achieving a co-operative effort through a unified CIS AD system. In practice, the only feasible solution would be to reintegrate the facilities, sites and equipments of the old SADF under a modified organizational structure, which would make C^2 responsive to a unified inter-state system: easy to propound but extremely hard to put into effect, not least because in some cases, notably the Baltic States, co-operation with Russia is simply not on the agenda.

A complicating factor is that attempts to resolve the crucial issues affecting strategic security have often led to AD facilities becoming bargaining chips in wider political games played in the region. For example, another of the FSU's major missile attack warning stations – the Gaballa facility in northern Azerbaijan – has provided Azerbaijan with a useful tool for manipulating Russian military negotiators. The Gaballa complex, being located only 40 km from Nagornyy-Karabakh, symbolizes the fragility of Russian efforts to attain aerospace defense. Additionally, it highlights security concerns regarding the Caucasus region, traditionally seen as Russia's weakest flank and certainly where the Russians appear to have been concentrating their greatest efforts.

Similarly, the case of Russian–Ukrainian military relations whilst complicated by political and economic factors could benefit greatly from the sharing of assets. Indeed, there has been some co-operation over space-based systems and the purchase of strategic SAM systems from Ukraine. None the less, while the Ukrainians may have benefited from the acquisition of AD equipment and facilities, they have also suffered from the Soviet legacy. For example, Ukrainian SAM and AAA units have been unable to conduct live training exercises because there are no suitable ranges in Ukraine. This highlights a general problem facing all of the former republics; no single country really benefits because while they may have inherited valuable equipment and facilities, vital components are often located beyond their borders. The problem is exaggerated in the Russian case simply because so many facilities were located along the periphery of the Soviet empire. There is no doubt that if Russia worked together with these countries they could provide an effective AD system that would be of benefit to all and could be a pivotal part of any CIS system. Problems concerning AD equipment and the introduction of new technologies – most of the R & D is conducted in Russia – are common to both countries. Moreover, there would be mutual profit in sharing radar and associated information in order to compile a first-rate RAP.

THE BILATERAL APPROACH

Given the enormity of the task facing Russia in its quest to establish a viable AD system, it should be expected that any number of separate talks would be required before it is possible to reach an agreement which could be signed by all the CIS states. In this way the Russians may be attempting to construct an overall CIS agreement using building blocks based on bilateral agreements. An example of the success of this Russian policy was the signing, in October 1994, of agreements with Armenia allowing Russian AD aircraft to be based at Yerevan. This has been described as a 'prototype of coalition armed forces', and signposted the first stage in the construction of the unified AD system, which had been agreed by CIS Heads of State at the Alma-Ata summit in 1995. In similar vein the military department of Kyrgyzstan signed an agreement for the leasing of territory occupied by the Russian Defense Ministry's seismic service. All this activity, while being part of the rebuilding process and evidence of AD co-operation, remains relatively insignificant in comparison to the wider problems associated with AD matters. Even when co-operation is agreed, there are insurmountable problems in finding the resources and finance for collaborative projects.

The basing of Russian AD aircraft in Armenia is an obvious example of the effort to strengthen the southern boundaries of the Russian Federation through a bilateral agreement, although it was presented as a joint enterprise involving both finance, resources, personnel and equipment for the benefit of the CIS as a whole. However, a more convincing example of the Russian commitment to funding joint ventures is clearly demonstrated by the agreement with Georgia to develop a unified AD system for the two countries. This widening of military co-operation in the region had the added advantage for the Russians of putting pressure on other republics to follow the Georgian example by making closer ties with Russia.

Another of Russia's closest, strategically important allies, Belarus, has been most accommodating in allowing Russia to retain the benefits of the old Soviet AD system. Under the auspices of the joint AD system of the CIS, Russian and Belorussian AD units began joint duty on 1 April 1996 to ensure the protection of the western boundaries. The majority of the resources and practical help needed to restore the AD facilities in Belarus came from the Russians and, in common with the Georgian case, this joint action can be viewed as a vital step towards achieving the aim of a unified AD system for the CIS.

Whether the contributions of funds, facilities and manpower promised at Alma-Ata will be forthcoming in the near future remains a matter of conjecture. Undeniably there are still large gaps in the system and, to date, both Moldova and Azerbaijan have resolutely refused to participate in AD co-operation under CIS guidance. Moreover, the Baltic states have severed all links with the CIS. In one assessment of the impact of these factors, Russian VPVO officers pointed out in late 1992 that if the Baltics, Moldova, Ukraine, Belorus and Georgia were to completely withdraw from a unified AD system, Russia would lose 1,000–1,500 km of extended air surveillance coverage – not an insignificant loss!

AIR DEFENCE OF THE RUSSIAN FEDERATION

Unfortunately, for those in the Russian Air Force who have inherited responsibility for AD, the Russian General Staff, which basically replicated the old Soviet General Staff, has adopted the traditional tendency to downplay the role of air power. This mindset effectively blocks any efforts or incentive to restructure along the lines envisaged and proposed by a number of forward-thinking military reformers in the years following independence. Russian General Staff thinking leans towards an evaluation that attributes the value of aerospace assets in its support of ground-force operations, rather than towards the development of an independent aerospace role. To that end it is unlikely, despite a general clamor for reform, that the proposed restructuring of the Russian Armed Forces for the year 2000 and beyond will be realized on time. This will have a serious impact on the effectiveness of any future AD system because of the essential role played by air force assets in such a system. It is true that the restructuring is intended to make major organizational changes to the aerospace forces: for example, in various proposals put forward by the General Staff, the once mighty and multi-faceted VPVO will be carved up, suggesting that fighter interception aircraft will be amalgamated into the Russian Air Force (RuAF). However, it is unlikely that such a major reorganization can be conducted with any degree of success given the myriad of problems facing the military leadership.

It has always been a widely held view in Russian military circles that the initial period of war is all-important and, in the era of precision weaponry, the outcome will be essentially determined in the first five to ten days of aerospace operations. It followed that great emphasis was placed on speed, surprise and the need for war to be concluded quickly before escalation into a nuclear phase. Despite these perceptions, the shortcomings of Russian air defenses are characterized by a lack of attention to issues which have greatest importance during this critical aerospace phase of future warfare. Little effort has been made to address the development and organization of a command, control and communication system with any capacity for survival. Nor has much attention been given to redressing the failure to upgrade AD facilities and equipment or improve training. To some extent this can be blamed on financial and resource restraints, but there are doctrinal problems which could have serious repercussions for the long-term future of Russian air defenses. There is a perception among some Russian analysts that the picture of future war may not have been adequately interpreted and, as a consequence, any reorganization that may be conducted will not necessarily be appropriate to the demands of the likely battlefield of the future. These views are endorsed by some Western analysts who argue, for example, 'that ground operations will continue to be the main emphasis of Russian military thought and that Russian aerospace power will not become a co-equal or independent actor in the foreseeable future'.

It should be noted that whatever future needs of Russian AD are determined there will be any number of permutations for force structure. The function of fighter aircraft could be split between the competing demands of offensive and defensive counter-air operations. For example, if attack from the air on the Russian homeland is perceived to be the major threat, then we would expect to see fighter aircraft organized according to the demands of defensive counter-air operations. On the other hand, it could be argued that the future battlefield will make more demands on the ability of fighter aircraft to conduct offensive counter-air operations. Any duality of roles or the ability to deploy air assets in a multi-role capacity would no doubt be welcomed, but flexibility of this kind tends to generate heavy demands on resources, training, equipment and facilities. Unfortunately, these essential ingredients are unlikely to be available either at the present time or in the foreseeable future. At the same time, fundamental questions of policy and doctrine need addressing before any decisions about structure, manning, equipment, training, tactics, etc. can be taken. There is little evidence to demonstrate that the Russians are directing enough attention to these matters and, as a consequence, the future for aerospace defense remains uncertain.

CUTS IN MANPOWER, EQUIPMENT, TRAINING, AND FACILITIES

By the mid-1990s, it was widely accepted, even at the highest levels, that Russia was unable to provide cover for about 40–50 per cent of its air space when utilizing its own resources. Russia's southern borders were effectively open and defenseless with no modern SAMs stationed in interior regions of the Volga, the Urals, and the Central Economic Region surrounding Moscow. Certainly there have been a number of critics who have argued that command and control of AD forces is less than satisfactory: one such critic even went so far as to suggest that the effectiveness of Russian air defenses in mid-1993 was no better than that of Iraq during the Gulf War.

As a result of the problems arising from the breakup of the Soviet Union, budget restrictions necessitated a reduction of manpower at a time when air staffs had to battle with any number of problems: the wide dispersal of units across the FSU; depressed living standards; the rising cost of equipment; low morale, etc. The morale of aircrew is already at rock bottom due to appalling living conditions, loss of status, low or non-existent pay, poor training programs and a general decline in job satisfaction. Shortages of manpower at the beginning of 1992 meant that most units were at the limit of their capability, with some not even being able to provide two shifts to maintain continuous operations at vital radar installations. This assessment of RuAF capabilities suggests that the rising costs of equipment and reduction of budgets will inevitably degrade the already low level of combat readiness well into the future.

Since 1993, the manpower strength of the RuAF has been reduced by 40 per cent. But much as manpower reduction has been a key issue during the military reform process, when it is considered alongside other factors, a sorry picture is painted. For example, it is accepted that about 50 per cent of all second- and third-generation aircraft will be at the end of their service life by the year 2005; a point exacerbated because the share of old and repaired aircraft of these generations has now reached 70 per cent of the full RuAF strength. Yet the Defense Ministry does not have the money to purchase the combat aircraft already ordered and awaiting delivery due to an 800 billion rouble budget shortfall. In 1995, only 32 new aircraft out of a planned 280 were made available to the whole of the Russian Armed Forces. Moreover, at any one time, only 50 per cent of aircraft are serviceable due to a chronic shortage of spare parts, engines, lubricants and aviation fuel – reserves are about 40 per cent of the actual need. It is not surprising, therefore, to learn that fighter aircrew now fly between 20–30 hours a year representing 17–20 per cent of the time, considered by even the most optimistic Russian commentators, necessary to maintain combat readiness. Lack of finance and resources means practically no training takes place across the full breadth of AD Forces and, as each month passes, fewer and fewer aircraft are fit to fly.

Aircraft deficiencies are having a particularly critical effect on AD Forces because the withdrawal from service of MiG-25, SU-15, and MiG-23 continues regardless of the erratic delivery of fourth generation fighter aircraft - many MiG-31s and SU-27s lie idle on airbases due to chronic shortages. Prospects for the future generation of aircraft is not rosy either. Despite having developed a fifth-generation combat aircraft (generally referred to as the MiG 1-42) there is great doubt as to the possibility of it ever entering operational service. With no money even to complete the final test cycle its serial production is in serious doubt. The same fate awaits the new SU-35 and MiG-33; both are fifth-generation fighters which require further technical trials before they can enter service.

Another matter of great concern is the lack of availability of force enhancers such as air-to-air refueling (AAR) tankers and airborne early warning (AWACS) aircraft. The tanker fleet has about 20 IL-78 (Midas) tankers; barely enough to support the strategic bomber fleet and transport operations, let alone interceptor aircraft. Some AAR has been conducted as part of a training program to co-ordinate MiG-31 interceptors with the AWACS aircraft, A-50 (Mainstay) and AAR support. A flight of four MiG-31s working with a single A-50 is capable of defending a front line extending to several hundred kilometers. Similarly, the range and endurance of the Su-27, another of Russia's latest fighters, is increased to 6,000 km or seven hours. Unfortunately, this type of development of interceptor tactics with force enhancers is severely affected by the debilitative condition of all AD forces.

AIRFIELDS AND FACILITIES

It is Russia's legacy that almost the whole of the former aerospace infrastructure has been subjected to considerable upheaval and in many instances practically destroyed. The division of territory of the USSR led to most of the AD facilities (airfields, weapon dumps, aircraft reserves, radar sites, command and control centres, air operations centers, etc.) now being situated beyond the borders of the Russian Federation. For example, it is believed that over 1,000 former Soviet aircraft, mostly fighter interceptors, are in storage in Turkmenistan alone. Furthermore, despite the fact that the Russians ended up with about 60 per cent of the Soviet Union's AD weapons and equipment, most are of poor quality because the first-rate equipment was located in the peripheral border areas.

The issue of airfields graphically illustrates Russia's dubious inheritance. Of all the available airfields spread across the FSU the Russians acquired about 210 (in the region of 35–48 per cent) but of those located in European Russia only about 65 have concrete runways. An estimated 140 airfields are in need of major repairs either to runways or ancillary facilities; many of the remainder are maintained in a usable state only with great difficulty. Russian military planners accept that of over 400 military and economic installations, over 100 of these are of strategic importance, and are vulnerable to air strikes. Without functional airbases from which to launch and operate fighter aircraft, it is inevitable that AD tasking, even in peacetime, will rapidly exceed the capabilities of the supporting services. Fighter interceptors are a crucial part of the AD network designed to protect many of the strategic installations in Russia because of 'holes' in the SAM system. Although the aim has been to have an integrated AD system comprising interceptors and long-range SAMs, there are any number of strategic sites that do not have SAM coverage.

CONCLUSION

Despite inheriting a sizeable proportion of the seemingly awesome might of SADF, by the turn of the century, Russian AD Forces will be unable to provide adequate aerospace protection. We have seen that the dispersal of facilities, equipment and personnel – albeit in various states of disrepair – across the FSU, has directed Russian attention towards the creation of a new integrated AD system. Faced with the enormity of going-it-alone, the obvious solution, at least to the Russians, would be a system based on co-operation between CIS states with Russia the major shareholder. While undoubtedly a logical conclusion, there are many in the FSU who regard this notion with great suspicion. The complexity surrounding the establishment of a unified AD system to protect the vast expanse of Russian airspace, has inevitably compelled the Russians to take the lead during negotiations; action that merely fuels suspicion among her neighbours and potential partners. Whether viable agreements can ever be reached remains a matter of conjecture but, even if this is possible in the future, the decay and decline of the remaining AD facilities, both in the Russian Federation and the former Soviet republics, will obviously need to be addressed without delay. When consideration is also given to the sorry state of equipment, lack of training and poor morale of AD personnel, then the prospect of a workable, unified Russian/CIS AD system is, to say the least, improbable. Without such co-operation and given the poor state of Russian AD forces, the ability of those forces to protect the integrity of Russian aerospace, both now and in the foreseeable future, must remain highly questionable.

Air Combat on the Periphery: The Soviet Air Force in Action during the Cold War, 1945–89

MARK A. O'NEILL

The United States Central Intelligence Agency (CIA) reported on 23 April 1948 that in the Soviet Air Force (SAF) 'first priority' was 'being given to the developments of an interceptor fighter force based on jet aircraft'. The report went on to state further that the Soviet military would focus on creating a force capable of stopping strategic bombers. The CIA believed the Soviets were developing their own long-range bombing capabilities as a secondary option. The author of this report concluded that the 'solution of both of these problems currently is being given high priority ... but the success achieved cannot be demonstrated by anything short of actual combat'.[1] The introduction of Soviet-piloted MiG-15 jet interceptors onto the world stage during the Korean War proved the accuracy of this CIA projection. In addition, the Tu-4 (a reverse-engineered copy of the US B-29) entered service with the long-range Air Force the same year as the CIA report and the Soviets successfully exploded an atomic bomb less than a year later.[2]

Once the Soviets demonstrated their ability to compete with US aviation and atomic technology, the stage was truly set for superpower competition. As Tony Mason has stated, 'Air power was a central feature in the 40 years confrontation between East and West. It was frequently prominent in conflict between client states beyond the central European arena.'[3] The use of Soviet pilots in combat beyond the borders of the USSR was another trend which would reoccur in various countries throughout the world until the late 1980s. The USSR not only transferred the technological capability to wage modern war to a large number of client states, but also backed its foreign policy commitments in several instances by fighting alongside its erstwhile customers. In the case of the People's Republic of China

(PRC) and the People's Democratic Republic of Korea (PDRK), the Soviets provided air defenses for these nascent communist regimes. After the death of Stalin, the USSR did not limit its support to Marxist-Leninist governments; its pilots flew combat missions to defend Nasser's Egypt as well as helping India develop an advanced aviation industry. The capabilities the Soviets made possible in many Third World countries became something more than a part of that nation's defenses. In Egypt's case, 'a large and effective air force was seen as a concrete example of the country's ability to use modern technology', and the Egyptian Air Force (EAF) 'became a symbol of Egypt's political aspirations, both at home and abroad'.[4]

The military standoff in Europe between the Soviet Union and the United States during the Cold War never erupted into the Third World War. While the two opponents faced each other across heavily defended borders from 1945–91, their competition in the rest of the world was much more animated. Both sides extended the Cold War to confrontations in Asia, Africa, and Central America, arenas where the military action was decidedly 'hot'. In this struggle to spread political and economic influence the Soviet Air Force played an important role. The SAF became the leading edge of the Soviet military commitment to many of the world's countries extending beyond merely supplying aircraft and training aircrews. In the People's Republic of China (1949–53), North Korea (1950–53), Cuba (1962), Yemen (1967), Egypt (1970), Vietnam (1965–70), and Afghanistan (1979–89), Soviet pilots participated in actual combat to support client states.

There was, of course, an historical precedent for Soviet pilots flying combat missions in foreign wars extending back to the Spanish Civil War and the Japanese invasion of China. The risks inherent in the Cold War battles were much greater and the scale of these involvements grew as the conflict progressed. In the Korean and Vietnam wars, Soviet pilots were directly engaged in combat against US pilots, thus risking a dangerous escalation to general war. Soviet air power played an important role in the Caribbean crisis of 1962. Khrushchev created perhaps the most dangerous of all the confrontations during the Cold War when he sent nuclear missiles and medium bombers to Castro's Cuba. The USSR began supplying weapons to Egypt in 1955; by 1970 Soviet pilots flew air defense missions against the Israelis, and by 1973 Sadat expelled the Soviets from the country altogether. Afghanistan was the largest and longest sustained Soviet military action since the Second World War and contributed to the final collapse of the entire Soviet system. In each of these conflicts, the SAF played a pivotal role within a unique geopolitical setting. The complexities of transferring advanced jet technology to societies ill-prepared to absorb them are a common thread throughout most of these cases.

Soviet foreign policy had distinct differences under late Stalinism (1945–53), Nikita S. Khrushchev (1954–64), Leonid I. Brezhnev (1966–82), and finally under Mikhail S. Gorbachev (1985–91). Each of these regimes built on the precedent set by the previous, but each was also influenced by the actions of the US and by their own often unpredictable clients. The SAF expanded its global presence throughout the late 1960s and into the 1970s, but fought its largest battle just across its own border in Afghanistan by the close of that decade. It was at the end of the 1940s and further to the east, however, that the Cold War began in earnest. These earliest aerial clashes between the US and the USSR are the best documented to date. Access to Soviet archival material is only slowly increasing, but the greatest number of documents declassified so far have dealt with the Soviet defense of North Korea and Manchuria during the Korean War.[5]

ORIGINS OF THE COLD WAR IN THE AIR

The Soviet Union did not emerge from the Second World War completely prepared to fulfill the role of a superpower. Stalin's regime did not have the atomic bomb and its economy was in ruins. Despite the size of the Red Army at the end of the war and the fact that it had occupied most of eastern Europe, Stalin still felt insecure in the face of US economic strength and military capability. This sense of insecurity, as Vojtech Mastny sees it, set the stage for the Cold War in Europe and shaped Stalin's policies along his immediate borders. The Soviet dictator was more likely to take risks when he concluded the advantages were on the side of the USSR.[6] Khrushchev put it more colorfully when he said that for Stalin 'foreign policy meant keeping the anti-aircraft units around Moscow on a twenty-four-hour alert'.[7] The air-defense analogy is an apt one and accurately reflects the defensive nature of Soviet air deployment during the early years of the Cold War.

Andrei A. Zhdanov set the ideological framework for that stage of the conflict during his secret speech on 22 September 1947. The Soviets called the meeting at the Polish resort town of Szklarska Poreba in order to frame their response to the Truman Doctrine and the Marshall Plan. Zhdanov announced that the world was divided into two openly hostile camps, the 'socialist' and the 'capitalist'. The socialist camp could count on support from countries such as Indonesia, Indochina, India, Egypt, and Syria, because these nations were resisting Western colonialism.[8] Although Stalin was not prepared to extend much more than moral support to these 'friendly' countries, each of these would later assume greater importance under Khrushchev and Brezhnev.

Stalin's primary focus was, and would remain, on his satellites in eastern Europe. Throughout 1948 and 1949 he struggled to tighten his grip on this most critical arena. In Czechoslovakia Stalin succeeded in installing a compliant regime; in Yugoslavia he failed to displace the upstart Tito; in Greece he abandoned the communists to avoid a confrontation with the US and Great Britain; and by May 1949 he was forced ignominiously to abandon the Berlin blockade. Before the end of 1949, however, Stalin's luck changed, but that change did not come in Europe.

Asia was an area of secondary importance to the USSR. The United States managed to keep the Soviets out of the postwar settlement in Japan. Stalin's own political miscalculations had contributed to that foreign policy failure. As late as 1948, the Soviets backed the Nationalist Chinese to win the civil war in China and thought the US would likely intervene if the Chinese Communists began to win. In Moscow's view the Indian Communist Party seemed more likely to lead the socialist camp in Asia.[9] Mao Zedong's forces began to prove that prediction wrong by 1949 at the same time that Soviet science produced its first atomic bomb. The combination of a communist victory in China and the sudden end of the Western monopoly on atomic weapons gave Stalin an opportunity to redress the foreign policy failures in Europe with success in the Far East.

Soviet participation in the air defense of Mao's forces began even before the signing of the Sino-Soviet Treaty of February 1950. In the summer of 1949, months before their victory over Jiang Jieshi (Chiang Kai-Shek's) Nationalist forces, several top Chinese Communist leaders were sent to Moscow to ease Stalin's suspicions that Mao might be an Asian Tito. Liu Shaoqi led the small delegation during meetings with the Soviet dictator in late June. The Chinese received assurances of friendship, support, and 360 anti-aircraft artillery (AAA) pieces to help with the defense of China's coastal cities. Early the next month several hundred pilot candidates and ground crew departed for training in the USSR. The Chinese asked for 200 Yak fighters, up to 80 bombers, and for Soviet advisors. At the same time, Liu Yalou, the newly designated commander of the embryonic Chinese Air Force (CAF), met with Soviet Defense Minister Aleksandr M. Vasilevskii to finalize the details of the sale of 434 military aircraft to China. On 5 October 1949 Stalin accepted the Chinese terms, and by 21 April 1950 the first CAF unit, the 4th Mixed Air Brigade, entered service at Nanjing.[10]

During Mao's first visit to Moscow, in a conversation with Stalin on 22 January 1950, the Chinese leader thanked the Soviet dictator for sending an air transportation regiment to China the previous year. Mao requested that this Soviet regiment be allowed to remain to airlift supplies to Liu Bocheng's forces massing for the impending campaign against Tibet. Stalin agreed to ask his military people if it was feasible to keep the transport planes in China. This is the earliest reference to the open participation of Soviet aviation in the PRC.[11]

Following the signing of the formal Sino-Soviet Treaty, Mao requested help in stopping the Nationalist bombing attacks on Shanghai and other important coastal cities. The 52d AAA Division, commanded by SAF Colonel S. Spiridonov, and a radar battalion were already deployed to the PRC at the time. This force was soon supplemented by MiG-15s of the 29th Guards Fighter Air Regiment (FAR), commanded by Hero of the Soviet Union (HSU) SAF Lt-Col. Pashkevich, and 45 La-11s of the 351st FAR, commanded by another HSU, V. Makarov, which were deployed to the Liadong Peninsula. SAF Lt-Gen. Pavel F. Batitskii set up an Air Defense district (*Protivovozdushnaia oborona* – PVO) headquartered in Shanghai. A mixed Bomber Air Regiment (BAR) of Tu-2s and Il-10s joined the fighter force and conducted combat training until the end of March. The Soviets scored their first kills in April 1950, when La-11 pilots P. Dushin and V. Sidorov shot down a Nationalist B-26 over Xuzhou and N. Gushev shot down two P-51 Mustangs during a single mission. The MiG-15s made their combat debut on the night of 11 May, intercepting two Nationalist B-24 Liberators. I. Shinkarenko became the first MiG-15 pilot to shoot down an enemy aircraft in combat. The acting regimental commander, Major Keleinikov damaged a P-38 Lightning not long thereafter.[12] The USSR had taken the first steps in utilizing its air forces as an active adjunct to its foreign policy mission.

The Soviets would help the CAF become the world's third largest air force in a dramatically short time. The USSR had more than just a military interest in the Far East and the Sino-Soviet Treaty contained an economic element as well. The Chinese and the Soviets engaged in joint ventures in Manchuria, including aircraft production facilities in Shenyang, to expand the productivity of this backward and war-torn region. Kim Il-Sung's North Korean regime was a decidedly junior partner in this arrangement. Stalin designated Mao to assume the leading role of the Communist revolution in Asia and afforded him a level of respect the North Korean leader never received. Although Kim Il-Sung had signed a treaty with the USSR a year before Mao, there had been no mention of military assistance and the relationship outlined therein was one of economic exploitation rather than co-operation. Stalin supplied arms and advisors to the PDRK, but in exchange extracted lead, gold, and other raw materials in a manner closer to colonialism than socialism.[13]

As advantageous as the Soviet position was in East Asia, Stalin took it too far. His foreign policy manipulations rarely produced the desired result for the USSR. By finally agreeing to allow the North Korean leader to invade the South in April 1950, Stalin unwittingly set the stage for one of the defining, and most dangerous, moments of the Cold War. Stalin had studiously avoided direct military confrontation with the US, but through his attempts to pit Mao and Kim Il-Sung one against the other he ended up committing his forces into combat against the US Air Force before the end of 1950. Stalin did not believe the United States would fight to defend South Korea, but his miscalculation, coupled with Kim's dream of a unified Korea and Mao's support for revolution in Asia, led to the Korean War beginning on the rainy Sunday morning of 25 June 1950.[14]

Truman's decision to intervene in Korea took the Soviets by surprise and the concurrent move to place the US Seventh Fleet in the Straits of Formosa irritated the Communist Chinese a great deal. Following the National Defense Conference of 7 July 1950, Mao ordered the People's Liberation Army (PLA) to deploy troops around the city of Andong on the Chinese side of the Yalu. These concentrations were to be completed by the end of the month.[15] On 21 July, the Soviet government ordered the 151st Fighter Air Division (FAD), equipped with 62 MiG-15s, to fly to Shenyang to provide air cover for the PLA's Thirteenth Army Group. The 151st also began retraining Chinese pilots in the new jets. By 10 August, while the United Nations Command (UNC) was still bottled up inside the Pusan–Taegu perimeter, the 151st was in place and ready to fly.[16]

The commander of the 151st FAD was Ivan M. Belov, former chief of staff of the Soviet First Air Army during the closing months of the Second World War. Belov, a highly capable staff officer, helped plan the devastation of Königsberg by Soviet bombers in 1945.[17] Belov's command, with the later addition of the 28th and 50th FADs, became the Sixty-fourth Fighter Air Corps (FAC). The Sixty-fourth FAC was the umbrella unit for all of the Soviet pilots, radar operators, and AAA gunners who fought the US Air Force in the northwestern corner of Korea better known as 'MiG Alley'. Other Soviet air divisions served in the Northeastern and Eastern Military Districts of the PRC. The central administration point for the SAF (*Tsentral'nyi punkt upravleniya*) was in Beijing under the direction of Major-General Prutkov. Major-General Golunov served in a similar position for the Eastern Military District headquartered in Nanjing. Shenyang, however, was the most critical command post for operations in Korea and served not only as host to the Sixty-fourth FAC but as the organizational and administrative center (under Major-General Gorlachenko) for the Northeastern Military District.[18]

Soviet participation in the air war over Manchuria and northwest Korea was a carefully orchestrated ballet in which political and economic considerations dominated purely military concerns. Stalin took great pains to keep the level of his commitment to his new Chinese allies large enough to satisfy Mao, but, more importantly, limited enough to avoid a general war with the United States. The history of the SAF's mission in China and Korea is the history of Soviet foreign policy during this defining engagement of the Cold War. The confrontation with the US and the transfer of modern technology to the PRC set the pace for an expanded Cold War arms race and eventually the introduction of a third superpower to the mix. The MiG-15, the USSR's most advanced conventional weapon at the time, was emblematic of the Soviet position in the world order. It was an indication to the West that the Soviet Union was capable of intercepting and destroying atomic bombers. To the Chinese, and to a lesser extent the North Koreans, it represented a rapid means to becoming a world-class military power and the political and economic leverage that that position implied.

Stalin and Mao's primary concern at the beginning of the Korean War was to defend Manchuria and the Soviet Far East against possible attack by the B-29s of the US Far East Air Force (FEAF). The four-engined bombers spent most of the first month of the war bombing targets close to the UNC frontlines around Pusan. At the beginning of August the two original B-29 groups were joined by two more to begin Interdiction Campaign No. 1 in an attempt to destroy North Korea's transportation system. On 12 August FEAF's Bomber Command incurred the wrath of US military and civilian leaders by attempting to bomb Rashin which was only 17 miles from the Soviet border. The bombers missed their target but did not violate Soviet airspace. The US State Department objected to attacks on targets so close to such a sensitive area and the Joint Chiefs of Staff (JCS) ordered a halt to raids on Rashin.[19] The show of restraint by the JCS probably did little to assuage the concerns of the leadership in Beijing and Moscow.

Top-secret reports from Soviet Defence Minister Alexander M. Vasilevskii, and his deputy German Malandin, to Nikolai A. Bulganin in late August 1950 revealed Belov's division requested permission to protect the Thirteenth Army Group from enemy air attacks. The Chinese notified the Soviets that on 27 August UNC planes had crossed the border into Manchuria and bombed population centers and railroad stations. The 151st FAD was in position and prepared to carry out its PVO mission.[20] The Soviet political leadership was not quite ready to allow Belov's pilots to take direct action against FEAF, even when they violated the Manchurian border. In their next report to Bulganin, Vasilevskii and Malandin stated that they did not believe it 'advisable' to fly missions to defend the Thirteenth Army Group. They were more concerned that the 151st had not started retraining Chinese pilots. According to plan, conversion to MiG-15s was to be completed by 25 January 1951. The report reiterated that Belov's division was not to violate the border with the PDRK.[21] Stalin was being extremely cautious in trying not to provoke the USAF. He was more concerned with preparing the Chinese to shoulder the burden of the war in Korea than he was in testing his new MiG-15s in combat.

Stalin appeared to lose his composure to some extent following MacArthur's attack at Inchon. In reports dated 21 and 23 September addressed to Stalin and circulated to Malenkov, Beria, Mikoyan, Kaganovich, Bulganin and Khrushchev, Vasilevskii informed the Politburo of the progress in dispatching two FARs directly to defend the North Korean capital. The Soviet dictator wanted 40 Yak-9s of the 34th FAR and an identical number of La-9s of the 304th FAR dispatched to the front as quickly as possible. Vasilevskii, while carrying out Stalin's orders, had serious reservations about this attempt to defend Pyongyang including the following statement at the end of his last report: 'At the same time I believe it necessary to report, it is unavoidable that the work of our pilots over Pyongyang will be discovered by the Americans after the first air battle, since it will be conducted by pilots speaking Russian over their radios.'[22] It seems very odd that Stalin appeared ready to risk Soviet pilots so close to the front lines in piston-engined aircraft that stood little chance in combat against FEAF's F-80s. There is no evidence that this deployment was actually carried out, but the orders indicate that Stalin was quite concerned with the Inchon landings and was unprepared to deal with them.

Soviet concerns with US air power received another strong shock shortly after MacArthur began his advance past the 38th Parallel. At 16.17 hours local time on 8 October, two USAF F-80 Shooting Stars dropped out of a low cloud cover to strafe the Soviet Fifth Fleet airbase at Sukhaya Rechka. The American jets destroyed one P-39 Aircobra and damaged six more at a facility less than 25 miles southwest of Vladivostok and 66 miles from the Soviet–Korean border. The piston-engined Aircobras, including a regiment from the Fifty-fourth Air Army which had landed only the day before, were relics of US Lend-Lease aid and would not have fared any better had they been aloft. The Soviets suspected that the two raiders had been launched by an aircraft carrier because of the F-80s' limited range, even with drop tanks. The base commander reported all of this directly to Stalin by the end of that day.[23] It must be assumed that the poor weather or the inability of Soviet radar to detect low-flying aircraft kept the two regiments of MiG-15s (from the 303d FAD) stationed at Vladivostok from intercepting and destroying the raiders.

By the second week of October the Soviets had begun to ship more jets and tanks to the Chinese. This was during the time when Chinese sources alleged that Stalin had reneged on his promise of air support for the PLA if it intervened in Korea.[24] The Soviet documents indicate that Stalin signed a directive from the Council of Ministers ordering Vasilevskii to send four MiG-9-equipped FADs to China. He also dispatched three regiments of Il-10s and ten independent tank regiments at the same time. These aircraft and tanks were to be used to defend Chinese industrial cities and to train Chinese crews. Section 3 of this directive states: 'The use of our aviation at the front and to cover troops and targets near the Chinese–Korean border is categorically forbidden.' The same restriction also applied to the tanks and their crews.[25]

The concern for defending Manchurian industrial assets found in the Soviet documents was reflected in the news coming from China during October 1950. Reports of panic in, and evacuation of, Chinese cities, air-raid defense measures, preparation for treating bombing victims, and the movement of industrial equipment deeper into China continued throughout the remainder of 1950. There was a concurrent increase in air activity, base construction, and PVO activity. Shenyang was placed under a state of emergency and its major factories were evacuated to Shanxi Province. A witness reported seeing a large steel manufacturing furnace being sent by rail from the Manchurian city of Anshan to the interior. The Chinese were removing every possible economic asset from the exposed coastal regions and areas near the Korean border to places that could be defended by the expanding PVO force.[26]

On 25 October, Mao's troops, the Chinese People's Volunteers (CPV), intervened in the fighting in North Korea. These first battles were merely skirmishes compared with the devastating offensive the CPV would launch at the end of November, but the entire character of the Korean War had changed decisively. The Soviets had used the month of October to prepare for the air defense of Manchuria and on 1 November 1950 Soviet pilots changed the entire nature of Cold War. When Russian MiG-15s attacked US aircraft over Sinuiju on the Korean side of the Yalu, they dramatically altered the US perception of the military capacity of the USSR.[27]

If the detonation of the Soviet A-bomb had come as a shock, then the appearance of the MiG-15 and its ground control and intercept (GCI) radar systems provided the final blow to US reliance on atomic domination as a military and diplomatic trump card. Although the Sixty-fourth FAC's mission to defend the bridges across the Yalu at

Andong and the vital Suiho hydroelectric facilities was limited in scope, the success of the MiG-15 threatened not only the UNC mission in Korea, but it also led the USAF Chief of Staff Hoyt Vandenberg to question the ability of the Strategic Air Command to launch successful atomic attacks on Soviet or Chinese territory. In a single day the MiG-15 completely outclassed every other aircraft in Korea. The CPV and the Sixty-fourth FAC did not co-ordinate their missions on the battlefield, but the existence of each completed the outline of a potential Third World War.[28]

On the afternoon of 1 November, Belov's 151st FAD and elements of the 28th entered the war from airfields at Anshan and Shenyang in Manchuria. According to the handwritten combat log kept by the Chief of Staff of the 151st, Colonel Merzelikin, the air division flew 16 sorties between 12.50 and 16.05 hours and claimed to have destroyed two F-80s without a loss. The Soviet MiGs were vectored onto their targets from an auxiliary control point (*vspomogatel'nyi punkt upravleniia* – VPU) at Andong, just across the Yalu from Sinuiju.[29] The Soviet intervention may have been a response to a UNC air raid that morning against a North Korean regiment of Yak-9s stationed at Andong airfield. The Soviet advisor attached to this unit, Colonel Petrachev, reported that one of his planes had been destroyed and three more damaged in the raid.[30] The USAF reported that it had attacked the North Korean Yaks after they were spotted by an RF-80 at midday in the revetments at Sinuiju airfield.[31] Andong is on the Chinese side of the Yalu and a UNC air raid would have been of great concern to the Chinese and the Soviets. It is unlikely that Petrachev would not know the location of his own regiment, or that the MiG-15s would have been scrambled from deeper into Manchuria without cause.

The airspace above Sinuiju was the focus for the epic air battles between Soviet-piloted MiGs and Bomber Command B-29s until the fall of 1951. The struggle for air superiority against UNC F-86 Saber jets stretched from early 1951 until July 1953. The B-29s failed to destroy the railroad bridge between Andong and Sinuiju and Bomber Command had to abandon daylight operations after its last, and most severe, mauling at the hands of the Soviets on 23 October 1951, 'Black Tuesday'.[32] The MiG-15 did not prove as adept an air-superiority fighter as did the F-86, but 'MiG Alley' remained a dangerous place for UNC, Soviet, and Chinese pilots for the duration of the conflict.

The disaster that hit MacArthur's forces following the CPV intervention caused a wave of panic among US political and military leaders. Truman mentioned the possibility of using atomic weapons in Korea at a press conference at the end of November 1950. The US president incorrectly stated that the use of A-bombs was a decision for 'the military commander in the field'.[33] The presence of Soviet air forces in Manchuria meant that UNC forces in Pusan and Japan could be threatened by air attack. Soviet air power appeared to have played a major role in the decision not to use atomic weapons against targets in North Korea and China. Truman went as far as authorizing the release of A-bombs from the Atomic Energy Commission in early April 1951. While Stalin's manipulations helped create the nightmare of Korea, his defensive posture in Manchuria helped prevent an escalation to general war.[34]

By the end of 1950, the Soviets had transferred the 50th FAD to Andong to defend the bridges and hydroelectric facilities. This action allowed the 151st and the 28th FADs to stand down from combat and complete training Chinese pilots.[35] Even as Soviet pilots fought the USAF over North Korea, Stalin continued to place his highest priority on supplying and training the Chinese and North Koreans. The Soviet Deputy Chief of Administration, Lt-Gen. Dratvin, filed a report on the number of military advisors, instructors, translators, and service and supply personnel assigned to the PRC and the PDRK. At that point the USSR had 1,487 people in China and another 176 in North Korea.[36] Dratvin also forwarded figures detailing the Soviet personnel involved in the transfer of the aviation and tank units to the PRC. There were 4,340 officers, 6,531 NCOs, and 5,944 enlisted men staffing the aviation divisions being transferred. An additional 815 officers, 1,647 NCOs, and 1,960 enlisted were in the independent tank regiments on station in northeastern China. A grand total of 21,237 military men and 9 civilians were in the PRC, or on their way, by December 1950.[37] It is unknown if these totals include the men of the Sixty-fourth FAC or only those men involved in the transfer of weapons and training. Stalin was determined to increase the military capabilities of Mao's regime as quickly as possible, but the Soviets were obviously not preparing to join the Chinese in large-scale combat operations in Korea. The Soviet dictator ensured that the major part of the dying would be left to the CPV and the North Korean People's Army.

As January 1951 got under way, Stalin set about creating a diversion for Truman and European allies by ordering the Soviet military to prepare for wargames in Budapest. The Soviets announced their intention to rehearse for an invasion of Tito's Yugoslavia. This only served to strengthen US suspicions that Stalin was preparing to take advantage of the preoccupation with Korea to launch an attack on western Europe. The dominoes could begin to fall in Europe and that by itself would lead to a rapid UNC withdrawal from Korea.[38] Truman considered Berlin, West Germany, Indochina, Yugoslavia, and Iran as areas under greater Soviet threat than Korea. The US president was concerned that Stalin might 'have decided that the time was in fact ripe for a general war with the United States'.[39] However, the US military attachés in Moscow found no evidence that the USSR was expanding its economy to support any large-scale aggression in 1951. They went a step further and projected that Soviet military capacity relative to the West would drop throughout the year, and tail off precipitously in the following years.[40]

On 3 January 1951, four Chinese pilots from the CAF's 28th FAR joined 16 Soviet MiGs from the 50th FAD on an intercept mission from Andong. The CAF had officially joined the air war against FEAF, but the Soviets continued to do most of the fighting in the air.[41] The fact that the USAF's F-86s had departed for Japan on the previous day, thus leaving air superiority over North Korea in Soviet hands by default, may have spurred the commitment of the newly trained Chinese pilots to combat.[42] The quality of the Soviet air defenses over the Yalu increased markedly in the early spring with the arrival at Andong of the 324th FAD commanded by three-time winner of the Hero of the Soviet Union medal and leading Allied ace from the Second World War, Ivan Kozhedub.[43] The 324th arrived just in time for one of the most dangerous periods of the Korean War.

While Stalin and Mao were busily planning to move MiG-15 units into North Korea, Truman was contemplating using atomic bombs in Korea again. After meeting with powerful members of Congress and reading MacArthur's latest slap at the commander-in-chief's authority, the president met with the Chairman of the JCS, Omar N. Bradley. Bradley showed Truman the most recent intelligence reports on Soviet concentrations at Vladivostok, the southern Sakhalin Islands, Manchuria, and the Shandong Peninsula. After consulting with CIA Director Walter Bedell Smith, the president called the head of the Atomic Energy Commission to the White House for a noon meeting. Truman convinced N. Gordon Dean to authorize the transfer of nine complete atomic bombs to the Air Force. The next day, 7 April 1951, the 99th Medium Bombardment Wing (atomic-capable B-29s) was ordered to pick up the weapons and proceed to Guam.[44] By moving the United States and the world closer to nuclear war than ever before, Truman also set the stage politically for the dismissal of MacArthur on 11 April. The USAF never dropped the nine atomic bombs, but the firing of MacArthur was nearly their political equivalent.

On 12 April, US and Soviet pilots fought the largest air battle of the war to that point over Sinuiju.[45] Sergei Kramarenko, Boris Abakumov, and Evgenii Pepeliaev participated in this battle. Kramarenko's FAR, the 176th, was the first group to hit the B-29s through the protective screen of F-84 Thunderjets. The escorting fighters, both F-84s and

F-86s, proved unable to stop the speedy MiG-15s from closing with the lumbering B-29s and the Soviet pilots set several of them on fire forcing many of the rest to jettison their bombs short of their target.[46] Abakumov was with Pepeliaev's regiment when it arrived to complete the task begun by the 176th.[47] The FEAF daily report listed three B-29s shot down and seven more damaged, along with an absurdly high number of MiG kills claimed. Several of the damaged B-29s crash-landed, either at Okinawa or at fighter bases in Korea, and were written off, while the Soviets reported no losses for the day. The report further stated that the 'enemy pilots reported very aggressive and pressed the attack to the bomber formation through flak. Most all attacks on B-29s made singly and in pairs.'[48] The Yalu bridges were off limits for Bomber Command until the even greater defeat at the hands of the 324th and the 303d FADs in October.[49] If Truman had ordered the B-29s on Guam to deliver their bombs, there was no guarantee that Ivan Kozhedub's pilots or the other Soviet pilots in the PRC would have let the bombers pass unmolested. The Soviet pilots could claim victory in their mission to keep the railroad bridge across the Yalu open, but they were never able to spread the air defense umbrella further south than 'MiG Alley' and in the summer of 1952 they were unable to prevent the destruction of the Suiho hydroelectric facilities by FEAF fighter-bombers.

The Korean War continued after the death of Stalin on 5 March 1953, but the most dangerous phase of the conflict had passed by the summer of 1951. As the ground combat slowed to a bloody stalemate along the 38th Parallel, the Soviets continued to cycle air divisions through Andong. Neither the PRC nor the USSR committed any significant air resources to the battlefield, preferring to conserve their expensive jets and under-trained aircrews for a future conflict. 'MiG Alley' became more of a training ground for Soviet, Chinese, and North Korean pilots, than a decisive arena of action. Stalin had managed, before his death, to alter the balance of the Cold War for good. Mao's PRC became the third superpower as a result of Soviet aid and Korean War experience. Stalin's successors would quickly rue his support of the Chinese comrades because of the increased expense of the Cold War that Korea triggered and the short-lived nature of Sino-Soviet co-operation.

THE COLD WAR UNDER KHRUSHCHEV

Immediately following Stalin's death, a group of Kremlin figures divided political power in the USSR among themselves. Georgii Malenkov and Lavrentii P. Beria made the first moves to put themselves in the positions of greatest power as chairman of the Council of Ministers and chief of the newly created Ministry of Internal Affairs respectively. In the remix that followed, the newest member of the Soviet élite, Nikita S. Khrushchev, watched his power base erode as his protégés Leonid I. Brezhnev and N. G. Ignatov were removed from the Secretariat. Khrushchev was also replaced as the Moscow party first secretary. Two of the most influential men of the next three decades of Soviet foreign policy appeared to be in political trouble as the new post-Stalin era dawned.[50] The struggle for power in the USSR had repercussions in eastern Europe with three major problems arising between 1953 and 1956.

Strikes and riots in East Germany during June 1953 were quickly crushed by the Soviet Army. West German military power was revived and brought under NATO command in May 1955. Poland and Hungary were both in an uproar by 1956. During this time, with Malenkov ostensibly in charge following the arrest and execution of Beria, Soviet foreign policy was limited to eastern European concerns. Khrushchev began to come into his own politically, by his own admission, during the negotiations with Austria in April 1955.[51] In September Khrushchev, accompanied by Bulganin, Molotov, and the foreign trade expert Anastas Mikoyan, went to Beijing to conclude new agreements with Mao to re-establish Sino-Soviet relations on a more equal basis. The USSR extended new loans to assist the technical development of Chinese industry.[52]

This was Khrushchev's first trip abroad, but it would not be his last as he prepared to launch the Soviet Union on a mission to support the new wave of revolution in the Third World. In April 1955 he and Bulganin journeyed from Yugoslavia to India, Burma, and Indonesia. Khrushchev was also pivotal in persuading the Politburo to supply weapons to Gamal Abdel Nasser's government in Egypt. Since Egypt was still struggling to free itself, as was India, from British colonial influence, supporting Arab nationalism was seen as the best answer to John Foster Dulles' Baghdad Pact. Dulles was hard at work building a series of treaty relationships ('pactomania' some called it) with Middle Eastern countries designed to help contain the USSR. Dulles was Khrushchev's *bête noire* in foreign policy.[53]

Khrushchev consolidated his power at home by deposing Malenkov, who had advocated reducing defense spending to levels of 'minimum deterrence' as an appropriate response to the destructive power of nuclear arsenals. The other members of the Presidium called for increased military spending and a buildup of conventional forces. Khrushchev courted the support of the Soviet military leadership by supporting increased military spending in open opposition to Malenkov. Khrushchev, in fact, adopted a position very close to the one Malenkov espoused in 1954 earning the enmity of the very military leaders who had originally supported him.[54]

After the successful launch of Sputnik in 1957, Khrushchev, by then in personal control of the USSR's political structure, became enamored with 'missile-nuclear' weapons. He was willing to reduce the size of the conventional armed forces and arms production in order to build inter-continental ballistic missiles (ICBMs). He backed the production of surface-to-air missiles instead of conventional fighter planes for the PVO forces.[55] It is somewhat ironic that Khrushchev's rise to power paralleled the design and construction of one of the best-known aircraft in the world. The USSR mass-produced the MiG-21, a Mach 2 point defense interceptor, for 28 years in 15 primary versions. The MiG-21 fought in more battles than any other aircraft in history. Forty-nine other countries have flown the type and three countries, Czechoslovakia, India, and China, were licensed to build it. If the MiG-15 was the emblem of Stalin's early Cold War commitment, the MiG-21 signified Soviet commitment to a substantial part of Asia, Africa, and Cuba.[56]

In looking at the use of the SAF in Cold War conflicts Khrushchev's era, like the man himself, is rather contradictory. On the one hand, he nearly single-handedly propelled the USSR onto the world stage as a major economic and political power. On the other hand, his Seven-Year Plan (1959–66), a departure from the usual Soviet Five-Year Plans, reduced the number of aircraft available for sale outside the USSR. He helped open the Middle East and Asia to Soviet economic influence through ventures such as the Bhilai Steel Plant in India and the Aswan High Dam project in Egypt.[57] His campaign of de-Stalinization helped undermine the already shaky Sino-Soviet relationship, but by the fall of 1956 the first Chinese-produced aircraft, the F-5 (a licensed copy of the MiG-17) had rolled off the production line at Shenyang and the Soviets had agreed to build a factory to manufacture Tu-16 bombers (the Chinese B-6).[58] That year Khrushchev ruthlessly crushed the Hungarian uprising after negotiating a settlement with the Poles of Poznan and threatened the Western allies with a missile attack if they did not stop the Suez war in Egypt.[59]

It should come as no surprise that this impulsive and contradictory Soviet leader played the leading role in precipitating the Cuban missile crisis. In embracing Castro's Cuban revolution, Khrushchev fulfilled his romantic notion of supporting Third World revolutions with the very missile technology that symbolized his military policies. By 1961 the USSR had ensured that Cuban sugar had a market and that Castro received Soviet weapons virtually for free. Khrushchev's shoe-banging behavior at the UN and his pursuit of revolutionary romanticism in Cuba caused serious concern among the Soviet political elite. Khrushchev had already survived one attempted coup in 1957, but as it turned out, he would not remain in power after the next.[60] Along with supplying weapons to Castro, the Soviet Union also contracted with India's Hindustan Aeronautics Ltd. (HAL) to produce the MiG-21, a prospect which did not please the PRC. The Soviets delayed delivery of the fighters until after the Sino-Indian War in 1962, but soon had another satisfied Third World customer as the Indians adapted and modified the Soviet jet to fit their needs.[61] Nasser's EAF also ordered MiG-21s in 1961 to replace the MiG-17s and MiG-19s which were outclassed by Israeli aircraft. The first group of Egyptian pilots transitioned to the MiG-21 in the USSR. These planes would serve as the backbone of Egypt's air defenses until the 1980s.[62] Even the Chinese were given approval to produce MiG-21s (F-7s) at Shenyang, until the Soviets canceled the contract as the Sino-Soviet split grew wider.[63]

MiG-21s were probably the least offensive weapons shipped to Cuba during Operation 'Anadir' beginning in the summer of 1962. Soviet Defense Minister Rodion Malinovskii and Chief of the General Staff Marshal Matvei Zakharov signed the final order sending fighters, bombers, 42,000 men and missiles with nuclear warheads to Fidel Castro. Lt-Col. N. Shibanov commanded the FAR of MiG-21s which was part of the PVO system Khrushchev provided to defend the Cuban revolution from the United States. Lt.-Gen. S. Grechko was in charge of the air defense system under the overall command of General of the Army Issa A. Pliyev. The Soviets also sent a regiment of Il-28 medium bombers which were soon spotted by U-2 spy planes as the bombers were being assembled on Cuban airfields.[64] The most interesting photographs the US spy planes took were of the launchers for elements of the strategic missile division. This division was the centerpiece of Khrushchev's reckless gamble to make his previously anemic missile threat more meaningful. The division contained 24 R-14 intermediate-range (2,200 miles) missiles and 48 R-12 medium-range (1,100 miles) missiles armed with 3-megaton warheads. Coupled with the nuclear-capable Il-28s and the two regiments of tactical nuclear rockets (25 miles), these Soviet weapons posed a considerable threat to the United States and to world peace.[65]

During Operation 'Anadir', Khrushchev sent more men to Cuba than the USSR sent to the PRC at the most dangerous stage of the

Korean War. Soviet pilots were ready to fly air defense and bombing missions from Cuba, but equipped with weapons of mass destruction under the very nose of the United States. Small wonder many in the Kremlin believed Nikita Sergeich may have lost his mind. The most critical factor in all of this was that these weapons were never used and were almost completely withdrawn (save for a single Soviet brigade) before the year was out. The Soviet and Cuban documents, thus far declassified and released, indicate that Castro was not nearly as pleased with the abrupt Soviet withdrawal as was much of the rest of the world. Khrushchev sent his closest advisor, Anastas Mikoyan, to try to soften the blow. The Soviets withdrew their missiles without consulting Castro. Even worse, they had also removed Castro's Il-28s after first promising they would never do such a thing. The Soviets had transferred the bombers to the Cubans, but not the missiles. Mikoyan's protestations that removing all offensive weapons was for the best and that Cuba was safe from US invasion fell on unbelieving ears and Soviet–Cuban relations were never the same thereafter.[66] As Castro himself explained, in reference to the Il-28s, in a secret speech given in 1968: 'They were useful planes; it is possible that had we possessed Il-28s, the Central American bases [from which Cuban exiles were launching Mongoose attacks] might not have been organized, not because we would have bombed the bases, but because of their fear that we might.'[67]

Between 1954 and 1964 the USSR spent roughly $2.7 billion (US dollars) in arms transfers to the Third World. Afghanistan was the first Third World nation to receive Soviet economic credits (when Third World is defined in political rather than economic terms) in 1954 and Nasser signed the first overt arms deal between the USSR and a Third World client in 1955. Khrushchev pushed a return to Lenin's vision of creating a link between nationalistic revolutions in the developing world and the Soviet Union's own revolutionary objectives. This policy changed at the 1956 Twentieth Party Congress where Khrushchev publicly announced the Soviet Union's openness to a 'vast peace zone' of nonaligned states at the same time he 'secretly' announced de-Stalinization. The Soviets added Syria, Yemen, Iraq, Indonesia, and Guinea to their trading list by 1959. During the last five years of Khrushchev's political career, Laos, Morocco, Algeria, Sudan, Ghana, Mali, Cambodia, Somalia, Tanzania, and Zaire (India has already been mentioned above) joined the nonaligned states doing business with the Soviets.[68] Although Leonid Brezhnev led the Kremlin faction which 'retired' Khrushchev from government service in 1964, the new Soviet leadership exploited the worldwide trade established by Khrushchev and Anastas Mikoyan to an even greater degree.

THE BREZHNEV ERA

The Central Committee plenum of February 1964 was one of the last times Nikita Khrushchev played a significant role in Soviet politics. His opposition in the Secretariat centered around Nikolai Podgorny, Brezhnev, Aleksandr Shelepin, and Dmitrii Polyansky. At the Presidium meeting of 13–14 October 1964, Khrushchev was officially retired.[69] He was ousted for primarily domestic reasons, but in his absence Soviet foreign policy began to change almost immediately. With Brezhnev installed as First Secretary of the Communist Party and Alexei N. Kosygin appointed as Soviet premier, the new leadership quickly faced a difficult situation in the Democratic Republic of Vietnam (DRV).

Ho Chi Minh's regime expected the aid of all 'fraternal socialist countries' in its military struggle against the United States. The Gulf of Tonkin incident indicated that there would be no diplomatic solution likely. The Chinese were supporting the Hanoi government to a much greater degree than was Moscow and Brezhnev and Kosygin did not want to risk losing influence in the socialist camp or pass on an opportunity to spread Soviet influence in Southeast Asia.[70] Once again the Soviets were dealing with a revolution deeply in need of air defense. The US punctuated the point by conducting bombing raids on North Vietnam during an announced visit by Kosygin in early 1965. Although the Soviets were concerned about damaging their relationship with the United States, they quickly decided to aid the DRV. US intelligence reported the first 15 Soviet MiG-17s along with 100 mobile AA guns arriving on 25 May 1965. On the next day US Intelligence spotted Il-28 bombers, the same kind which had been considered offensive weapons during the Cuban missile crisis, in North Vietnam.[71]

The United States opened a sustained air war over North Vietnam with the Rolling Thunder strikes beginning in 1965 and lasting until 1968. The Soviets increased their arms shipments to the DRV so that during the period 1965–74 they supplied almost 70 per cent of Hanoi's military needs. The USSR walked a fine line between giving the Vietnamese enough weapons to defend themselves properly and trying to interest the United States in a negotiated settlement to the conflict. With the help of the Soviet Union, North Vietnam's air force doubled in size by 1967.[72]

Oleg Sarin, a journalist with the military paper *Krasnaia Zvezda*, states unequivocally that, 'Soviet pilots were participating actively in flights of the Vietnamese Air Force (VAF) in countering American raids into North Vietnam.'[73] The Soviet pilots lived in small groups close to their airfields and their aircraft, just as in Korea, carried local national markings. Soviet engineers and technicians usually maintained the aircraft. Colonel Alexei Vinogradov, a Soviet pilot in Vietnam, reported that, 'Sometimes we participated in two to three sorties each day.' Another pilot, Anatoli Gribov, was apparently shot down over South Vietnam and repatriated in exchange for a downed USAF crew.[74] Given the nature of past Soviet Air Force ventures it is likely that a major part of the air war against the USAF in Vietnam was conducted by Soviet pilots. The number of pilots reported in the country between 1965 and 1970 ranged from 102 (1965) to 171 (1969). In 1971 and 1972 there were 84 and 64 pilots respectively.[75] The Soviets reported losing 13 men in battle and another three to non-combat causes between July 1965 and December 1974. Fifteen out a total of 16 were officers, so this may indicate pilots killed in combat and non-combat-related accidents.[76] More research in the Ministry of Defense archives needs to be conducted before anything definitive can be written on this subject, but it does seem to be a very promising area of future research. :

The final two SAF engagements of the Cold War occurred in countries that Khrushchev had initiated contact with in the early 1950s, Egypt and Afghanistan. Soviet involvement with the Egyptian Air Force continued to grow after the 1956 Suez Crisis and by mid-1967 the EAF had some 450 combat aircraft and 350 support planes. Many of Egypt's 700 pilots had seen combat in the Yemeni Civil War. The Egyptian pilots were still in the process of converting to the new MiG-21 fighters and Tu-16 bombers. This was a complex process which often took more than a year to accomplish. The operational ready rates for the new aircraft were somewhere around 65 per cent. The Soviets had supplemented the EAF's fighter force with early warning radar, AAA, and SA-2 missiles very similar to the defenses they had installed in North Vietnam. The Soviets designed their PVO system around the concept of high-altitude, high-speed interception, and missile defense. Soviet radar could not detect low-flying targets and EAF pilots were not trained in low-altitude dogfighting, where their MiG-21s were less maneuverable.[77]

When the Israelis launched their surprise air attack against Nasser's forces on 5 June 1967 they almost destroyed the EAF in a relatively short time. Most of the aircraft were hit on the ground since the low-flying Israeli jets were not spotted by radar. During the next six days the Israeli Air Force (IAF) pounded the Syrian and Jordanian air forces out of existence as well. The IAF provided ground support to Israeli tank columns as they overran the Sinai, the Golan Heights, and the West Bank of the Jordan. Soviet training and equipment were defeated in a matter of days and Nasser's Egypt was nearly defenseless against IAF raids.[78]

Much like the SAF in the opening phases of Barbarossa during the Second World War, the EAF had lost many more aircraft than pilots. Nikolai Podgorny visited Nasser after the débâcle and the Egyptian leader offered to place his country's air defenses in the hands of the Soviets. Podgorny declined for the moment, but did arrange for a massive influx of replacement aircraft and new advisors. The Soviets flew in all manner of military supplies during an 'air bridge' lasting for 40 days and soon 15,000 advisors were in the country. Soviet influence soared as Egypt plunged even further into the superpower's debt, but this did not limit Egyptian nationalism and the EAF chafed at Soviet caution.[79]

The pilots got an opportunity to avenge their humiliation as President Nasser began a 'war of attrition' against the Israelis after the USSR had provided enough military support to make good Egypt's losses from the Six Day War. In March 1969, the Egyptians began air raids and artillery bombardments across the Suez Canal. IAF retaliation raids hit a variety of vital targets throughout Egypt.[80] The War of Attrition escalated throughout the summer of 1969 and by September the IAF had commenced destroying Egypt's air defense system, neutralizing the SA-2 and SA-3 batteries, and attacking Egyptian morale. By the end of 1969 the IAF had air superiority over most of the country and the Egyptians had to delay their plan to retake the Sinai. They were engaged in the War of Attrition as an end in itself. Israeli air power alone could not score a decisive victory and so Egypt continued to attack despite the pounding it was taking from the IAF.[81]

Since November 1967, Soviet pilots had been flying combat missions in support of the Republican faction in the Yemeni Civil War. The Soviets replaced Egyptian pilots who returned home after the Six Day War débâcle. They flew combat missions in MiG-17s, MiG-21s, and Il-28s against Royalist forces, and flew patrols over the Red Sea and the Gulf of Aden. The Soviets fought until 1969 when the civil war ended. Their presence in Yemen was confirmed by British intelligence who intercepted radio traffic in Russian between ground control stations and airborne MiGs.[82] In January 1970, Nasser flew to the USSR to plead for direct Soviet participation in Egypt's air defense. The Soviets were at first hesitant to commit their pilots into combat with the IAF, but eventually they acceded to Nasser's plea. As a result, the USSR began shipping more advanced surface-to-air missile systems, later model MiG-21s, and trained Russian crews to operate the new systems.[83]

The SAF pilots operated from Egyptian airfields in March 1970, following a surge in IAF deep-strike bombing raids in February. New surface-to-air missiles arrived to help plug the low-altitude gap in the air defense screen. During this early phase, the IAF tried its best to stay away from the airfields where Soviet air regiments were posted. Initially, the Soviet pilots patrolled near Cairo and the Nile Delta but did not approach the front lines along the Suez Canal. By April the Soviets were operating off of at least five major bases and it was getting increasingly difficult for the IAF to avoid Egypt's new defenders.[84] In addition to the five or so squadrons of new MiG-21s, the Soviets supplemented Egypt's defenses with Yak-26 intelligence overflights of Israel from bases in Soviet Armenia.[85] In order to reduce the chances of an Israeli surprise attack, a MiG-25 group was sent to Egypt to conduct reconnaissance missions along the Suez Canal. These intelligence overflights continued even after the end of the War of Attrition.[86]

The Soviets and the Egyptians increased the pressure on the IAF by positioning their SA-2 and SA-3 screens closer to the Suez Canal. They augmented the density of the defensive zone and, in June, began firing the AA missiles in salvos of five rather than the usual two. The Israelis lost five F-4 Phantoms to these new tactics. The Soviets ambushed and damaged an A-4 Skyhawk on 25 July. The Israelis retaliated by ambushing three flights of MiG-21s which had been scrambled to intercept a decoy group of IAF Mirages. In the subsequent action five Soviet MiGs were shot down much to the delight of the Egyptian pilots who had endured all they could from the arrogant Russian pilots. As a result, the PVO chief paid an inspection call on his troops in Egypt to investigate the cause of this disaster.[87]

The Soviets were not significantly better than the Egyptians in combating IAF air attacks, but Soviet participation raised the diplomatic stakes significantly. The US pressured Israel to avoid escalating the conflict to general war. With the strategic situation stalemated and losses rising, both sides agreed to a diplomatic solution to the expensive War of Attrition and the action ended by 7 August 1970. In Egypt from 18 October 1962 until 1 April 1974, the USSR lost 11 officers in action and ten to non-combat causes. An additional 35 died during roughly the same period in Syria and another officer was lost in combat in Yemen.[88]

A little over a month after the end of the War of Attrition, President Nasser died and Vice President Anwar Sadat was chosen to lead the country. Egypt's relationship with the USSR deteriorated over the next two years. During 1971 the Soviets flew high-speed reconnaissance missions in MiG-25s to photograph Israeli defenses in the Sinai. When the EAF asked for more pictures of the Sinai defenses, the Soviets refused since they were primarily concerned with watching the US Sixth Fleet. The USSR also refused repeated requests for Tu-22 bombers and MiG-23 fighter-bombers to give the EAF a better chance against the more advanced Israeli types. When they were refused for the third time, the Egyptian government expelled the Soviet advisors, technicians, and pilots with ten days' notice. With this expulsion, the USSR lost its influence in a strategic area of the Middle East and could no longer base aircraft in that critical corner of the eastern Mediterranean. Although the USSR did help resupply the Egyptian military after the horrendous losses of the Ramadan War in 1973, they never again played a dominant role in internal Egyptian affairs.[89]

The last major battle fought by the SAF before the effective end of the Cold War in 1989, began near the end of the Brezhnev era and did not cease until the close of the Mikhail Gorbachev era. The Afghan War (1979–89) differed markedly from the engagements mentioned previously in this chapter, since there was no battle for air superiority or any need for PVO forces. Afghanistan is often called the Soviet Union's Vietnam and in many ways the experiences of the SAF in that country were closer to those of the United States in Southeast Asia than to those of the Soviet pilots who fought in Korea and Egypt. In Afghanistan, the Soviets were not fighting a defensive battle against a numerically superior or even a technologically advanced foe. Instead, helicopters played a greater role in the combat than did fighter-bombers. The USSR attempted to prop up a regime which was incapable of defending itself against a more popular guerrilla movement. The Soviets discovered, as did the US in Vietnam, that massive firepower did not guarantee victory and that it was far easier to advance into one of these nasty civil wars than it is to withdraw from one. The casualty figures were much higher for Afghanistan than they were for all of the previous Cold War conflicts combined. For the 110 months of the war the USSR sustained 14,751 killed and a further 469,685 medical casualties. The air forces lost 118 planes and 333 helicopters. The highest casualty rate occurred during the second period of the war from March 1980 to April 1985.[90]

Just how the USSR arrived at the decision to invade its southern neighbor is still not completely clear. A top secret memorandum from the meeting of the Politburo on 12 April 1979 signed by Andrei Gromyko, Yurii Andropov, Dmitrii Ustinov, and Boris Ponomarev indicated that the 'revolution' of the previous April (1978) occurred in an 'economically weak, backward feudal country with primitive economic forms and limited domestic resources'. The Democratic Republic of Afghanistan was in serious trouble and the reporting members of the Politburo decided that: 'our decision to refrain from satisfying the request of the leadership of DRA to send Soviet military units to Afghanistan was correct and this policy should be continued further because the possibility of new rebellions against the government cannot be excluded.'[91]

Despite this display of good sense, the USSR was ready by Christmas Eve to launch a military coup of its own to overthrow Prime

Minister Hafizullah Amin. Amin had risen to power in a bloody coup
in September of 1979 and had murdered his fellow communist and
Moscow's favorite, Nur M. Taraki. The Soviets used paratroopers to
seize strategic points in Kabul and other important towns in
Afghanistan.[92] The success of the first phase of the operation belied the
ten years it would take to finally get Soviet troops out of Afghanistan.

The initial land invasion of Afghanistan by the Fortieth Army had
relatively few helicopters and ground-support aircraft. By the summer
of 1980, the Soviet commanders realized that air power greatly
increased their odds of defeating an elusive enemy in some of the
harshest terrain in the world. A year after that, hardly a ground operation was mounted without support by the SAF. Determining the exact
numbers of fighters, bombers, and helicopters the USSR had available
at any given time is difficult because many of the fixed-wing planes
were based on Soviet territory and the rotary-wing units moved very
often in country.[93]

Aside from the use of helicopters for rapid troop movements, close
fire support, and forward air control missions, the rest of the SAF
inventory acted as flying artillery. Bombing missions were conducted
either to support troops already in battle or to area-bomb Afghan
villages and agricultural targets in a campaign of aerial 'scorched
earth'. Despite its age and lack of suitability for the role, the MiG-21
was the primary bomb hauler for the SAF. These Mach 2 high-
altitude interceptors were forced to fly low and slow to drop bombs
and fire rockets at the Mujahedin. Tu-16 and Su-24 medium and light
bombers were used 'almost daily' to carpet-bomb suspected guerrilla
areas.[94] The Soviet equation for successful operations against the local
guerrillas included delivering powerful, and accurate, air and artillery
preparation, making several trips to ferry soldiers into the combat
zone, and using spotter helicopters to direct air attacks and artillery
bombardment.[95]

Even after ten years of combat service, ground units continued
having difficulty calling in artillery and air attack. The lack of intelligence preparation and the paucity of maps for most of Afghanistan
resulted in only 25 per cent of ordnance being delivered on target.
Frontal aviation pilots flew up to four sorties per day during offensives, but the effectiveness of these sorties was limited.[96] Until 1985–86
Soviet pilots only had to contend with the usual kinds of light anti-
aircraft weapons, such as 12.7mm machine guns and 20mm AAA.
When the US Department of Defense authorized the release of the
Stinger shoulder-launched AA missile to the Mujahedin, Soviet loss
rates rose sharply and the effectiveness of air strikes and helicopter
support dropped dramatically.[97]

In general, the Soviet pilots found that they were poorly prepared for the kind of missions they were expected to fly in Afghanistan. They had to learn the ropes as they went. The usual Soviet training program closely paralleled the training the Soviets gave their Third World clients. Ground attack had not been a priority of the SAF since the Great Patriotic War and did not become a priority for most of the SAF outside of Afghanistan. Rigid training under unrealistic conditions left the Soviet pilot nearly as poorly prepared for combat as his Egyptian or Syrian counterpart. A pilot's combat skills improved with combat experience, if he lived. Throughout the entire period of the USSR's 'bleeding wound', the SAF delivered an impressive tonnage of munitions but with negligible effect in the end. The analogy of striking a flea with a sledgehammer is an apt one that US Vietnam veterans may find familiar. The Soviet Air Force played a major role in covering the Fortieth Army's slow withdrawal out of Afghanistan. New MiG-27s and Tu-22 bombers flew 100 sorties a day in an effort to keep the Kabul regime viable and cover the Soviet retreat during October 1988.[98]

The SAF experience had other similarities with the US experience in Vietnam, such as the indiscriminate use of air power causing death, injury, and homelessness to a massive number of civilians. The SAF also spread millions of mines by air, which continue to kill Afghanis even though the Soviet Union is no more. Millions of Afghanis were forced to become refugees as their homes and villages were bombed, rocketed, and shelled into rubble. In the end, the Soviets were no more successful than the US was in carrying out its policy objectives. Air power alone could not counter the problems of terrain, weather, and lack of concentrated targets. The Mujahedin, like the Vietcong and the North Vietnamese Army, received military aid from another superpower and utilized tactics to lessen the impact of air power. As devastating as the war in Vietnam had been to the US politically, the retreat from the ten-year failure in Afghanistan helped speed the destruction of the USSR. By 1991 the Cold War, like the Soviet Union, had come to an end.

CONCLUSION

We have looked at only a small element of US–Soviet military competition over the past 40 years. By focusing on the times when the Cold War turned hot, we can see some of the incredible risks that were taken during our recent past. The Soviet Air Force was committed, after the Second World War, to developing advanced jet technology to defend itself against atomic bombing. It accomplished that goal, however imperfectly, but ballistic missiles changed the nature of the game. The

Soviet concept of air defense changed very little between the Korean War and the 1973 Ramadan War. The Soviet pilots and those who used the Soviet system seemed always to have been a generation or two behind in technology and a step behind in aircrew preparation. This is a common theme in Soviet/Russian history and when viewed in the historical context of the last two centuries, the Soviet weapons systems during the Cold War represented some of the highest scientific achievement in the world. However, the success of Soviet science could not, in the end, overcome the economic, political, and diplomatic mistakes committed by the Soviet leadership.

NOTES

1. ORE-19-48. CIA Report on 'Soviet Military and Civil Aviation Policies', 23 April 1948, 6; cited in Mark O'Neill, *The Other Side of the Yalu: Soviet Pilots in the Korean War,* unpublished dissertation (Tallahassee, FL: History Department, Florida State University, August 1996), 1.
2. David Holloway, *Stalin and the Bomb: The Soviet Union and Atomic Energy, 1939–1956* (New Haven, CT: Yale University Press, 1994), 243.
3. Tony Mason, *Air Power: A Centennial Appraisal* (London: Brassey's (UK), 1994), 198.
4. Lon O. Nordeen and David Nicolle, *Phoenix Over the Nile: A History of Egyptian Air Power, 1932–1994* (Washington, DC: Smithsonian Institution Press, 1996), 150.
5. During a dissertation research trip to the Old General Staff building Moscow in the spring of 1995, I was given access to the unit histories, or fragments of those histories, for all of the MiG-15-equipped Soviet FADs that flew in defense of the Yalu River bridges. Kathryn Weathersby, my dissertation director, had been instrumental in getting the Russians to declassify many of the documents relating to the Korean War. Many of the documents she found in the Ministry of Foreign Affairs and the Presidential Archive of the Russian Federation pertained to the air war. These documents formed the basis of my dissertation cited in note 1. Much of what follows in this section is derived from that research.
6. Vojtech Mastny, *The Cold War and Soviet Insecurity: The Stalin Years* (New York: Oxford University Press, 1996), 191.
7. Nikita Khrushchev, *Khrushchev Remembers,* Strobe Talbott, trans and ed. (Boston: Little, Brown, 1970), 393; cited in Walter LaFeber, *America, Russia, and the Cold War, 1945–1992,* seventh edition (New York: McGraw-Hill, 1993), 256.
8. Mastny, *Soviet Insecurity,* 30–1.
9. Ibid., 55.
10. Shu Guang Zhang, *Mao's Military Romanticism: China and the Korean War, 1950–1953* (Lawrence, KS: University of Kansas Press, 1995), 39, 49–50.
11. Danny Rozas, trans., 'Stalin's Conversations with Chinese Leaders: Talks with Mao Zedong, December 1949–January 1950, and with Zhou Enlai, August–September 1952', *Cold War International History Project Bulletin,* 6–7 (Winter 1995/96), 9.
12. Aleksandr Kotlobovskii, 'Malaia vozdushnaia voina za Velikoi stenoi' ('A Little Air War behind the Great Wall') *Mir Aviatsii* (March 1993), 28. Thanks to George Mellinger for showing me this source.

13. Mastny, *Soviet Insecurity*, 92–4; Vladislav Zubok and Constantine Pleshkov, *Inside the Kremlin's Cold War: From Stalin to Khrushchev* (Cambridge, MA: Harvard University Press, 1996), 62–3.
14. Ibid.
15. Sergei N. Goncharov, John W. Lewis and Xue Litai, *Uncertain Partners: Stalin, Mao, and the Korean War* (Stanford, CA: Stanford University Press, 1993), 270–1. Mao's telegrams to Nie Rongzhen and Gao Gang.
16. Central Archive of the Ministry of Defense (Tsentral'nyi Arkhiv Ministerii Oboroni – TsAMO); fond 16, opis 3139, delo 16, list 1. All Soviet-era documents hereafter cited in O'Neill, *The Other Side of the Yalu* (see Acknowledgements and Further Reading).
17. M. N. Kozhevnikov, *Soviet Army Air Force Command and Staff in the Great Patriotic War, 1941–1945* (Moscow: Nauka, 1977), 174, 177.
18. TsAMO-A; f. 16, op. 3139, d. 15, ll. 10–11.
19. Robert F. Futrrell, *The United States Air Force in Korea, 1950–1953*, revised edition (Washington, DC: Office of Air Force History, USAF, 1983), 129, 192.
20. TsAMO-A; f. 16, op. 3139, d. 16, l. 1.
21. TsAMO-A; f. 16, op. 3139, d. 16, ll. 4–5.
22. Archive of the Foreign Policy of the Russian Federation (AVPRF); fond 3, opis, 65, delo 827, listi 79–82.
23. AVPRF; f. 3, op. 65, d. 827, ll. 102–3.
24. Goncharov *et al.*, *Uncertain Partners*, 183–4, 279.
25. TsAMO-A; f. 16, op. 3139, d. 16, ll. 162–5.
26. Mark A. Ryan, *Chinese Attitudes Toward Nuclear Weapons: China and the US during the Korean War* (Armonk, NY: M. E. Sharpe, 1989), 7–108.
27. Goncharov *et al*, *Uncertain Partners*, 199.
28. T. R. Fehrenbach, *This Kind of War: A Study in Unpreparedness* (New York: The Macmillan Company, 1963), 558, 560; A. W. Jessup, 'MiG-15 Dims USAF's A-Bomb Hopes', *Aviation Week* (4 February 1952), 16; Roy E. Appleman, *Disaster in Korea: The Chinese Confront MacArthur* (College Station, TX: Texas A&M University Press, 1989), 212; Futrell, *USAF in Korea*, 219.
29. TsAMO-A; f. 151gv. iad, op. 152688ss, d. 7, ll. 1(ob)-2.
30. AVPRF; f. 45, op. 1, d. 335, ll. 71–2.
31. Futrell, *USAF in Korea*, 219; David Rees, *Korea: The Limited War* (New York: St Martin's Press, 1964), 130.
32. Daniel T. Kuehl, 'Refighting the Last War: Electronic Warfare and USAF B-29 Operations in the Korean War', *The Journal of Military History* 56, 1 (January 1992), 94.
33. Clay Blair, *The Forgotten War: America in Korea, 1950–1953* (New York: Times Books, 1987), 522–3.
34. Bruce Cumings, *The Origins of the Korean War, Volume II: The Roaring of the Cataract, 1947–1950* (Princeton, NJ: Princeton University Press, 1990), 751.
35. TsAMO-A; f. 50 iad, op. 174045ss, d. 39, l. 55.
36. TsAMO-A; f. 16, op. 3139, d. 16, ll. 133–4.
37. TsAMO-A; f. 16, op. 3139, d. 16, l. 145.
38. William Stueck, *The Korean War: An International History* (Princeton, NJ: Princeton University Press, 1995), 147–8; Melvyn P. Leffler, *A Preponderance of Power: National Security, the Truman Administration, and the Cold War* (Stanford, CA: Stanford University Press, 1992), 417–18.
39. Harry S. Truman, *Memoirs, Volume II: Years of Trial and Hope, 1946–1952* (Garden City, NJ: Doubleday & Company, Inc., 1956), 420–1.
40. Leffler, *Preponderance of Power*, 441–2.
41. TsAMO-A; f. 50 iad, op. 174045ss, d. 39, l. 58.

42. Futrell, *USAF in Korea*, 276–81.
43. TsAMO-A; f. 324 iad, op. 152706ss, d. 3, ll. 10–11.
44. Roger Dingman, 'Atomic Diplomacy During the Korean War', *International Security*, 13, 3 (Winter 1988/89), 69–74.
45. D. Clayton James, *The Years of MacArthur, Volume III: Triumph and Disaster, 1945–1964* (Boston, MA: Houghton-Mifflin, 1985), 573.
46. Yefim Gordon and Vladimir Rigmant, *MiG-15: Design, Development, and Korean War Combat History* (Osceola, WI: Motorbooks International Publishers and Wholesalers, 1993), 127–8.
47. Ibid., 131–2.
48. Air Force Historical Research Center (Maxwell, AFB, AL); Daily Operations Summary Far East Air Forces, 12 April 1951.
49. TsAMO-A; f. 303 iad, op. 174133ss, d. 1, l. 13.
50. Amy Knight, *Beria: Stalin's First Lieutenant* (Princeton, NJ: Princeton University Press, 1993), 181.
51. Zubok and Pleshkov, *Kremlin's Cold War*, 180.
52. Joseph L. Nogee and Robert H. Donaldson, *Soviet Foreign Policy Since World War II*, 4th edn (New York: Macmillan, 1992), 123.
53. Zubok and Pleshkov, *Kremlin's Cold War*, 186, 190.
54. Nogee and Donaldson, *Soviet Foreign Policy*, 124–5.
55. Paul J. Murphy, ed., *The Soviet Air Forces* (Jefferson, NC: McFarland, 1984), 170–71.
56. R. A. Belyakov and J. Marmain, *MiG: Fifty Years of Secret Aircraft Design* (Annapolis, MD: Naval Institute Press, 1994), 275.
57. Nikita S. Khrushchev, *Khrushchev Remembers: The Last Testament*, Strobe Talbott, trans. and ed. (Boston, MA: Little, Brown, 1974), 303, 344.
58. Kenneth W. Allen, Glenn Krumel, and Jonathan D. Pollack, *China's Air Force Enters the 21st Century* (Santa Monica, CA: RAND, 1995), 222, 228.
59. Zubok and Pleshkov, *Kremlin's Cold War*, 191–2.
60. Ibid., 207–8.
61. Ramesh Thakur and Carlyle A. Thayer, *Soviet Relations with India and Vietnam* (New York: St Martin's Press, 1992), 93–5.
62. Nordeen and Nicolle, *Phoenix over the Nile*, 184–5.
63. Allen *et al.*, *China's Air Force*, 223.
64. Oleg Sarin and Lev Dvoretsky, *Alien Wars: The Soviet Union's Aggressions Against the World, 1919 to 1989* (Novato, CA: Presidio Press, 1996), 143–53.
65. Yuri Pavlov, *Soviet–Cuban Alliance: 1959–1991* (Coral Gables, FL: University of Miami North-South Center, 1994), 40.
66. Vladislav M. Zubok, ' "Dismayed by the Actions of the Soviet Union:" Mikoyan's talks with Fidel Castro and the Cuban leadership, November 1962', *Cold War International History Project Bulletin*, 5 (Spring 1995), 90.
67. Philip Brenner and James G. Blight, 'Cuba, 1962: The Crisis and Cuban-Soviet Relations: Fidel Castro's Secret 1968 Speech', *Cold War International History Project Bulletin*, 5 (Spring 1995) 82.
68. Bruce Porter, *The USSR in Third World Conflicts: Soviet Arms and Diplomacy in Local Wars, 1945–1980* (Cambridge: Cambridge University Press, 1984) 16–19, 241.
69. Sergei Khrushchev, *Khrushchev on Khrushchev: An Inside Account of the Man and His Era* (Boston, MA: Little, Brown, 1990) 45, 55.
70. Ilya V. Gaiduk, *The Soviet Union and the Vietnam War* (Chicago: Ivan R. Dee, 1996), 17–19.
71. Ibid., 27–8, 40.
72. Thakur and Thayer, *Soviet Relations*, 116–17.

73. Sarin and Dvoretsky, *Alien Wars*, 107.
74. Ibid., 108.
75. Ibid., 94.
76. G. F. Krivosheev *et al.*, *Grif Sekretnosti Sniat: poteri vooruzhennikh sil SSSR v voinakh, boevykh deistviiakh i voennykh konfliktakh, statisticheskoe issledovanie* (Moscow: Voennoe Izdatel'stvo, 1993), 396.
77. Nordeen and Nicolle, *Phoenix over the Nile*, 193–5. The Soviet involvement in the Angolan Civil War (1974) and the Ogaden War between Somalia and Ethiopia (1977–78) were significant, but most of the air combat in those conflicts was conducted by Cuban pilots and does not fit the scope of this work.
78. Ibid., 205–15.
79. Ibid., 222–5.
80. Alex P. Schmid, *Soviet Military Interventions since 1945* (New Brunswick, NJ: Transaction Books, 1985), 94–5.
81. David Klubes, 'Air Power Paradox: Israeli Bombing during the War of Attrition, 1969–1970', *Air Power History*, 43, 2 (Summer 1996), 26–7.
82. Murphy, *Soviet Air Forces*, 296.
83. Klubes, 'Air Power Paradox', 29.
84. Nordeen and Nicolle, *Phoenix over the Nile*, 248–9.
85. Victor Flintham, *Air Wars and Aircraft: A Detailed Record of Air Combat, 1945 to the Present* (New York: Facts On File, 1990), 62–3.
86. Murphy, *Soviet Air Forces*, 297; Nordeen and Nicolle, *Phoenix over the Nile*, 253.
87. Nordeen and Nicolle, *Phoenix over the Nile*, 253; Klubes, 'Air Power Paradox', 31.
88. Krivosheev, *Grif Sekretnosti*, 396.
89. Nordeen and Nicolle, *Phoenix over the Nile*, 256–7.
90. Krivosheev, *Grif Sekretnosti*, 405–7.
91. Mark Kramer, 'Soviet Foreign Policy During the Cold War: A Documentary Sampler. Document Six: Soviet Policy in Afghanistan, 1979: A Grim Assessment', *Cold War International History Project Bulletin*, 3 (Fall 1993), 67–8.
92. Scott R. McMichael, *The Stumbling Bear: Soviet Military Performance in Afghanistan* (London: Brassey's (UK), 1991), 1–5.
93. Ibid., 80.
94. Ibid., 83.
95. Oleg Sarin and Lev Dvoretsky, *The Afghan Syndrome: The Soviet Union's Vietnam* (Novato, CA: Presidio Press, 1993), 111.
96. Ibid., 117–19.
97. McMichael, *Stumbling Bear*, 90–1.
98. Tom Rogers, *The Soviet Withdrawal from Afghanistan: Analysis and Chronology* (Westport, CT: Greenwood Press, 1992), 45.

The Rise and Fall of Aeroflot: Civil Aviation in the Soviet Union, 1920–91

DAVID R. JONES

INTRODUCTION

By 1981 Aeroflot, the Soviet Union's state monopoly of civil aviation, was the world's largest civil airline in terms of the mileage of its scheduled routes and numbers of passengers carried. Since 1945 the Soviet Union had won recognition as a 'superpower' and the aircraft of Aeroflot, the only official Soviet carrier, proudly flew into the major cities of 87 countries in Europe, the Middle East, Sub-Saharan Africa, South and Southeast Asia, and operated direct flights from Moscow to Havana and Montreal. Within the USSR it served some 3,600 cities and 'populated centers' and overall, it carried over 109 million passengers and 2.6 million metric tons of freight. Equally impressive, these figures represented 15 per cent of all passengers carried on scheduled air services throughout the world.

On the surface these figures marked one major success of the Soviet regime. Russian and Western observers alike agreed in concluding that the state airline had played a significant role in modernizing the Russian empire's social and economic system, in strengthening the centralized authority of the Soviet leadership, and in legitimizing its authority in the eyes of the USSR's Russian and non-Russian citizens alike. Yet a mere decade later matters appeared very different. As was the case with the Soviet system as a whole, the coming of *glasnost* under Mikhail Gorbachev brought revelations of deep-seated problems within Aeroflot. By 1988 these were the subject of bitter articles in the Soviet press that in turn spawned a host of much publicized projects for improvement. But the decline continued so that by 1991 Aeroflot was on the verge of bankruptcy, service on many routes was inadequate and even irregular and, before further remedial measures could be attempted, the airline became a casualty of the collapse of the Soviet Union.

When the USSR dissolved, so too did its state monopoly of civil aviation. Although aircraft of 'Aeroflot-Russian International Airlines' (ARIA) remain in service on both domestic and international routes today, it is only a shadow of the company's once mighty Soviet predecessor. Whereas the latter had ranked first in the world in terms of total (international and domestic) passenger-kilometers and tonne-kilometers performed in 1985, by 1994 it had dropped to sixth and twelfth places, respectively. Indeed, Aeroflot now lies fragmented into some 50 major and over 400 smaller separate companies that continue to service, or attempt to service, the numerous domestic routes within the vast territory that made up the former USSR, and to provide the aircraft used for agriculture, forest protection, flying doctor and ambulance services, aerial surveys, and such ancillary services as the operation of aerodromes and hotels. Outside the Russian Republic these responsibilities devolved to the successor states' so-called 'Babyflots' which although their safety and other standards often leave much to be desired, have joined other Russian and foreign air services in competing internationally with their former parent line.

Despite Aeroflot's sudden collapse, the real past services and accomplishments of the Soviets' civil aviation monopoly merit close attention from students of modern Russia. Its expansion since the 1920s reflects the growing importance the Soviet authorities accorded air transport as a solution to the daunting problems they faced in the realm of what has been called 'spatial management'. Given the vast distances and other problems traditionally faced by Russia's economic and transport planners, it is hardly surprising that Igor Sikorsky first envisaged his large, multi-engine aircraft as being air-transports rather than as the Il'ya Muromets bombers that brought him initial fame during the First World War. As for the later Soviet Union, after 1945 it stretched some 6,000 miles (9,660 km) from the Baltic in the west to the Bering Sea in the east, 3,000 miles (4,830 km) from the Arctic in the north to the Pamir Mountains in the south, and had an area of over 8,736,000 square miles (22,400,000 sq. km), over 75 per cent of which is in Asia. Within this land mass resources are scattered widely in regions remote from the main centers while on both sides of the Urals, much of the area is arid, frozen or rugged in nature, or otherwise unsuited for large-scale habitation. Taken with the extreme variations in climatic conditions, this means that large sections have been very thinly populated and that as a consequence, the creation of an effective and integrated transport system has been a major problem both for political administrators and economic developers.

While Russia's rivers long provided one essential means of transport and communication, the creation of overland routes also became an early priority. But, despite continuing programs of road-building and repair, highways are notoriously difficult to maintain and of necessity, are sparse in many regions. Thus by 1 January 1978, the Soviet Union still had only 201,250 miles (322,000 km) of paved highway as compared with the much smaller United Kingdom's (Northern Ireland excluded) 214,570 miles (343,315 km). True, after the 1860s, railways had assumed tremendous importance and had improved matters considerably. None the less, like Sikorsky, the leaders of the young Soviet state saw great possibilities in the development of aerial transport as a means of promoting economic growth and helping them to modernize the culturally backward regions of their vast country. They further believed that under conditions of central planning, this mode of transport should be integrated fully into a 'uniform transport network', and so be only one specialized component that complemented rather than competed with the other modes of transit. Yet as with airlines elsewhere, and despite the earlier efforts made by the State Planning Commission (Gosplan) and other agencies, it was only after the Second World War that technological advances made possible the design of more powerful aircraft, and so permitted Aeroflot to make a really substantial contribution within the overall transport system of the USSR (see Table 10.1).

Table 10.1

INTER-URBAN PASSENGER TRAFFIC IN THE USSR (PERCENTAGES BY MODES)

Year	Rail	Road	River	Sea	Air
1950	91.1	1.9	3.8	1.6	1.6
1955	89.1	4.5	2.9	1.2	2.3
1960	79.3	9.9	2.6	0.8	7.4
1965	66.8	13.3	2.2	0.7	17.0
1969	60.8	14.3	1.8	0.5	22.66

Source: Hugh MacDonald, *Aeroflot: Soviet Air Transport since 1923* (London, 1975), p. 247.

SOVIET CIVIL AVIATION BEFORE AEROFLOT

Although the name 'Aeroflot' was adopted for civil aviation only in 1932, its origins stretch back to the birth of Russian aviation. It was in fact a revised form of the obsolete term *vozdushnyi flot* (aerial or air fleet) that in late Imperial and early Soviet times referred to the air forces in general and denoted 'the totality of piloted aerostats [airships] and aeroplanes which a state has at its disposal'. In this form the term survived until the military reorganization of 1924 and the appearance of the Administration of the Military Air Forces of the Workers' and Peasants' Red Army (*Upravlenie Voenno-Vozdushnykh Sil RKKA*). If the term 'air forces' (VVS) then replaced 'aerial fleet' for Soviet military aviation, the older usage survived in the forms of 'Civil Air Fleet' (*Grazhdanskii vozdushnyi flot*) and 'Aeroflot'. On 26 March 1932 the latter name was adopted formally in the title *Upravlenie vozdushnykh soobschenii 'Aeroflot'* (Administration of the 'Aeroflot' Air Lines), a step that capped a decade of uneven growth in the field of aerial transport.

The Soviet regime had established an All-Russian College for the Administration of the Air Fleet in the wake of the October Revolution of 1917, but it was only in April 1918 that this agency turned its attention to civil as opposed to military aviation. If a decree of 24 May 1918 then established the Main Administration of the Aerial Fleet as a branch of the People's Commissariat of Defense, its attention necessarily focused on the problem of raising an air service to meet the needs of the Civil War. Only when that conflict began winding down in 1920 was an effort made to develop civil aviation, but temporary military services already had appeared to carry mail and important officials. One such was that which operated briefly from 16 March of that year along the Smolensk–Gzhatsk–Moscow route. This provided a model that was followed by a second military mail service that opened on 13 January 1921 along the route Kharkov–Kiev–Ekaterinoslav–Sevastopol, which on 29 January was extended to Moscow. This service's pilots made only 15 flights, but a third route was of greater significance. Served by two detachments of converted Sikorsky Il'ya Muromets bombers, this comprised the two connected routes of Kharkov–Kursk–Orel and Orel–Tula–Moscow. Its passenger and mail flights began on 1 May and continued two to three times weekly until 10 October 1921. Finally, the Red air services' Fourth Reconnaissance Detachment operated weekly flights from August to October 1922 along the Tashkent–Alma-Ata and Pishpek–Alma-Ata routes in Turkestan.

If these efforts did not result in a regular service, others did. During 1920 Soviet officials discussed the opening of a Berlin–Lithuania–Moscow route with the German Albatross Aircraft Company, and of a Stockholm–Leningrad route with the Swedes. Further negotiations with Berlin led to the formation of the joint German–Russian company Deruluft (*Deutsche–Russische Luftverkehrgeschellshaft*) on 24 November 1921. This line formally opened scheduled service along the Königsberg–Kovno–Smolensk–Moscow route with Fokker F.IIIs on 1 May 1922. The route was soon extended westward as far as Berlin and complemented later by a service from Königsberg to Leningrad *via* Tilsit, Riga and Tallinn. By 1932–33, Deruluft was flying 754 flights annually along 1,613 miles (2,645 km) of routes, and carrying 5,195 passengers, 89,970 pounds (40,810 kg) of freight, 171,782 pounds (77,920 kg) of luggage and 165,345 pounds (74,000 kg) of mail over a total of 1,292,209 miles (1,970,835 km). Subsequently, it used a variety of German and Soviet aircraft that included Dornier Merkurs, Rohrbach Rolands, Junkers Ju-52s, Tupolev ANT-9s, and by 31 March 1937, when the service terminated, a chartered Douglas DC-2 as well.

Important as were Deruluft and a much shorter-lived German *Junkers Luftverkehr* service (1922–25) as predecessors of Aeroflot's later international system, the real basis for a domestic service was laid on 9 February 1923. On that date a resolution of the Council of Labor and Defense USSR established a Council of Civil Aviation attached to the military's Main Administration of the Aerial Fleet. Immediately thereafter an Inspectorate of Civil Aviation was created beneath the new body to serve as its executive agency. In that March these organizational measures found practical expression in the creation of the Soviets' first national civil air organization in the form of the joint-stock company known as the All-Russian (from September 1926, All-Union) Volunteer Association of the Aerial Fleet, or 'Dobrolet'. On 15 July 1923 the new line opened a regular air service using German Junkers F-13s on the 262-mile (420 km) route between Moscow and Nizhnii Novgorod (later Gorkii) with a flight by the experienced pilot Ya. N. Moiseev. Meanwhile, sister organizations appeared in 1923 in the form of *Ukrvozdukhput* on 26 March, as well as *Zakavia* and *Azdobrolet* for the Caucasus.

During 1923 Soviet civil aviation carried a total 229 passengers and 4,190 pounds (1,900 kg) of cargo. Thereafter Dobrolet continued to expand its services throughout the European part of the USSR, extended its routes into Siberia and the Far East, and sought to open routes into Central Asia and Outer Mongolia. When these efforts brought only mixed results, steps were taken to reorganize and centralize the domestic services and on 10 December 1929 the Ukrainian–Caucasian services (which had been merged in 1925) were absorbed by Dobrolet. Although the last continued to acquire new aircraft and expand its routes during 1930, in the view of the Soviet authorities the First Five-Year Plan necessitated that all concerns be state owned.

Determined to end private participation in this aspect of their transport network the Council of People's Commissars (Sovnarkom) resolved on 29 October 1930 to combine Dobrolet and the Main Administration of the Civil Air Fleet into a single, state-owned All-

Union Combine of the Civil Air Fleet (*Vsesoiuznoe ob'edinenie Grazhdanskogo vozdushnogo flota* or VO GVF). Equally committed to avoiding the use of imported and foreign-built transports, the Russians already were designing and building their own machines. To this end the authorities had established a 'scientific research institute' (*nauchno-issledovatel'skii institut*) for civil aviation (the NII GVF) in October 1930, and on 3 October 1931 they created *Grazhdaviastroi*, 'All-Union Trust for the Construction of Civil Aviation'. This last was to direct the work at aircraft factories Nos. 81–89 and simultaneously supervise the construction of airfields and other facilities. In addition, Soviet specialists also experimented widely with the use of transport gliders and, in the early 1930s, with lighter-than-air dirigibles. For constructing these latter they brought the Italian expert Umberto Nobile to Moscow and with his aid, built several large airships. As elsewhere, the use of flammable hydrogen led to several disasters but despite this, talk of creating dirigible routes persisted throughout the decade.

In the interim, Dobrolet had formally ceased to exist on 1 November 1930 and its assets, which included some 200 aircraft, passed to the new Combine and its nine subordinate, regional air line administrations (*upravleniia vozdushnykh linii* or UVL). But because this agency was no longer autonomous in terms of ownership, it had to be separated formally from the command of the Red Air Force, with which the civil air authorities had been competing throughout for both aircraft and other resources. This administrative emancipation was confirmed with the creation of GU GVF in 1932, and again when this agency was transformed into the Ministry of Civil Aviation (MGA) of the USSR in 1964. None the less, the relationship of civil aviation with the People's Commissariat (later Ministry) of Defense remained close throughout the 1930s, and again became one of formal subordination from 1941–48. Although legally autonomous thereafter, Aeroflot was always considered to be a reserve for the Air Forces' Military Transport Aviation (VTA) and until the 1980s was invariably headed by a military officer.

THE CREATION OF AEROFLOT

In spite of some successes during 1931, both the civil air fleet and *Grazhdaviastroi* failed to meet the goals set by Moscow's central planners. On 25 February 1932 the Council of People's Commissars therefore reorganized the Combine and transformed it into the Main Administration of the Civil Air Fleet (*Glavnoe upravlenie grazhdanskogo vozdushnogo flota* or GU GVF). As noted above, GU GVF then adopted the name 'Aeroflot' for its air transport fleet with the creation of an administration for its air lines on 26 March 1932. It also continued to maintain its own research, development and test facilities, and to have overall control over the output of civil aircraft at the factories operated by Grazhdaviastroi, which itself became the Administration for Capital Construction (*Upravlenie kapital'nogo stroitel'stva* or UKS). As for the Soviets' determination to equip their airline with Russian-built aircraft, in 1935 this was underlined by a decision that made the use of domestic transports mandatory by all branches of the GVF. And indeed, at this time Soviet production did increase, new models did appear, and by mid-decade K. A. Kalinin's K-5 and A. N. Tupolev's ANT-9 (PS-9) were used widely on Aeroflot's lines. Even so, the foreign machines imported earlier continued in service and despite production of Tupolev's ANT-35 (PS-33) after 1937, by 1941 it was the American DC-2/DC-3 – license-built in Russia under B. P. Lisunov as the PS-84 (Li-2) – that was becoming the workhorse of both the civil and military transport fleets.

Despite its new structure, by 1933 civil aviation had failed to meet the planners' goals for the First Five-Year Plan. The government now sought further to mobilize the energies of Aeroflot's personnel, and to guarantee their political loyalties, through the activities of a network of Communist Party cells and committees. Established throughout all the GU GVF's various administrations and branches, after 1933 these were headed by a 'Political Administration' similar to that found in the Armed Forces. Apart from Marxist-Leninist educational and propaganda activities, this agency carried out its own investigations, imposed disciplinary measures, and expelled Party members in its efforts to enforce compliance with Communist and governmental directives. Again as in the Armed Forces, Aeroflot's Political Administration made full use of these powers during the Stalinist purges of the latter 1930s when, according to reports in the Soviet press, the airline lost an assistant director, its construction chief and some 22 engineers, technicians and pilots. As elsewhere in Soviet society, these represented only the tip of the iceberg and by instigating these arrests the Political Administration undoubtedly diminished the very efficiency it was to promote.

Meanwhile, in 1934 GU GVF as a whole had undergone another substantial internal restructuring in accord with the Communist Party's decision to replace agencies with single functions with multifunctional authorities formed on a regional basis. On 19 May the Council of People's Commissars ordered the dissolution of the transport and agricultural trusts, and of the repair and construction bodies, as well as of the local agencies created in 1932. These were replaced by

12 territorial administrations for Moscow, the Ukraine (Kiev), Central Asia (Tashkent), Transcaucasus (Tiflis), Kazakhstan (Alma-Ata), Azov–Black Sea (Rostov), the Volga (Saratov), the Urals (Sverdlovsk), West Siberia (Novosibirsk), East Siberia (Irkutsk), the Far East (Khabarovsk), and the North (Arkhangelsk [Archangel], later Leningrad). Each of these was charged with organizing and supervising all categories of civil aviation within its region so as to make the most effective use of the available aircraft. As in the case of the political administration, the Armed Forces again provided the model for these civil aviation regions. Thus both the geographical distribution and the duties of the new GVF administrations mirrored those of the military districts and their VVS commands, and within each region Aeroflot's assets were structured along paramilitary lines with 'detachments' (*otriady*) being established for regular transport duties, for special duties, and for training. At the same time, responsibility for constructing and equipping airdromes now passed to a new Central Office for Research and the Planning of Air Routes, or *Aeroproekt*.

Over the next seven years this regional structure was modified with some districts being abolished (the Volga and Central Asian) or renamed (the Transcaucasian into the Georgian), and others being established in Central Asia (the North and South Kazakh, the Uzbek, the Tadzhik, and the Turkmenian). None the less, the structural essentials remained intact and provided the basis upon which Soviet civil aviation entered the Second World War. Otherwise, on 7 August 1935 the Aerial Code of the USSR gave precise definition to the activities of GU GVF and Aeroflot. This code assigned the new agency three areas of responsibility: (1) operating an air transport system; (2) providing over 60 types of special services (aerial surveys, forest-fire fighting, agricultural spraying, medical evacuation, etc.); and (3) promoting the educational, recreational and athletic activities connected with aviation. Consequently, the Code of 1935 (revised in 1962) consolidated GU GVF's monopoly over all non-military aerial activities within the USSR.

FREIGHT AND PASSENGER SERVICES TO 1941

On the surface the first of these missions would seem the primary one. In 1932 some 27,000 passengers and about 900 tonnes of freight were carried over 19,940 miles (31,900 km) of routes. In 1933 the main trunk route eastward from Moscow to Kazan was extended into Siberia to Novosibirsk, Krasnoiarsk, Irkutsk, and eventually to Vladivostok. During the mid-1930s, regional air services began serious development in Siberia and the Far East, in the Soviet republics of

Central Asia (Kazakhstan, Kirgizia, Tadzhikistan, Turkmenistan, and Uzbekistan), and they were extended in the European regions of the USSR as well. In 1937 the network of airline routes reached 52,000 miles (93,300 km), and in 1940, 90,000 miles (144,000 km). This represented an increase in routes of more than twelve times the 6,575 miles (10,778 km) recorded in 1928, and by the time of the German invasion in 1941 the mileage of scheduled routes had risen still further to 91,250 miles (146,000 km).

One Soviet source insists that during 1940 the Civil Air Fleet carried over 400,000 passengers. Others, however, give lower figures (see Table 10.2) and figures for the first half of 1941 list 127,000 passengers, 24,900,000 metric tons of freight and 3,000 metric tons of mail. As these and similar figures demonstrate, if Aeroflot's route network had expanded over the previous decade, the carriage of freight and mail was of much greater importance than that of passengers. In fact, by 1939 the Soviet Union finally surpassed the United States in terms of its volume of air-freight, and in that year freight and mail comprised some 85 per cent of the total volume carried. Furthermore, even the so-called 'passengers' often were anything but paying customers. Rather, they were usually government, economic or military officials. In this regard Aeroflot provided an important link between the central government and remote areas in both the interior and far-flung borderlands. Apart from helping service developing economic enterprises in these regions, during the late 1930s civil air transports played a significant role in transferring troops to threatened frontiers in the Far East and Mongolia, and to the newly 'liberated' or occupied Baltic Republics in 1940.

Table 10.2

DEVELOPMENT OF CIVIL AVIATION IN THE USSR (SELECTED YEARS, 1923–40)

Year	Passengers (000,000)	Cargo/Mail (000 tonnes)	Route Network (000 km)
1923	——	——	0.4
1924	——	——	3.5
1928	0.007	0.2	9.3
1932	0.03	0.9	31.9
1937	0.21	44.4	93.3
1940	0.36	58.4	143.9

Source: Hugh MacDonald [Klaus Vomhof], *Aeroflot: Soviet Air Transport since 1923* (London, 1975), p. 275.

None the less, in practice very real obstacles meant that Aeroflot had great difficulty in maintaining regular schedules and in many areas of the USSR, winter flights were few and far between. To some extent this simply reflected atrocious climatic and weather conditions, but is also partly explained by the still low level of aviation technology in general, and of Aeroflot's technology in particular. True, Moscow's Khodnyka Field, later known as the Frunze or Central Airdrome, was relatively well serviced. Again, after 1928 a Moscow–Novosibirsk mail service became possible thanks to night-flying facilities set up at Moscow and Kazan. But before 1941 such examples were exceptional. Regardless of plans drawn up by Grazhdaviastroi and Aeroproekt for the construction of airports at all major cities, and despite the opening of Moscow's Vnukovo Airport on 2 July 1941, progress elsewhere was slow, and especially so in the Far East. True, by 1940 the USSR boasted that it maintained some 150 airports, a network of operational aerodromes for local lines, and numerous other airdromes and airports in various stages of construction or reconstruction. Yet the overwhelming majority were primitive fields with unsurfaced runways, and in fact, planners at times attempted to route new lines along coasts and rivers so as to permit the use of float planes and so avoid building expensive fields for landplanes. Consequently, by 1941, Aeroflot's ground facilities and capabilities for instrument flying and air traffic control in European Russia already lagged behind those found in the West, and all were still more rudimentary east of the Urals. For instance, construction of an airport at Vladivostok had begun only in March 1932 while Khabarovsk, the hub of the GVF's Far Eastern network, only received its first grass field in 1938 and continued to maintain a primitive seaplane base until 1953.

Despite the public praise heaped on Soviet pilots, Moscow's economic planners still showed little interest in providing the funds needed to improve the situation. And given the comparatively low levels of state investment allocated to civil aviation, the latter's failure to raise revenues by developing commercial passenger services may seem surprising. Yet apart from the problems just mentioned and the resulting unavoidable scheduling uncertainties, the development of commercial passenger services was also hindered by other factors. These included the inhibitions placed on private travel by the regime, the low quality of service in general, the lack of private enterprise in Stalin's Russia which restricted significantly the base from which the line could draw paying customers, and the high cost of individual tickets for private travellers. At the opening of the 1930s, for instance, a single Khabarovsk–Okha ticket cost 175 rubles during the brief, three-month season of fair-weather flying, and 350 rubles (or roughly half of an average worker's monthly salary) during the rest of the year. This situation would remain basically unchanged until the late 1950s–early 1960s. It was only then that technical advances and the availability of more capable transport aircraft lowered the investment required to develop passenger routes and made this an attractive alternative in the eyes of Soviet planners.

What was true for domestic passenger services was even more so for the foreign routes. To begin with, under Stalin the generally xenophobic Soviet regime had little interest in expanding the contacts between its citizens and the external world, or in encouraging a tourist industry. Although some effort was made to open routes with Asian allies such as Mongolia and the Chinese warlord in Sinkiang, in the West Moscow remained satisfied with the contacts provided by Deruluft. This situation began to change after Adolf Hitler assumed power in Germany in 1933 and the Soviet Union responded by entering into formal alliances with France and Czechoslovakia in its pursuit of a 'United Front' against Fascism. Consequently, Aeroflot opened a Moscow–Prague route on 31 August 1936. But the need for more extensive Soviet-operated services became pressing when Deruluft ceased operations at the end of March 1937. Consequently, on 19 July Aeroflot set up an Administration for International Routes and quickly began scheduled flights between Moscow and Stockholm. In 1939 plans were made to expand Aeroflot's services with Chiang Kai-shek's Nationalist China and in 1940, after the signing of the Nazi–Soviet Pact, both Aeroflot and Lufthansa began flying along the Berlin–Moscow axis. None the less, by 1941 international operations consumed only a fraction of the Soviet Union's still limited resources in the field of civil aviation.

AERIAL PHOTOGRAPHY AND MAP-MAKING

Apart from its transport functions, one of the primary missions of the Soviets' civil aviation was to use aerial photography as a means of collecting accurate information regarding the USSR's terrain. As in the case of transportation its wide expanses, rugged mountain ranges, inhospitable deserts, and severe climates create major problems for map-makers. Topographic surveys by ground crews are difficult, time-consuming and very costly, and topographic surveying by means of photogrammetry and metric or mapping cameras was an obvious if only partially effective alternative. And of course, the high quality topographic maps so produced have both civil and military uses.

The Soviets' use of aerial photographs in civilian topographic surveying and mapping began before the First World War. Efforts to develop aerial photographic equipment aside, a number of early Russian publications dealt with the uses of this technology. In 1907, for example, the Military Engineering Academy published *Metric*

Photography and Its Applications in Flight and the Imperial air services used these techniques during the First World War. Then, in 1918, the Red Army's Military Topographic Service (VTS) began making use of aerial photographs in designing new and updating old maps. In March 1919 V. I. Lenin signed a decree creating the Higher Geodetic Administration (VGU), which subsequently became the Main Administration of Geodesy and Cartography (GUGK). In that same year the Workers and Peasants' Red Air Force Higher Aerial Photogrammetric School opened as well.

In the civil field, the Forestry Administration began making use of aerial photographs on an experimental basis in 1922, and on 1 March 1924 Dobrolet set up an Aerophotographic Surveying Section (*Otdel aerofotos'emka*) under the direction of veteran revolutionary M. O. Bonch-Bruevich. Although the VGU still employed fewer than 500 engineers and technicians in 1924, by 1925 maps were appearing that were based on aerial photographs. Matters soon improved still further thanks to the creation of the State Technical Bureau 'Aerophotographic Surveying' (*Gostekhbiuro 'Aerofotos'emka'*), which shared work with the existing agencies, and to the availability of more accurate devices. In particular, N. M. Aleksapol'skii developed a photograph rectifier, the production of which began at the Moscow Geodetic Institute in 1926, and the PPS rectifier (named for its designer P. P. Sokolov) entered serial production in 1927. By using these rectifiers cartographers and photogrammetrists could make use of the 'composite method' of topographic mapping. Put simply, this meant that map-makers now could avoid being misled by the erratic flight characteristics of the day's aircraft when drawing contour lines from mosaics of rectified images, and so could use field surveys to provide elevations.

By 1927–28 aerial photographs were being used in experimental land-use planning in the Fergana Valley, the Ukraine, and the province of Moscow, and these images were used in laying out the new collective and state farms. Since this procedure grew in importance with the First Five-Year Plan, on 1 January 1930 the earlier agencies were abolished and merged into *Gosaerofotos'emka*, the State Technical Bureau for Aerial Photography and Surveying (*Gosudarstvennyi tekhnicheskii biuro 'Aerofotos'emka'*), which is also known simply as GAFS. During its first year *Gosaerofotos'emka* undertook as many projects as its two predecessors had done over the previous five-year period. Equally important, on 27 December 1931 the authorities organized the Administration of Agricultural Aerial Photography. The 1930s also brought further advances in the equipment used by Aeroflot and VVS flyers for producing maps. In particular, between 1934 and 1938 F. V. Drobyshev designed several models of stereometers, including the widely used STD-1. These devices permitted the creation of stereoscopic (or three-dimensional) models from rectified images, and so increased the accuracy of map-maker's contour lines. Furthermore, the stereoplanographs developed at this time permitted elevations as well as contours to be obtained from aerial photographs. As a result, by 1940 Aeroflot's topographic and geologic survey activities were playing an increasingly important role in the work of Soviet economic as well as military planners.

OTHER 'SPECIAL' DUTIES

Apart from map-making, both Dobrolet and Aeroflot contributed to the Soviet economy by aiding agriculture through pest control. In 1918, this became the responsibility of the Plant Protection Section of the People's Commissariat of Lands and Forests (*Narkomzem*); by 1921, Soviet specialists were urging that aircraft be used in combating locusts, and in 1922 a special commission was established and provided with two aircraft to carry out the relevant trials in crop dusting. Fiscal constraints, as well as the need to design and test the necessary equipment, delayed the formation of a dusting and spraying detachment until February 1925. Thereafter, this task assumed growing significance and in October 1930, at the height of the drive to collectivize agriculture, these missions became the responsibility of the All-Union Combine for Agricultural Pest Control. Although it initially had a mere 13 aircraft at its disposal, the Combine acquired an additional 59 during 1931 and established five main bases in Leningrad, Moscow, the North Caucasus, Transcaucasia, and Central Asia.

After additional administrative tinkering, on 21 November 1932 the Council of People's Commissars resolved to replace the Combine with the All-Union Agricultural and Forest Aviation Trust, placed the latter under GU GVF, and assigned it a total of 197 aircraft. The Trust remained in charge until it was disbanded and incorporated into Aeroflot's regional administrations during 1934–35, by which time it had a strength of 393 specially equipped aircraft. Despite this decentralized structure, the number of machines assigned to agriculture and forestry continued to grow and by September 1941 a total of 886 reportedly were in active service. As for their activities, during the first six months of that year Aeroflot maintained that its pilots had spent 11,340 hours in spraying or dusting 500,000 hectares in the struggle with locusts and sugar-beet weevils, and another 687,000 hectares in an effort to suppress malaria mosquitoes.

Aeroflot also contributed to Soviet life through the development of its Flying Ambulance Service (*Sanaviatsiia*). This use of aircraft also had been pioneered during the First World War by the French in particular, and by others elsewhere as well. Given its vast and sparsely populated expanses, the Soviet Union obviously would benefit greatly from such a service, both to evacuate patients and to airlift medical personnel, equipment and supplies into remote areas. The first Soviet ambulance aircraft – the Kalinin K-3 – was delivered in 1928, by 1931 an ambulance brigade had been formed in co-operation with the Soviet Red Cross, and in August 1933 such services were centralized within an Administration of Ambulance Aviation (*Sanaviatsia*). Under its direction a nation-wide network was slowly established to serve both civilians and the military and again, despite limited resources, equipped with better aircraft. In 1937 this branch of civil aviation too was integrated within the system of GU GVF, and during that year the air ambulances were credited with aiding 8,129 people, 1,616 of whom actually had been evacuated. Although the service still had only some 70 aircraft in 1938, a total of 550 air ambulances had been registered by June 1941. The growing scope of their activities is indicated by figures claiming that in the first six months of that same year, the air ambulances had made 11,356 flights, evacuated 3,399 patients and airlifted 10,352 medical personnel into areas of need.

PROPAGANDA FLIGHTS

Aircraft had been employed as a means of delivering propaganda in both the First World and Russian Civil Wars. But it was only in 1925, when two aircraft toured the Caucasus, and three others traveled from Moscow to Peking, that the Soviet government began to exploit the full potentialities of aerial agitation. As might be expected, during the 1930s, Aeroflot's contribution in promoting aviation in particular, along with the legitimacy of the Soviet regime in general, became as important a mission as its more obvious services to the transport system and the economy. Indeed, during that decade special 'propaganda' squadrons were formed to carry the Communist message to the populace. The first such appeared in the form of the *Pravda* squadron of 1931. This was responsible for delivering matrices for the Party's central daily newspaper from Moscow for printing in Kharkov, Kiev, Leningrad, Groznyi, Odessa, Kazan, Rostov, Tiflis, and Sverdlovsk. Although this special unit was disbanded after a year's operation, Aeroflot continued to perform this service throughout its existence. Meanwhile, on 17 March 1933 the GVF began organizing a much more important propaganda squadron that was based in Moscow and named after the eminent writer Maxim Gorky. Its flagship was the single example of A. N. Tupolev's five-engine ANT-14, which later was named the Pravda, as well as his giant ANT-20 Maxim Gorky, which joined the squadron in 1934. Built especially as an 'agitation aircraft', the second was a large, eight-engine machine that housed a movie theater, a photographic dark-room, a print shop, a telephone system for internal communications, and external loudspeakers.

Apart from these two aircraft, by the end of 1934 this squadron included 25 aircraft of various types. After the crash of the ANT-20 in May 1935, the squadron was reorganized the following October with a strength of 40 aircraft, the ANT-14 included, and it continued active until being disbanded in February 1939. During its existence it reportedly brought Moscow's message to some 10 million people through films, loudspeakers and leaflets. The presses carried by its larger aircraft could produce up to 8,000 leaflets per hour, and these were helpful in promoting literacy as well as in the campaigns carried out during the springtime sowing seasons, and those aimed at explaining shifts in the Party 'line' or urging that greater efforts be made to raise levels of production, public health, and so on.

At this time other pilots, including many from Aeroflot, sought to increase the USSR's prestige at home and abroad by setting world records in speed, endurance, altitude, and distance. Throughout this period Soviet flyers had gained considerable experience in flying in northern and arctic conditions, experience that later stood them in good stead in both war and peace. Among their feats was the first landing on the North Pole and the charting of a north Pacific route to North America. Although honors were heaped on these pilots, who became known as 'Stalin's Falcons', and although many of their feats were spectacular, the practical significance of these exploits remained limited at best.

THE PROMOTION OF 'AIR-MINDEDNESS'

Before 1914 national aeroclubs had promoted the cause of aeronautics throughout Europe, and they had combined to form the International Aeronautics Federation, or FAI, which had established its offices in Paris in 1905. At that time the Imperial All-Russian Aeroclub had served as that nation's representative. Although that link was broken in 1917, after 1923 its duties were assumed first by the various aviation clubs and 'voluntary' associations formed to support aviation. These quickly developed into mass organizations to prepare individuals for work in military and civil aviation, to channel the general populace's donations for the development of aviation within the USSR, and to carry out the required propaganda. From March 1923 this was the task of the Society of Friends of the Air Fleet (*Obshchestvo druzhei vozdushnogo flota* or ODVF) which, after a series of mergers with other defense organizations, emerged in 1927 as the Society of Friends of Defense and the Aviation-Chemical Industry of the USSR (*Obshchestvo druzhei oborony i aviatsionno-khimicheskogo stroitel'stva* or *Osoaviakhim SSSR*).

Although the new umbrella society had some 15 million members by 1934, aviation was only one of its concerns and even this area included parachuting and gliding as well as pilot training. For this last task Osoaviakhim still had only two such flying schools with fewer than 900 student pilots and relatively few training technicians in 1932. Even so, its efforts to expand aerial training and sports already had received a major boost on 25 January 1931. Then delegates to the Ninth Congress of the Young Communist League (VLKSM or Komsomol) had enthusiastically become the patrons of aviation under the slogan of *'Komsomolets – na samolet!'* ('Komsomol members – to an airplane!') with the aim of providing Soviet aviation with 150,000 trained pilots from the service schools and volunteer aeroclubs. Thanks to the close co-operation that immediately developed between the Komsomol and Osoaviakhim, the number of available aeroclubs expanded to 20 during 1932. Similar motives lay behind the designation of 18 August as the 'Day of the Red Aerial Fleet' which since 1933 has served as the occasion for air shows and other events that publicize the achievements of the Soviet Union's military, civil and sports flyers.

This impetus continued over the next few years and was reinforced still further by the decree *On Osoaviakhim*. Adopted jointly by the Party Central Committee and the Soviet government on 8 August 1935, it demanded that the society strengthen its efforts in military training. The result of this so-called Komsomol 'pass' or 'ticket' (*putevka*) was a large influx of men and women under 24 years of age into the expanding network of aviation schools, air clubs and circles being established by Osoaviakhim, as well as the entry of 30,000 communists and 'Komsomoltsy' into the Air Fleet's regular schools from 1931 to 1936. Consequently, by 1934, the number of flying clubs had risen to 115 and, depending on the source consulted, to from 115 to 140 by 1936. In any case, by mid-decade, student numbers had increased seven-fold, and between 1931 and 1936 the aircraft provided to Osoaviakhim's aircraft park, largely by GU GVF, grew by nineteen times. Despite the disruption of Stalin's purges, this growth continued so that by 1 January 1941 the Osoaviakhim network included from 180 to 190 aeroclubs and 46 glider schools. In all, from 1930 to 1941 the society claimed to have trained 121,000 aircraft and 27,000 glider pilots for the civilian and military air services.

Meanwhile, in 1933, work had begun at Tushino Field at Moscow on facilities for the Central Aeroclub of the USSR. Opened in 1935, it immediately federated with the FAI as the Soviet Union's official representative. In 1936, the Soviet government assigned this club responsibility for registering aviation records, and it rapidly became the main methodological center, as well as one of the best training institutions, in the Soviet Union. In December 1938 it was renamed *Tsentral'nyi Aeroklub SSSR imeni V. P. Chkalova* (TsAK SSSR im. V. P. Chkalova), or Chkalov Central Aeroclub, after the famous long-distance flyer and Hero of the Soviet Union of the mid-1930s. Among the club's graduates are such well-known air aces as A. I. Pokryshkin and I. N. Kozhedub, both of whom were three-time Heroes of the Soviet Union, as well as numerous other prominent pilots, designers and scientists such as S. V. Ilyushin, S. P. Korolev and A. S. Yakovlev.

Although GU GVF was forced to devote significant resources to both its propaganda flights and the creation of the network of air clubs and associated groups, Aeroflot derived significant benefits from these activities in particular and the public's enthusiasm for aviation in general. First Aeroflot, like the Imperial Air Services before 1914, was able to supplement the limited resources received from the central authorities with funds raised by bonds sold to, and from gifts received from, the public. Then, second, the air club training network provided the pilot schools and training wings of Aeroflot, as well as those of the VVS, with a steady stream of candidates in all fields. For like the latter, the GU GVF maintained its own establishment of research and engineering institutes, aero-technical and pilot schools, and practical flyer-training centers and training detachments (*otriady*). To ensure that these establishments fulfilled the tasks assigned them, the GVF provided them with several thousand aircraft during the 1930s.

AEROFLOT AT WAR, 1941–45

On 23 June 1941, in the immediate wake of the German attack of the preceding day, Aeroflot was mobilized to provide direct support for the military forces, and the People's Commissar of Defense assumed operational control of the Civil Air Fleet. Although certain vital airline routes to Siberia and the Arctic continued in regular service, a subsequent meeting at GU GVF led to most of Aeroflot's crews, technical personnel and aircraft being formed into 'special air groups' on 26 June. Of these the most important were the Special Liaison Services Air Group, the Moscow Special Purposes Air Group, and the Northern, Kiev and North Caucasian Air Groups. Throughout the conflict, these and other Aeroflot units, subunits and detachments provided general support for the front, maintained service along the essential routes for the movement of key personnel and special freight throughout the country, took part in such special missions as the dropping of parachutists and supplies to partisans, and evacuated wounded and other personnel.

During the four years of war Aeroflot crews distinguished themselves in all the great battles of the Eastern Front. At Leningrad and Moscow, Kiev and Odessa, and Stalingrad and Berlin, Aeroflot aircrews and their aircraft flew alongside regular VVS units. Over a three-day period in October 1941, for example, 60 aircraft of the Moscow Aviation Group and units of Long-Range Aviation (ADD) transferred nearly 5,500 men and their equipment to plug a hole in Moscow's defensive line. Again, during the prolonged battle of Stalingrad of 1942–43 the Civil Air Fleet undertook more than 46,000 sorties and transported 31,000 troops with 2,587 tonnes of military cargo, and this despite persistent enemy opposition and appalling climatic conditions. The importance of Aeroflot's overall contribution is indicated by Soviet statistics that show that its pilots flew over 93 per cent of all Soviet transport missions, 1,622,384 flights involving the delivery of personnel and/or *matériel* within the zones of combat and in all, a total of over 5,400,000 combat flights. Furthermore, in 1942 the special air groups were transformed into regular air regiments to become the nucleus of the VVS' reorganized and expanding transport service (*Voenno-transportnaia aviatsiia* or VTA).

Aeroflot's pilots were especially important in sustaining 'aerial bridges', as Soviet and Russian military writers call the frequent air-supply operations or airlifts that were mounted whenever large Red Army formations, defensive positions or cities were encircled. Perhaps the most notable example of such a 'bridge' was the air supply service established for besieged Leningrad as a supplement to the transport vessels operating on Lake Ladoga, and the 'road of life' that crossed that lake's ice during the winter. The work of organizing this airlift's initial phase became official on 13 September 1941, the day on which the command of the Leningrad Front issued its directive 'On the Organization of Air Transport Communications between Moscow and Leningrad', and responsibility for maintaining the aerial communications and transport lines between Leningrad and Bol'shaia Zemlia was assigned to the *Osobaia severnaia aviatsionnaia gruppa* (Special Northern Aviation Group), or OSAG, of the Civil Aerial Fleet or GVF (Aeroflot). Commanded by A. A. Lavrent'ev, it began service with some 15 transport aircraft. These comprised four twin-engine Li-2s or Soviet DC-3s also simply known as the 'Lee DWA' or Duglus (Douglas); two four-motor G-2s (TB-3 heavy bombers); two twin-motor G-1s (TB-1 heavy bombers); three two-engine PS-41s (SBs medium bombers); and three to four single-engine R-5 postal aircraft that served to maintain communications between Leningrad and Moscow.

These were clearly insufficient for the task and aircraft of both the VVS and the Baltic Fleet were co-opted for supply flights. This force was reinforced further on 20 September, when the State Council of Defense (GKO) approved a proposal from the command of the Leningrad Front and Aeroflot by its resolution 'On the Organization of Air Transport Communications Between Moscow and Leningrad', and added three of the six squadrons of the Moscow Special Purpose Aviation Group (MAGON) to the airlift. Meanwhile, the Special Baltic Aviation Detachment had been assigned to the airlift as well. By the month's end a total of some 64 transports, serving under the command of the OSAG, were flying foodstuffs into and evacuating *matériel* and people from the besieged city. According to official statistics, by 31 December 1941 the transports had brought in 3,605 tonnes of foodstuffs, 1,273 tonnes of arms and 138 tonnes of mail, and removed 52,827 essential skilled workers, specialists and scientists, along with members of their families, and some 9,000 wounded. If figures vary from source to source, even Western scholars have concluded that Aeroflot's success in supplying Leningrad when ground communications with the beleaguered city had been severed, and in taking out civilians and wounded on the return journeys, has never been rivalled in any other wartime operation.

By 1943 the use of transport aviation had increased significantly. During that year more than 390,000 men and 29,000 tonnes of supplies were transported to the fronts in general, to advanced mechanized spearheads in particular, and to the partisans. Both landings to aid these last, and the parachute drops of agents and troops into enemy-occupied areas became increasingly common. As for the role of Aeroflot pilots, this is evident from the fact that aircraft of GVF regiments again flew 466,035 sorties with 352,674 of the total troops carried. GVF personnel also were able to make use of their other prewar skills as well. Its flyers joined with the military as well as the personnel of other civilian agencies in preparing maps for the military, the primary responsibility for which rested with the Military Topographic Service (VTS). As the Germans advanced eastward during 1941–42, the demand arose for 1:1,000,000 scale topographic maps of European USSR as far east as the Volga. With the help of GU GVF, teams from both military and civilian agencies involved surveyed 1,600,000 square kilometers. By the beginning of 1942 these maps had been readied by the VTS and civilians from the Main Administration of Geodesy and Cartography. In addition, maps with scales of 1:500,000 and 1:200,000 appeared for the same region during early 1942. Other Aeroflot pilots meanwhile provided the backbone for the military's air ambulance service which, during this conflict, transported some 340,000 wounded, 2,044 tonnes of blood and 1,678 tonnes of medical supplies. And despite the damage, if not complete destruction of its prewar facilities, GU GVF moved in to re-establish its domestic passenger and mail routes as soon as the Red Army had pushed the Germans out of European Russia during 1943–44.

During the war the 2nd, 6th, and 7th Air Regiments of the Civil Air Fleet were designated as 'Guards' units for the courage displayed by their pilots in battle and large numbers of decorations were awarded to their aircrews. More important still, by the end of this struggle Soviet transport pilots had gained considerable experience in flying in the most miserable conditions believable, had an expanded network of airfields and were receiving both more sophisticated instruments and more capable transport aircraft. The USSR's own aircraft production increased steadily after factories evacuated eastward in 1941 had reopened and, as a result, a growing stream of new aircraft had swollen the meager fleets available in 1941–42. It was at this time that the PS-84 (Li-2), the Soviet version of the DC-3, came into its own. If these Russian-built models were supplemented by those ferried from the USA via Alaska and Siberia, the bulk of the Li-2 fleet was provided and maintained by Soviet factories. Equally important, it also would continue to meet most of Aeroflot's needs for some years after the war as well.

POSTWAR EXPANSION, 1945–68

By 1945, Soviet air transport had proven to be a factor of major significance in modern warfare and had acquired the capabilities to begin playing an equally vital role in the postwar transport system. When hostilities ended, Aeroflot's first task was to rebuild its ruined facilities in the western USSR, and then to begin extending the route network far beyond that of its prewar system. Apart from the emphasis on transporting the personnel involved in the country's reconstruction, special attention was devoted to establishing direct services between Moscow and the capitals of the union republics. Also significant was the simultaneous introduction of services running from major industrial cities to the holiday resorts in the Black Sea and Caucasus regions. The Soviet authorities announced that the GVF was to receive 1,000 aircraft released by the VVS for these purposes. Equipped in this manner with large numbers of Li-2s, Aeroflot entered upon a period of rapid growth. During 1945, for instance, the airline increased the number of passengers carried from the 359,000 of 1940 to a total of almost 537,000. During the next five years numerous additional routes were opened and in 1950 its pilots carried some 1.5 million passengers over a route network that stretched 186,300 miles (300,000 km), of which 11,800 miles (19,000 km) were international.

Apart from the basis provided by the war, Aeroflot's expansion after 1945 was possible largely thanks to advances in aviation technology. Although the Li-2 (DC-3) bore the brunt of the traffic, these planes were soon being supplemented by more modern, Soviet-designed transports. The first postwar Soviet airliner, the twin-engine Ilyushin Il-12, had a higher performance and greater capacity than the Li-2. It entered Aeroflot's inventory on 22 August 1947 and by 1948 was in service along all of Aeroflot's trunk routes. This transport also saw service on Aeroflot's longest domestic route from Moscow to Vladivostok, which involved 33 hours and nine intermediate stops. Other Il-12s operated on the new international routes to eastern Europe, as well as on those from Moscow to Peking and to Kabul, and were employed to maintain scheduled cargo services. In 1954, this model was replaced as the standard carrier on main routes by the Il-14. A direct development of the Il-12, it also was powered by twin piston engines but had 28–32 seats as compared with the latter's 21–27. Two four-engine transports – the Tu-70 and Il-18 – also underwent tests. But despite their advantages, both needed airfield facilities that were still generally unavailable and so did not enter service. Pending modernization of the GVF's system of ground facilities, it was therefore left to the Li-2, Il-12 and Il-14, supplemented by the An-2 and Yak-12 on short-haul routes, to provide the bulk of Aeroflot's fleet on the eve of the jet era.

This last opened for Aeroflot with the introduction of the twin-jet Tupolev Tu-104 airliner in 1956. Developed in a remarkably brief time thanks to the availability of major components from the earlier Tu-16 bomber, its pressurized cabin initially had seating for 50 passengers. A Tu-104's arrival in London in March 1956 created a sensation and thanks to Britain's withdrawal of the 'Comet' (after two were lost) from service in 1954, from 1956 to 1958 the Tu-104 was left as the world's only jetliner in civilian service. Although its initial operations were uneconomical, its passenger capacity rose in successive versions to 70, to 85–100, and, over certain routes, as high as 115. Some 200 of these aircraft were built, and about half of this number were still operational in 1976. Its derivatives, the twin-jet Tu-124 (56 passengers) and Tu-134 (up to 76 passengers), meanwhile had entered service along Aeroflot's routes and in all some 112 and 700 of these types, respectively, were built.

Apart from the jets, aircraft with turboprop engines began entering service on Aeroflot's long-range and medium-range routes, as well as on the short-haul runs. A new Ilyushin Il-18 in the form of a 84-

passenger, four-turboprop-powered transport appeared in 1957 and proved to be so versatile that over 500 were built. More impressive still was the appearance of the four-engine Tupolev Tu-114 in October 1957. The largest Soviet turboprop aircraft, and the world's largest commercial transport of its day, during the 1960s this 400 mph (645 kpm), 220-seat intercontinental liner facilitated development of Aeroflot's long-range internal and international services. Meanwhile, O. K. Antonov's large-load An-8 cargo transport had been tested in 1955, and then entered service with both the GVF and VVS, while production of that designer's four-turboprop, 84-seat An-10 began in 1959. Although continuing problems brought its withdrawal from service in 1972, Antonov's An-24 and An-26 twin-engine, short-haul machines were of more lasting significance. The first began service along Aeroflot's Moscow–Voronezh–Saratov route in September 1963 and by 1978 1,100 had been built. The An-26, of which some 1,700 were completed, was a development of the An-24 that appeared in the early 1970s. With from 40–52 seats these transports carried the bulk of travellers along the GVF's complex network of local routes and by the mid-1990s 850 An-24s still operated commercially in the former USSR (171 with Aeroflot). Otherwise, during the 1960s, the single-engine An-2 biplane, with a capacity for seven to ten passengers, continued to operate on short hops and perform other duties that ranged from serving as air ambulances to agricultural work.

The introduction of these new machines was naturally reflected in the continued expansion of Aeroflot's activity. In 1955, the last year of the Fifth Five-Year Plan, it carried over 2.5 million passengers, 194,960 tonnes of freight and 63,769 tonnes of mail over a route network of 199,650 miles (321,500 km). By 1958 this last had grown to 216,850 miles (349,200 km), and the line had carried 8,321,500 passengers and 445,640,000 tonnes of freight and mail for a total of 941,500,000 tonne-kilometers, which was a 25.3 times that of 1940. Subsequent expansion continued apace so that by 1966 the network of internal routes had risen to some 294,975 miles (475,000 km), of which 114,265 miles (184,000 km) were classified as being of All-Union significance, to which international routes added another 31,000 miles (50,000 km). In that same year traffic involved 47 million passengers, who were carried an average distance of 621.6 miles (1,001 km), and 1.3 million tonnes of freight over an average haul of 8,135 miles (1,310 km).

It was the Il-18 turboprop and Tu-104 jetliners that were decisive with regard to the expansion of Aeroflot's international routes. Work on building a postwar network abroad had begun as early as 1944 with flights operating first to Rumania, and later to the other capitals of newly 'liberated' eastern Europe. Even so, Moscow held aloof from the International Civil Aviation Organization (ICAO) that emerged in that year in Chicago, as well as from the prewar International Air Transport Association (IATA) that was revived during 1945–47. Indeed, as late as 1957 Aeroflot's east European services, along with that serving Scandinavia via Helsinki and those operating to China, Mongolia and Afghanistan, comprised the totality of foreign routes. Furthermore, these remained subordinated administratively to the GU GVF's Moscow Administration of Transport Aviation (MUTA), and this last only created a special 'Aviation Group for International Air Services' as an autonomous subdivision in 1962. By that date the new aircraft entering service had permitted GU GVF to supplement the USSR's more aggressive diplomatic activity by negotiating its own bilateral agreements. As a result, in 1958 Tu-104s began regular flights from Moscow to Amsterdam, Brussels and Paris while KLM, Sabena and Air France operated reciprocal schedules. Service was extended on similar terms with British European Airways (BEA) to London in 1959, and thereafter Aeroflot's international expansion was little short of spectacular. In January 1963, for instance, regularly scheduled flights opened along the Moscow–Montreal–Havana route, in 1967 Aeroflot began serving Tokyo, in 1968 New York, and so on.

One consequence was that tourism became a component of growing significance for Aeroflot's international service. But if route expansion beyond the Soviet frontiers offered financial as well as political benefits, air transport became an increasingly attractive and cost-effective component of the Soviets' domestic transport network as well. True, operational costs were still often high thanks to the vicissitudes of weather and climate, as well as the high fuel consumption of many Soviet aeroengines. Even so, the introduction of more efficient aircraft, combined with Aeroflot's greater experience and relatively improved technology in all spheres to lower even these costs and increase efficiency. In 1929, for instance, a letter from Irkutsk to Moscow still required 36 hours to reach Moscow by air (as compared with six days by rail), but by the 1970s took only eight hours by jet transport. On the basis of such advances, proponents of expanded air transport and passenger services began justifying them as a time-saving means of cutting labor and other costs. For this and other reasons, in 1959, plans were announced for developing Aeroflot into a common transport carrier within the overall transport system.

This also reflected the fact that the initial costs of establishing an air route were attractive indeed in comparison with other forms of transport. In 1960, establishing a new air route cost an estimated 10,000 rubles per kilometer, but building a highway cost 800,000 rubles per kilometer, and a railway 1.3 million rubles per kilometer. Furthermore, a properly managed, large-scale airline should be able to fund much of its expansion from its own income and, despite continuing subsidies from the state budget for capital investment in aircraft and ground facilities, the GVF had been reporting a profit over its operational budget since 1952. From that time on, therefore, Aeroflot had transferred portions of its operational revenues to its capital expenditure and after the so-called 'Liberman reforms' of the early 1960s, sales and profitability became primary concerns of GVF's management.

One obvious way of increasing revenues was to increase Aeroflot's volume of business. After 1957 the airline's management therefore began channeling profits into an effort to lower fares. This process began modestly and took time but by 1961 the cost of a Khabarovsk–Petropavlovsk flight had been cut to 76 rubles from the 170 rubles of 1958. Overall, by 1967 fares had been cut in half and were competitive with those of the railways, and the majority of 'long-haul' passengers (those travelling over 930 miles or 1,500 km) chose Aeroflot. This was particularly the case in distant regions like the Far East, and of the some 27,500 who traveled along the Moscow–Khabarovsk route during August 1970, 25,000 (89 per cent) did so by plane rather than train. During the 1960s this success was promoted as well by a mass campaign initiated by the Komsomol to acquaint Soviet citizens with the new advantages offered by air travel, as well as by Aeroflot's growing role in serving the Black Sea 'tourist' resorts.

As a result of all these factors, by 1970 the airline was carrying 24.1 per cent of the USSR's intercity passengers (as compared to 1.6 per cent in 1950). At the same time, this expansion in scale also had demanded the construction of new facilities to handle the larger and faster aircraft in use, as well as to accommodate the larger volumes of freight and passengers carried. When Aeroflot's expansion began in the 1950s–1960s, most Soviet airports still remained little more than 'pens for passengers manned by teams of authoritarian women' (see Wilson and Bachkatov in Acknowledgements and Further Reading, p. 296). In accord with the decisions of 1959, plans were made to upgrade these facilities, as well as the airport hotels that the airline had begun operating after the Second World War and which numbered 89 by 1968. Throughout the 1960s therefore work went ahead on building a system with five classes of airdromes that could handle from 15 to 400 passengers per hour accordingly.

In the interim, Soviet air travellers had to endure conditions that their Western fellows would have found intolerable. Throughout the postwar period, foreign travellers on Aeroflot's domestic routes complained of an absence of even such elementary safety equipment as seat belts, life jackets and oxygen masks in aircraft that often relied on visual means of navigation and airports lacking even the most rudimentary systems of ground control. Even so, by 1969 Aeroflot flights reportedly flew with an average of some 79 per cent of their seats filled as compared to 50 per cent on American airlines. This clearly testified to the popularity of air travel among the Soviet citizenry, whatever the risks involved, but it also spoke volumes about the advantages Aeroflot gained from its monopoly position. For if complaints did surface occasionally, Soviet customers had no choice but 'to make the best' of the often mediocre services provided.

YEARS OF PROSPERITY, 1968–85

One aspect of Aeroflot's rapid postwar growth was a massive increase in the personnel employed – from some 37,000 in 1947 to roughly 400,000 in 1973 – and eventually a corresponding administrative reorganization became necessary. Although the Main Administration of the Civil Air Fleet once again had recovered its formal autonomy from the Ministry of Defense in 1948, its ongoing expansion initially was accommodated largely through the creation of additional territorial administrations and 'aviation groups'. As of 1955, for instance, there were sixteen of the former and nine of the latter, as well as occasional special units. These last subsequently included that mentioned above for 'international air services', and the once autonomous unit operating along the 'Northern Sea Route' (on the USSR's northern coast) which merged with Aeroflot in 1960 as the Administration of Polar Aviation. But by that date a more serious restructuring clearly was required. Steps in this direction came with the revised Aerial Code of 1962, and with the transformation of MUTA's special aviation group into the independent Transport Administration for International Air Services (TUMVL) in 1964. The process continued with GU GVF itself being transformed into the Ministry of Civil Aviation (MGA) by a decree of 27 July 1964. Yet despite its new name GVF (like DOSAAF, the successor to Osoaviakhim) long remained under the direction of the former VVS commanders E. F. Loginov and his successor V. P. Bugaev. Perhaps surprisingly, both of them proved to be relatively capable as managers of a civilian airline.

Within the new ministry operations remained geographically subdivided. If it initially retained the main territorial administrations and supplementary 'aviation groups', with the introduction of the timetable for 1966 the word 'territorial' was dropped and the existing 'groups' became administrations in their own right. Thereafter, however, new units were formed within some of the new, regionally based administrations and again termed aviation 'groups'. By 1976, there were 30 administrations or directorates, 13 of which represented subdivisions of the RSFSR and 14 the other Union Republics. The remaining three were the Moscow Transport Administration for long-distance links with the capital and central regions, the Polar Aviation Administration with its special responsibilities in the north added to coverage of the central Russian area, and the Central Administration of International Air Services (TsUMVS). Formed on the basis of the TUMVL in 1971, this last controlled the services operating abroad from Moscow and the other territorial administrations, and it formally used the trade name of 'Aeroflot-International Airlines.'

By the late 1970s Aeroflot's international connections had been extended to 69 countries and included destinations in most European nations, numerous states in Africa and Asia, the United States, Canada, Cuba, and Peru. This continuing expansion was stimulated by the introduction of new aircraft after the mid-1960s. One such was the 168-seat Il-62, a four-jet airliner which had begun operating on Aeroflot's transcontinental and other long-haul services in the late 1960s, and this provided a long-range workhorse for the following decade. Unfortunately, the same was not true of the Tu-144, the Soviet SST competitor to the Anglo-French Concorde. Although first displayed at the Paris Air Show in 1965, this aircraft only began mail services some ten years later, in December 1975, between Moscow and Alma-Ata. Despite Aeroflot's unhappiness concerning this machine, it then entered passenger service in 1976 but after one crashed in 1978, was retired. If this meant the considerable efforts devoted to developing the Tu-144 had been largely wasted, the airline's planners faced similar difficulties in obtaining all of the types required from the Ministry of the Aircraft Industry, the agency now responsible for the design and production of its equipment. Aeroflot had more or less held its own against such Western aircraft as the Boeing 707 and Douglas DC-8, but by the 1970s it badly needed 'jumbo' and 'airbus' equivalents to the Boeing 747 and the DC-10.

Soviet industry did somewhat better in providing aircraft suitable for use on Aeroflot's other domestic routes. First the Tu-154 trijet liner, which carried up to 158 passengers, provided a transport suitable for the medium-range, high-density routes. The prototype first flew in October 1968, scheduled cargo flights began in 1971 and passenger models began service on both Aeroflot's domestic and east European routes in 1972. By 1976 some 100 were reportedly flying with the GVF and versions of this durable liner still remained in production in 1997. By that date more than 1,015 had left the production line and although some had been exported, in 1990 Tu-154s still flew half of the passenger-kilometers logged by Aeroflot's pilots. Second, the shorter routes were served from 1968 by the trijet, 27–34 seat Yakovlev Yak-40. On entering service it became the first operational jet transport designed especially to operate from grass airfields and unpaved runways of 1,000 meters or less, and it significantly improved the service provided to towns in the more remote regions of the former Soviet Union. There may well have been over 500 Yak-40s in service by 1976, and in all some 1,000 were built. By 1976 Aeroflot also still reportedly flew 80 Il-62s as well as specialized, four-engine cargo air-craft in the form of some 250 Antonov An-12s and 40 very large An-22s. In addition, large numbers of An-24s and An-26s continued to serve on short-haul routes and by that date the airline had received some 200 Il-76 freighters, which had entered production in 1973 and already seen service with the VVS.

With these aircraft, in 1975 Aeroflot carried roughly 97 million passengers and 2.5 million tonnes of freight, which were double the figures of 1966. By 1975, the network of scheduled flights had been extended to reach a total of 496,800 miles (800,000 km), including 186,300 miles (300,000 km) of international routes. Those within the Soviet Union were classed as either *Linii Soiuznogo Znacheniia* (All-Union Services, or LSZs) or *Mestnyie Vozdushnyie Linii* (Local Air Services, or MVLs). The LSZs were listed in Aeroflot's central timetable and all the administrations contributed to maintaining their operations. If the majority of routes radiated from Moscow, they none the less served the capitals of all the union republics, the major indus-trial centers and the holiday resorts. By that decade, however, a grow-ing number of links were appearing that did not require passage through Moscow and there were direct flights from many cities, including smaller industrial towns of special significance (e.g. Bratsk in eastern Siberia), to the Black Sea and Caucasian holiday resorts. Services along the MVLs meanwhile offered regular flights to over 3,500 towns and villages, as well as to farming, forestry and mining camps or settlements. In many cases these were the sole means of transport to the larger centers, and all of the GVF administrations, with the International (TsUMVS) and the Moscow Transport (All-Union services) Administrations excepted, supported this effort.

If the postwar GVF devoted much less effort to mounting obvious propaganda extravaganzas or providing direct support to aerial sports and aviation clubs, its international flights none the less served a similar purpose. Apart from serving travellers and increasing revenues, they 'showed the Soviet flag' around the world and gave proof of the USSR's technical and industrial might. At the same time, Aeroflot continued to provide a number of 'services to the national economy' at home. As before 1941, civilian mapping and the use of other applications of metric photography remained a central growing concern. In 1947, for instance, the All-Union Ministry of the Coal Industry employed aerial photographs in compiling the large-scale maps it used for planning, and in 1948 the Council of Ministers USSR published a set of special 'Instructions on the Topographic-Geodetic Surveying of Populated Rural Areas'. By the mid-1950s production was under way of stereophotogrammetric instruments with automatic photograph rectification. In 1959 the SPR-2 stereoprojector of G. V. Romanovskii entered production and with this and other mechanisms, Aeroflot's pilots could carry out near complete topographic surveys by stereotopographic methods. Mapping cameras generally were placed in such relatively stable aircraft as the An-2, Li-2, Il-14, Il-20, An-24, and An-30, all of which were employed in aerial topographic surveys. When the An-30 appeared in 1974, it was the Soviets' first specialized aerial survey aircraft and quickly became one of the most widely used platforms for aerial photography.

As earlier, Aeroflot's other special services included operating air ambulances and the aircraft used in agricultural and forestry protection and, with the military, search and rescue missions. Overall responsibility for the first of these services belonged to the Ministry of Health but it was the GVF that provided the aircraft, pilots and maintenance personnel who served the some 200 stations dotting the USSR by the mid-1970s. The aircraft involved included An-2s, Yak-12s and a range of helicopters. By that date they reportedly made 100,000 flights, and carried over 200,000 passengers (medical personnel as well as patients) with 550 tonnes of medical supplies annually. Meanwhile, each of Aeroflot's territorial administrations also provided the specialized aircraft needed for crop dusting and spraying, the control of mosquitoes and other pests, and forest patrols and fire-fighting.

As just indicated, Aeroflot has acquired a large fleet of helicopters which are used in carrying out these functions and maintaining its MVLs cargo and passenger services. The first to be deployed widely was the Mil Mi-1. Next came the 16-passenger Mil Mi-4, which equipped an air taxi service that opened between Simferopol and Yalta in 1958, and which later saw service on a number of other routes in the Black Sea region and the Caucasus, as well as between Moscow and its airports. This craft in turn served as the basis for the 28–31 seat Mi-8. With two turbine engines, it entered Aeroflot's fleet in 1967 and rapidly began replacing Mi-4s on the scheduled services. The same design bureau was responsible for the large Mi-6, which flew first in 1957 and quickly set numerous helicopter records, and for the Mi-10. Although one model of the Mi-6 carried up to 80 passengers, its lift capability of up to 12,000 kilograms internally or 9,000 kilograms slung made it especially suitable for carrying heavy cargoes and performing lift work. As for the Mi-10, it was used largely as a flying crane in major construction work, and for lifting other heavy loads while the later Mi-26 can carry some twenty tons. Meanwhile, the twin-engine Kamov Ka-26 proved to be an extremely versatile and widely used machine thanks to its specialized and interchangeable modules that could be fitted to the basic airframe and engine unit. One of these provided seats for six passengers, but others were designed for ambulance, agricultural, mapping and geological prospecting, search and rescue, or other duties.

With regard to ground facilities, after 1968 the categories of airports had been revised so as to provide facilities capable of dealing with 1,000 passengers hourly. Consequently, by the early 1970s Aeroflot's construction schedule foresaw completion of 220 major installations for its national and 1,000 for its local lines by 1990, which figures were to rise to 1,600 and 2,000, respectively, by the year 2000. Similarly, the airline hotel chain was to be upgraded as well. Furthermore, in 1970 Aeroflot finally had joined the ICAO and two years later announced a ten-year plan to bring its ground control system up to international standards. For this purpose it opened negotiations for the purchase of advanced air traffic control computers from the West while spurring on its own developmental efforts. To meet the demands of the 300 per cent increase in air traffic projected for the years 1970–75, Aeroflot also introduced the Soviet-built Minsk-23 and Siren-1 computers for use in ticketing. If Aeroflot continued to stand aloof from IATA and instead still preferred bilateral agreements, it none the less avoided a price war by keeping its rates in line with the world's other airlines.

The Soviet Union also signed the convention on 'sky-jacking' of 1973. But meanwhile, Moscow's refusal to publish its accident statistics gave rumor the power to magnify the impression made by authenticated accidents. In addition, despite well-publicized improvements and a prohibition of genuine public discussion, during the early 1980s articles in the professional press made it obvious that problems still persisted in the areas of safety, the central ticketing system, the carriage of long-haul freight, the air-traffic control facilities, and so on. Matters had seemed to improve in 1980 when foreign contractors provided Moscow with a new international aerodrome, Sheremetevo-2, for use during the Olympics. Yet the capital otherwise was still served by four other outdated fields (Domodedovo, Sheremetevo-1, Vnukovo, and Bikovo), and other major cities were no better off. Similarly, only a few major fields were served by nearby hotels, and these were of dubious quality. Indeed, many domestic provincial fields comprised little more than a runway and collection of sheds, and in many cases the air-traffic control and other facilities were primitive at best. If this situation had been tolerated while the line operated in isolation, by the late 1970s the modernization and upgrading of both Aeroflot's ground support structure and tourist accommodation was an obvious condition for any future expansion of the airline's international and domestic traffic.

AEROFLOT AND THE COLLAPSE OF THE USSR

Although Aeroflot's expansion since 1959 had indeed been impressive, significant problems obviously still demanded resolution. Its fleet did contain a range of aircraft, but this was insufficiently inclusive to meet all the needs posed by the USSR's diverse route categories. In particular, the Soviet line still lacked equivalents to the American Boeing-747 'jumbo' jetliner, and to the DC-10 and Lockheed L-1011 Tristar wide-bodied transports. A remedy for the second deficiency seemed on hand when the first Il-86 wide-bodied airliner began scheduled service in December 1980 with its seats for 28 first-class and 206 economy passengers. None the less, a lack of suitably powerful engines and other problems had complicated its development for the last decade. When production halted in 1994, only 99 had been built, 88 of which remained in service. However, in January 1990 Aeroflot still felt it necessary to lease five of the more efficient French A-310 'Airbuses'. As for a Soviet 'jumbo', if a candidate at last seemed on hand when a prototype An-400 reportedly flew in 1983, this aircraft never entered production. Once again the short-haul routes seemed better served when, in 1980, the trijet Yak-42, which exists in 120-passenger and freight models, entered service with the GVF. But unfortunately, this machine rapidly proved incapable of meeting the demands made on it.

Behind the sparkling and modernized facade of Moscow's new Sheremetevo-2 airport, serious problems continued to plague Aeroflot's ground and support services as well. Throughout the early 1980s, occasional comments in the press indicated the frustration felt by other sectors of the Soviet economy over inefficiencies in Aeroflot's central planning and the scheduling of freight services, and a lack of effective co-ordination between the central and regional administrations. Other reports indicated that at most airports both the passenger and freight handling facilities were inadequate and congested, that fuel shortages were frequent, that the ground-control network was still backward thanks to the slow delivery of new equipment, and that it meanwhile had become dangerously overloaded by the ever increasing traffic. Equally troubling, despite continuing campaigns aimed at promoting safety, many of Aeroflot's passengers remained unimpressed. The servicing provided by ground crews was frequently sloppy at best, and planes were often spotted in conditions that would be unacceptable in the West (with bald tires, for example). Again, stewardesses often neglected to check seat-belts (assuming that these were available and in working condition) and themselves remained standing during takeoffs.

After 1986, full-scale public debate became possible thanks to Mikhail Gorbachev's policy of *glasnost* (openness). The resulting complaints rapidly demonstrated that Aeroflot had far to go before it achieved the standards of service, comfort and safety set by the world's other major carriers. By 1988 Deputy Minister of Aviation Boris Paniukov felt forced to respond to his critics in the press by announcing still one more program of modernization. In accord with this, by 1995, much of the existing roster was to be replaced with the much more modern and efficient long-range Il-96s and medium-range Tu-204s, and major programs meanwhile were to be implemented to improve the ground services provided at Soviet airports, hotels included. More often than not, these programs involved 'joint ventures' that brought the investment of Western capital and know-how. In 1988, for example, Moscow's Sheremetevo-2 (which handled 6.2 million passengers a year) finally obtained a duty-free shopping center that was operated jointly by Aeroflot with Ireland's Aer Rianta, and in 1990 that field's customs procedures were streamlined. In this last year the regime announced an Aeroflot–Lufthansa agreement on implementing a £273 million program for modernizing Sheremetevo in general, and for the rebuilding of Sheremetevo-1 in particular, so as to accommodate 15–18 million passengers annually. A similar plan that involved Canadian and other investors existed for Leningrad–St Petersburg, and the subsequent opening of other destinations brought in their train similar if smaller schemes in Kiev and elsewhere. Equally significant, in late July 1989 Aeroflot at last had become a member of IATA, and in this manner received access to the international ticketing system needed to expand its foreign currency revenues through advance bookings and increase the influx of tourists.

Once again fulfillment fell short of intentions. Many Aeroflot aircraft in fact still lacked such standard equipment as emergency oxygen masks for its passengers, the provision of which only become publicly announced policy in April 1990. Worse still, in that year news reports continued to pillory equipment shortages, maintained that up to 60 per cent of Aeroflot's fleet suffered from excessive wear and tear, and that the airline's limited capabilities were leaving some 15 to 20 million potential passenger without tickets! Again measures were ordered to rectify the situation but by that time, Aeroflot was on the verge of bankruptcy. Refused a government subsidy, it was forced to triple its international fares. Despite such emergency measures, shortages in foreign currency left Aeroflot without the funds needed to purchase fuel, 92 airports (roughly half of the total) had to close down, flights to eastern Europe and Africa were cancelled, and a leading Soviet admiral suffered the humiliation of being briefly stranded in New York. Worse still, in 1990 Aeroflot was faced with the loss of its international monopoly when a presidential decree authorized formation of a rival under the name of ASDA that was to begin operations from Vnukovo with Boeing 747s and 767s in June 1992, the same year in which Air Russia – a joint venture of Aeroflot and British Airways was to begin flying from Domodedovo Airport. If neither of these ended by posing serious threats, other rivals surfaced of which TransAero was the major contender.

Aeroflot's domestic monopoly was already disappearing as well in the months that preceded the Soviet Union's dissolution in the autumn of 1991. During that year a number of Western companies received permission to open international routes to various destinations (like Kiev) within the once-closed Soviet interior. With the breakup of the USSR in December, the Soviet airline's structure was completely shattered when the former republican administrations officially transformed themselves into new national 'Babyflots' to serve their own domestic and, increasingly, international routes with varying degrees of efficiency and levels of safety. As a consequence, the Aeroflot that emerged was significantly smaller in terms of both equipment and personnel. By 1994, for example, its employees numbered a mere 14,838 (911 pilots and co-pilots included) as compared with the comparable figures of 43,491 (3,064) for Lufthansa, 48,823 (2,865) for British Airways, 39,815 (1,191) for Air France, and 23,762 (1,179) for Iberian airlines. But at the same time, it also meant that the Russian line lost its responsibility for maintaining a number of unprofitable routes as well.

These developments marked the end to a long-lived monopoly and at last offered real choices to travelers in the former Soviet Union. By so doing, they marked the end of an era in the history of civil aviation as well. Yet during this enforced 'downsizing' Aeroflot's old international division provided a core around which its domestic components could rally to form a legal successor that survives to bear its once proud name as 'Aeroflot-Russian International Airlines' (ARIA). As such, it has continued to face its numerous difficulties head on so as to serve the new Russian republican regime by sustaining its domestic and international networks, and by operating the 'Russian Government Air Service' that transports President Boris Yeltsin and other top officials. If it now finds itself competing internationally with both foreign lines and domestic rivals such as the Moscow-based Trans-Aero Airlines, it has attempted to recoup its losses in other areas through subsidiaries such as the AJAX freight service.

Perhaps surprisingly given the state of contemporary Russia, or perhaps thanks to the airline's own leaner establishment and new competitive spirit, these efforts achieved considerable success. According to Valerii Okulov, Aeroflot's new acting general manager, by the end of 1996 its passenger traffic was up 9.2 per cent (to 3.82 million) over 1995, its traffic in the former USSR had grown by over 200 per cent, its turnover amounted to 7,441 billion rubles ($1.29 billion), and its profits amounted to 170 billion rubles. In 1997 Okulov also announced plans to relieve the pressure on Moscow's Sheremetevo terminal by diversifying Aeroflot's operations through the creation of a network of regional 'hub' airports at St Petersburg and Nizhnii Novgorod in European Russia, at Ekaterinburg in the Urals, at Yakutsk in Siberia, and at Khabarovsk or Vladivostok in the Far East. At the same time, the line's fleet of 115 aircraft was to be modernized and the types in service be reduced from nine to five through the future purchase or leasing of both domestic and foreign machines. But even without these measures, by 1997 Western observers were reporting that Russia's national airline at last compared adequately with other international flag carriers in such relevant areas as safety, punctuality, on-board service, and service record.

Epilogue

ROBIN HIGHAM

Scanning *Flight International* for the years since the decline and fall of the Soviet Union and the emergence of the Commonwealth of Independent States, we can observe a number of trends.

The switch to a capitalist economy created a shortage of funds in a society where costs had rarely been reckoned. Hard hit were the armed forces, whose king-pin role was rapidly reduced to the point that by 1997 the Chief of the Air Staff declared that he would have to ground the MiG-31, pride of the fighter forces, because pilots had too little time in the air to be able to fly it safely. Moreover, not only were the Russian planes being grounded, much of the strength had been siphoned off to the sister states in the CIS. The Soviet carriers were either in port or being sold off. And the Air Defense and Air Forces were heading for a merger.

The former satellite air forces were beginning to come to the end of the maintainable life of their aircraft such as the MiG-21, though the Israelis were coming forward with an inexpensive upgrade. The fact that the former Eastern Bloc air forces were going to need new aircraft placed the Russian aircraft industry and the former USSR factories in the Ukraine in a bind. The opportunity was there, but they lacked not so much suitable designs as the capital to certificate them and be able to offer competitive production bids. Instead, it seemed quite likely that early model US F-16 fighters would fill the void if not other Western designs or even indigenous machines from Czech, Polish and Rumanian sources. More sophisticated Russian aircraft such as the MiG-29 and the Sukhoi-27 suffered also from the long Soviet and Russian reputation for poor maintenance and a shortage of spare parts. Moreover, the less the Russian Air Force itself flies, the harder it is to sell aircraft.

The invisible infrastructure upon which the current Russian and Ukrainian air services are based is in turmoil. Both a state committee in Russia and the responsible minister were in 1997 demanding the merger of the design bureaux with their autonomous factories into a consolidated Western-style business structure in which the entire product would be produced within one organization. Moreover, as the realities of modern defense economics have their impact, further mergers may be expected just as has happened in the West with some of the smaller entities simply being driven out of business. Already some of the pressures have been felt in the airliner business, with new planes being fitted with both imported electronic suites and even with Rolls-Royce or Pratt & Whitney engines. In fact, of course, as noted earlier, this follows a long precedent in Russia. On the other hand, for the first time the world may see large, sophisticated airliners being exported because the combination of rugged Slavic airframes with Western engines and electronics will become financially attractive, due to the lower cost provided by cheap Russian wages. Just as Japan may have failed to conquer China militarily 50 years ago, so she is now doing it economically. Thus Russia may become a serious threat to the Imperialist powers in a decade or so.

One consequence of the breakup of the USSR was that Aeroflot lost its monopoly of civil aviation. This not only shrank the organization, but also saw it split into divisions for internal and external services. At the same time some Western airliners were acquired, to be followed by a trend towards operating links, pools, or whatever, with airlines such as the German Lufthansa This also is a trend that is likely to expand as one of the laws of economics is mergers for strength, whether formal or informal.

What we have seen portrayed in this book are Russian and Soviet patterns and actions which have had their parallels in the outside world, though often on a different timetable. That this has been so has been due to a combination of space and resources, the nature of the people, and the ideologies of their government. Now that the barriers are down and tourism is occurring, Russia is entering the modern unitary global world with consequent effects upon her aviation and air power.

Acknowledgements and Further Reading

INTRODUCTION

For the general background of the geographic area in which the Russians have operated, see Robin Milner-Gulland with Nikolai Dejevsky, *The Cultural Atlas of Russia and the Soviet Union* (New York, 1991) and the companion volume on China by Caroline Blunden and Mark Elvin (1991); both also provide historical information.

For general information and background on Russia, see the US Army DA Pam 550–95 *Area Handbook for the Soviet Union* (Washington, DC, 1971 and later).

Since 1977 David R. Jones has edited *Soviet Armed Forces Review Annual* (Gulf Breeze, FL) and is a useful source for command diagrams and other information, and see also *Transformation in Russia and Soviet Military History*, ed. Carl W. Reddel (USAF Academy, 1986).

On Soviet theories of war fighting see Harriet D. Scott's edition of V. D. Sokolovskiy, *Soviet Military Strategy* (New York, 1968, 1975), and David M. Glantz, *Soviet Military Operational Art in Pursuit of Deep Battle* (London, 1991). More particularly, on the Second World War, read Albert Seaton, *Stalin as Military Commander* (New York, 1976) and Harold Shukman, *Stalin's Generals* (New York, 1993). Otto Preston Chaney, *Zhukov* (Norman, OK, 1971, 1998).

David M. Glantz's US Army Combat Studies Institute Study provides an overview of *The Soviet Airborne Experience* (Fort Leavenworth, KS, 1984).

Mikhail Haller and Aleksandr M. Nekrich, *Utopia in Power: the History of the Soviet Union from 1917 to the Present* was published in Russia in 1982 and in the West (New York) in 1985.

On the USSR's post-1945 activities, see Oleg Sarin and Lev Dvoretsky, *Alien Wars: the Soviet Union's Aggressions Against the World, 1919 to 1989* (Novato, CA, 1996), and Jacques Lévesque, *The Enigma of 1989: The USSR and the Liberation of Eastern Europe* (Berkeley, CA, 1996), and Michael Dobbs, *Down with Big Brother: The Fall of the Soviet Empire* (New York, 1996).

John Morrow's *The Great War in the Air 1914–1918* (New York, 1994) is a broad scholarly overview of that conflict.

For eastern Europe, see John E. Jessup, *Balkan Military History: A*

Bibliography (New York, 1986). The services' own writings have been catalogued in Robin Higham, *Official Histories* (Manhattan, KS, 1970) with another updating volume in process.

For a general introduction to the history of aviation, see John W. R. Taylor and Kenneth Munson, *History of Aviation* (London, 1972), John T. Greenwood, *Milestones of Aviation* (New York, 1989), and Robin Higham, *Air Power: A Concise History* (Manhattan, KS, 1988).

Kenneth R. Whiting, *Soviet Air Power, 1917–1976* (Maxwell AFB, AL, 1976) (1986) provides an older overview as does Asher Lee's, *The Soviet Air Force* (London, 1962). Christopher C. Lovett's *Soviet Naval Aviation: Continuity and Change* (College Station, TX, 1984); David Woodward, *The Russians at Sea* (New York, 1965), and the series *Soviet Naval Developments* (Washington, DC, 1974–) are helpful.

The greatest Russian and Soviet operational period was the 1941–45 war. The older Soviet view is to be found in the official *Great Patriotic War of the Soviet Union, 1941–1945: A General Outline* (Moscow, 1974). The companion volume is *The Soviet Air Force in World War II: The Official History*, edited by Ray Wagner (New York, 1973), the more recent work in English being Von Hardesty, *Red Phoenix: The Rise of Soviet Air Power, 1941–1945* (Washington, DC, 1982 [paperback has no date and no copyright]).

The experience of the Great Patriotic War was the subject of a number of volumes, of which the following are still useful: Robert Jackson, *The Red Falcons: The Soviet Air Force in Action, 1919–1969* (London, 1970) and Robert P. Berman, *Soviet Air Power in Transition* (Washington, DC, 1978)

For a general guide to the literature on the Russian and Soviet air forces and aviation see Myron J. Smith, Jr, *The Soviet Air and Strategic Rocket Forces, 1939–1980: A Guide to the Sources in English* (Santa Barbara, CA, 1981); and Michael Parrish's three works, *Soviet Security and Intelligence Organizations, 1917–1990: A Bibliographical Dictionary and Review of the Literature in English* (New York, 1992), *Soviet Armed Forces: Books in English, 1950–1967* (Stanford, CA, 1970), *The USSR in World War II: An Annotated Bibliography of Books Published in the Soviet Union, 1945–1975*, 2 vols (New York, 1981), as well as the sources noted in Robin Higham and Jacob W. Kipp, *Soviet Aviation and Air Power* (Boulder, CO, 1977).

For information on hostilities since 1945, consult Victor Flintham, *Air Wars and Aircraft: A Detailed Record of Air Combat, 1945 to the Present* (New York, 1990); also Edgar O'Ballance, *Afghan Wars: 1839–1992* (London, 1993).

There are many books devoted to aircraft, ever a popular subject. Start with Jean Alexander, *Russian Aircraft since 1940* (London, 1975) and John Stroud, *Soviet Transport Aircraft since 1945* (London, 1968), Nico Sgariato, *Soviet Aircraft of Today* (Warren, MI, 1978), Hugo Hooftman, *Russian Aircraft* (Fallbrook, CA, 1965), and Bill Sweetman and Bill Gunston, *Soviet Air Power: An Illustrated Encyclopedia of the Warsaw Pact Air Forces Today* (London, 1978), and note how more and better information and photographs have become available. Also relevant is Gerald Howson, *Aircraft of the Spanish Civil War 1936–1939* (Washington, DC, 1990).

On airlines, R. E. G. Davies has an introductory volume, *A History of the*

World's Airlines (New York, 1964). There are now a number of histories of the various world's airlines, including several of PanAmerican, which had links to Aeroflot. Perhaps the best of these are Robert Daley, *An American Saga* (New York, 1980), and Marylin Bender and Selig Altschul, *The Chosen Instrument* (New York, 1982).

On the establishment of long-distance airlines see Sir Hudson Fysh, *Qantas Rising* (Sydney, 1965) and *Qantas at War* (Sydney, 1968) and Robin Higham, *Britain's Imperial Air Routes, 1918–1939* (London, 1960). For the more recent atmosphere, Dan Reed, *The American Eagle: The Ascent of Bob Crandall and American Airlines* (New York, 1993), *High Risk: The Politics of the Air, The Autobiography of Adam Thompson* (London, 1990), and for a very different approach Kevin and Jackie Freiberg, *Nuts! Southwest Airlines, Crazy Recipe for Business and Personal Success* (Austin, TX, 1996).

On the Russians' Japanese opponents see Shinji Kondo, *Japanese Military History: A Guide to the Literature* (New York, 1984) and René Francillon, *Japanese Aircraft of the Pacific War* (London, 1979) which includes introductory information on the aircraft industry. For the operations which so impressed the Kremlin, Kenneth P. Werrell, *Blankets of Fire: US bombers over Japan during World War II* (Washington, DC, 1996). The Soviets also got to view USAAF grand-strategic bombers on their own soil, for which see Richard C. Lukas, *Eagles East: the Army Air Forces and the Soviet Union: 1941–1945* (Tallahassee, FL, 1970).

For comparative studies of the various important armed forces of the world including Russia and its opponents in the twentieth century, see Allan R. Millett and Williamson Murray, *Military Effectiveness*, 3 vols (Boston, MA, 1988).

Bradley I. Smith, *Trading Secrets with Stalin* (Lawrence, KS, 1996) provides information on Allied–Soviet Intelligence, 1941–45.

The most important Soviet opponent in the two world wars was Germany. A start may be made with Dennis Showalter's overall view, *German Military History, 1648–1982* (New York, 1984). More detailed reading can be undertaken in the 1983 (London) reissue of the British Air Ministry's account *The Rise and Fall of the Luftwaffe, 1933–1945*, in Williamson Murray, *Luftwaffe* (Baltimore, MD, 1985), and in David Irving's *The Rise and Fall of the Luftwaffe: The Life of Field Marshal Erhard Milch* (Boston, MA, 1973) and, for the background, in James S. Corum's, *Luftwaffe: Creating the Operational Air War 1918–1940* (Lawrence, KS, 1997). For German aircraft see William Green, *Warplanes of the Third Reich* (London, 1970), and for background, Edward L. Homze, *Arming the Luftwaffe: the Reich Air Ministry and the German Aircraft Industry, 1919–1939* (Lincoln, NE, 1976). On the leadership of the German Air Force, R. J. Overy, *Goering – The Iron Man* (London, 1984) or David Irving, *Göring: A Biography* (Boston, MA, 1989), and David Baker, *Adolf Galland: The Authorized Biography* (London, 1996).

On the French Air Force, see *French Military Aviation: A Bibliographical Guide* by Charles Christienne, Patrick Facon, Patrice Buffotot, and Lee Kennett (New York, 1989) and Charles Christienne, Pierre Lissarague *et al.*, *Historie de l'aviation militaire française* (Paris, 1980) of which there is an abridged English version.

CHAPTER 1: EARLY FLIGHT IN RUSSIA

Early Russian aeronautical journals are an excellent source of information on the genesis of Russian aviation. The records of the Imperial All Russian Aero Club were published annually in *Vozdukhoplavatel'* [Aeronaut], beginning in 1908. One important illustrated periodical, *Avtomobil' i vozdukhoplavaniye* (The Automobile and Aeronautics), appeared briefly in 1910 and 1911 as a 'bi-monthly, illustrated, scientific-popular and sport' magazine. *Avtomobil' i vozdukhoplavaniye*, for example, recorded one of the most detailed accounts of the 'St Petersburg to Moscow Air Race' of 1911 (see issue 14, 1 August, 1911), pp. 401–11). *Zaria aviatsii* (Dawn of Aviation) and the *Sevastopolskiy aviatsionnyy illustrirovannyy journal* (The Sevastopol Illustrated Aviation Journal) reflected other, more regional, attempts to publish a Russian periodical devoted to aviation. For two years, 1916 and 1917, the Sevastopol Military Flying School published *Voyennyy letchik* (Military Pilot), a journal that reported on the war and provided Russian air units with descriptions of the latest aeronautical technology. The shortlived *Vestnik letchikov i aviatsionnykh motoristov obnovlennoi Rossii* (Messenger of Pilots and Aviation Mechanics of a Reformed Russia) represented the last independent aviation periodical. The first Bolshevik periodical, *Vestnik vozdushnogo flota* (Messenger of the Air Fleet), appeared in 1920.

Few histories of the early period have been written. P. D. Duz', *Istoriya vozdukhoplavaniya i aviatsiya v SSSR (Period 1914–1918)* (A History of Aeronautics and Aviation in the USSR. The Period 1914–1918), (Moscow: Oborongiz, 1944, followed by reprints in 1960, 1979, 1986, and 1995) represents a standard Soviet treatment, a work possessing considerable information, but marred by Soviet censors. Pre-revolutionary accounts are few in number, but always interesting for what they say and do not say about the formative period of Russian aviation: L. Vladimirov, *Sovremennoye vozdukhoplavaniye i yego istoriya* (Contemporary Aeronautics and its Historical Background) (Kiev, n.p., 1909); N. Borozdin, *Zavoyevaniye vozdushnoi stikhii* (Conquest of the Air) (Warsaw, n.p., 1909); M. L. Frank, *Istoriya vozdukhoplavaniya i yego sovremennoye sostoyaniye* (History of Aeronautics and its Current Situation) (St Petersburg: Izdatel'stvo 'Vozdukhoplavaniye,' 1910); *Vozdukhoplavaniye i letaniye, Russkiye letuny* (Aeronautics and Flight, Russian Aviators) (St Petersburg: Tipografiya A. S. Suvorina, 1911); A. A. Rodnykh, *Istoriya vozdukhoplavaniya i letaniya v Rossii* (History of Aeronautics and Flying in Russia) (St Petersburg: 'Gramotnost'', 1911); K. Ye. Veigelin, *Zavoyevaniye vozdushnago okeana, Istoriya i sovremennoye sostoyaniye vozdukhoplavaniya* (Conquest of the Air Ocean, History and Contemporary State of Aeronautics) (St Petersburg: Knigoizdatel'stvo P. P. Soikina, 1912); and *Vozdushnyy spravochnik, Ezhegodnik Imperatorskogo Vserossiiskago Aero-Kluba* (Yearbook of the Imperial All-Russian Aero Club), published in 1915–16 by the Imperial Russian Aero Club. An excellent documentary source for Russian aeronautics up to 1907, one that includes materials on Alexander Mozhaiskiy, is *Vozdukhoplavaniye i aviatsiya v Rossii do 1907, Sbornik dokumentov i materialov* (Aeronautics and Aviation in Russia to 1907, a Collection of

Documents), edited by V. A. Popov (Moscow: Gosudarstvennoye Izdatel' stvo Oboronnoi Promyshlennosti, 1956).

Newspapers provide occasional coverage of aeronautical developments, in particular air shows and races. For this article, see *Moskovskiye vedomosti* (Moscow Gazette) for 16 January 1910, o.s. (old style); 17 June 1910, o.s.; and 8 March 1911, o.s. Newspaper accounts with warnings about the destructiveness of air power are supplemented by more serious studies of the potential of air warfare such as A. A. Rodnykh's *Voyna v vozdukhe v byloye vremya i teper'* (The War in the Air, Past and Present) (Petrograd: Delo, 1915).

Specific technical subjects related to aeronautics are covered in a number of titles that appeared before 1914. For the question of air law, see L. I. Schiff, *Vozdukhoplavaniye i pravo* (Aeronautics and Law) (St Petersburg: Vozdukhoplavaniye, 1911). Pre-revolutionary Russian aviation possessed a number of specialized titles: *Opredeleniye vysoty poleta aeroplanov i nekotoriye meteorologicheskiye nablyudeniya* (The Measurement of Altitude of Airplanes and Some Meteorological Observations) (St Petersburg: Usmanov, 1910); and S. Mesentsov, *Aviatsionnaya karta* (Aviation Maps) (St Petersburg: Tipografiya Imperatorskoi Nikolayevskoi Voyennoi Akademii, 1913). During the First World War period there were numerous technical manuals published on selected aviation topics, for example, P. Yankevich, *Aerofotografiya rukovodstvo vozdushnoi Fotografii* (Aerial Photography) (Petrograd: A. N. Lavrov, 1917).

Among the titles available on Russian aircraft designs, the most authoritative remains V. B. Shavrov, *Istoriya konstruktsii samoletov v SSSR do 1938* (History of Aircraft Design in the USSR up to 1938) (Moscow, Mashinostroyeniye, 1969, 2nd edn, 1985). Shavrov also published a second volume for the years 1938 to 1950. For a brief survey of the early Soviet period, see G. V. Kostyrchenko, 'Otechestvennaya aviatsiya v 1918–1925 godakh' (Fatherland Aviation of the Early Period, 1918–1925) (*Voprosy istorii*, 3 (1994), 170–3). Igor Sikorsky's autobiography, *The Story of the Winged-S, An Autobiography* (New York: Dodd, Mead & Company, 1938), provides key insights into his pre-revolutionary years as part of a larger account of his career. For a contemporary account of this formative period, one can read in English, B. Roustem-Bek, *Aerial Russia* (London: John Lane, 1916).

K. N. Finne, a flight surgeon who served with Sikorsky's Il'ya Muromets squadron, wrote his own personal account of his service with Igor Sikorsky, in Russian, under the title, *Russkiye vozdushnyye bogatyri I. I. Sikorskogo* (Russian Air Knights of Igor Sikorsky) (Belgrade: n.p., 1930); this same book has been translated into English as *Igor Sikorsky: The Russian Years*, co-edited by Carl Bobrow and Von Hardesty (Washington, DC: Smithsonian Institution Press, 1987). For a recent work in English on the career of Andrei N. Tupolev, see L. L. Kerber, *Stalin's Aviation Gulag, A Memoir of Andrei Tupolev and the Purge Era*, ed. Von Hardesty (Washington, DC: Smithsonian Institution Press, 1996).

Many Russian historians in the post-communist context freely acknowledge the Grand Duke Alexander Mikhailovich as the 'Father of Russian Aviation'. The Grand Duke's role was central, first to build airmindedness and second to provide leadership for the Imperial Russian Air Force in the First World War. For his own autobiography, see Alexander, Grand Duke of Russia, *Once a Grand Duke* (New York: Farrah & Rinehart, 1932); the Russian-language version being *Kniga vospominaniya* (Memoirs) (Paris: *Izdaniye zhurnal illustrirovannaia Rossiya*, 1933). For an illustrated memoir by an *émigré* Russian pilot, read Alexander Riaboff, *Gatchina Days: Reminiscences of a Russian Pilot*, edited with an introduction by Von Hardesty (Washington, DC, 1986). To place these personal accounts in a wider perspective on the air war, one should read John H. Morrow, Jr, *The Great War in the Air, Military Aviation from 1909 to 1921* (Washington, DC: Smithsonian Institution Press, 1993).

Selected historical monographs provide detail and nuance on early Russian air power: 'Znacheniye aeroplanov dlya artilleriiskoi strel'by', (Application of the Airplane for Artillery Spotting), *Vozdukhoplavatel'*, 8 (August 1912), 586–7; Robert A. Kilmarx, 'The Russian Imperial Air Force in World War I,' *Airpower Historian*, 10, 3 (July 1963), 90–5; David R. Jones, 'The Birth of the Russian Air Weapon, 1909–1914', *Aerospace Historian*, Fall (September 1974), 169–71; and Carl Bobrow, 'The Beginnings of Air Power: Russia's Long Range Strategic Reconnaissance and Bomber Squadron, 1914–1917', *Bulletin of the Russian Aviation Research Group of Air Britain*, 31, 112 (December 1992), 18–29. One of the most richly illustrated volumes on the czarist air force (if lacking in consistent accuracy with details) is the large format book, *Imperial Russian Air Service*, edited by Alan Durkota, Thomas Darcey, and Viktor Kulikov (Stratford, CT: Flying Machine Press, 1996. For the role of women, see Christine White, 'Gossamer Wings: Women in Early Russian Aviation, 1910–1920', *Proceedings of the Second Annual National Conference on Women in Aviation* (Parks College of Saint Louis University, March 21–23, 1991). George L. Stamper, Jr, 'The Sikorsky S-16 and Russian Aviation During the Great War' (MA Thesis, University of Georgia, 1995) is one of the few detailed accounts of the S-16 in English. Vadim Mikheyev has published a Russian language history of this same aircraft, *Sikorskiy S-16, Russkiy Skaut* (The Sikorksy S-16, Russian Scout) (Moscow: Polygon, 1994). Mikheyev (with G. I. Katishev) also published *Aviakonstruktor Igor I. Sikorskiy, 1889–1972* (Aircraft Designer Igor I. Sikorsky, 1889–1972) (Moscow: 'Nauka', 1989) and *Kril'ya Sikorskogo* (Wings of Sikorsky) (Moscow: Voyenizdat, 1992), both milestone publications on Sikorsky to be published in Russia in modern times.

The technical side of Igor Sikorsky's pre-revolutionary designs has prompted considerable interest. Harry Woodman has emerged as one of the most authoritative writers on Igor Sikorsky, in particular the Il'ya Muromets; a representative sampler of his work would include 'The First Airbus' (Sikorsky's Grand), *Air Pictorial* (December 1989 and November 1990); 'Les Bombardiers géants d'Igor Sikorsky', *Le Fanatique de l'aviation*, 150 (May 1982), 10; 151 (June 1982), 16; and 'Il'ya Muromets Type "B" of WW 1', *Airfix Magazine* (May 1985), 351. Woodman has published a three-part feature containing sketches of the variants of the Sikorsky four-engine designs in *Windsock International*, 6, nos 3, 4, and 5 (May–October, 1990). Carl Bobrow provides data on the design of the Il'ya Muromets in his article,

'A Technical Overview of the Evolution of the Grand and the Il'ya Muromets', *W.W.I Aero*, 127 (February 1990), 40–55.

Russian *émigrés* played an influential role in Western aviation development. A recent Russian-language book by D. A. Sobolev, *Nashi sootechestvenniki v zarubezhnom aviastroyenii* (Our Compatriots in Foreign Aircraft Design) (Moscow: Izdatelstvo 'Libri', 1996), represents an effort to recover the story of how Russian-born aviation figures pursued aviation careers abroad. The story of Boris Sergievsky, a Russian fighter pilot who emigrated to the West, may be found in two sources: *Boris Vasilievich Sergievsky, 1888–1971, Sbornik statei* (Boris Vasiliyevich Sergievsky, A Collection of Articles) (New York, 1975), and 'Sergievsky', *The New Yorker*, 9 (November 1940), 12–13. Frank J. DeLear in his *Igor Sikorsky: His Three Careers in Aviation* (New York: Dodd, Mead, 1969) provides coverage of Igor Sikorsky's American years.

CHAPTER 2: FROM CHAOS TO THE EVE OF THE GREAT
PATRIOTIC WAR, 1922–41

General works on Soviet military aviation include Alexander Boyd, *The Soviet Air Force Since 1918* (New York, 1977); Asher Lee, *The Soviet Air Force* (New York, 1962); and Kenneth R. Whiting, *Soviet Air Power* (Boulder, CO, 1986). See also John Erickson, *The Soviet High Command 1918–1941* (New York, 1962); Heinz Nowarra and G. R. Duval, *Russian Civil and Military Aircraft 1884–1969* (London: Fountain Press, 1970); Jean Alexander, *Russian Aircraft Since 1940* (London, 1975); and Lennart Andersson, *Soviet Aircraft and Aviation, 1917–1941* (Naval Institute, 1994).

Works on Soviet Air Force theory of the 1920s and 1930s include numerous books and articles by A. S. Algazin, Ya. Alksnis, V. V. Khripin, A. N. Lapchinsky; and B. L. Teplinski; the key journal of the period was *Vestnik Vozdushnogo Flota*. See also V. V. Anuchin and O. N. Zdorov, 'Genesis and Development of the Theory of Combat Employment of Air Forces (1917–1938)', *Voenno-istoricheskii zhurnal* (August 1988). On VVS leadership, see G. P. Skorikov, 'Komandarm 2 Ranga Ia. I. Alksnis', *Voenno-istoricheskii zhurnal* (January 1987); I. Svetlichnyi, 'Komkor Smushkevich', *Aviatsiia i kosmonavtika*, 9 (1982), 38–9; Dmitrii Yakovlevich Zil'manovich, *Na orbite bolshoi zhizni* (Vilnius: Izd. Mintis, 1971); Grigori A. Tokaev, *Betrayal of an Ideal* (London: Harvill Press, 1954) and *Comrade X* (London: Harvill Press, 1956); Dmitrii Antonovich Volkogonov, *Stalin: Triumph and Tragedy* (New York: Grove Weidenfeld, 1991). On the Soviet–German co-operation of the 1920s and 1930s, see Yuri Dyakov and Tatyana Bushuyeva, *The Red Army and the Wehrmacht* (Amherst, NY: Prometheus, 1995), as well as Lennart Andersson and Kenneth Whiting (cited above).

On the VVS in the 1930s, see Kenneth E. Bailes, 'Technology and Legitimacy: Soviet Aviation and Stalinism in the 1930s', *Technology and Culture*, 17, 1 (1976), 55–81. Interesting information on VVS theory and training is contained in sections of I. V. Timokhovich, *Operativnoe Iskusstvo Sovetskikh VVS v Velikoi Otechestvennoi Voine* (Moscow: Ministry of

Defense, 1976); M. N. Kozhevnikov, *Komandovanie i shtab VVS Sovetskoi Armii v Velikoi Otechestvennoi voine 1941–1945 gg.* (Moscow: Nauka, 1977); and P. N. Pospelov, ed., *Istoriia Velikoi Otechestvennoi Voiny Sovietskogo Soiuza 1941–45* (Moscow: Voenizdat, 1960–65). See also Walter S. Dunn, *The Soviet Economy and the Red Army, 1930–1945* (Westport, CT: Praeger, 1995); A. E. Golovanov, 'Long-Range Bombing Aviation', *Soviet Studies in History* 23, 3 (1984–85), 34–82; William E. Odom, *The Soviet Volunteers* (Princeton, NJ: Princeton University Press, 1973); A. S. Yakovlev, *Tsel' zhizni*, (Moscow: 1966); M. V. Zakharov, 'On the Eve of World War II', *Soviet Studies in History* 23, 3 (1984–85), 83–122.

On the VVS in the Far East, see *Japanese Special Studies on Manchuria* (Washington, DC, 1956); G. K. Zhukov, *The Memoirs of Marshal Zhukov* (New York: Delacorte, 1971); M. V. Novikov, 'V nebe Khalkin-Gola', *Voprosy istorii* (March 1974); Claire Chennault, *Way of a Fighter* (New York: 1949), and Gordon Pickler, 'United States Aid to the Chinese Nationalist Air Force, 1931–1949' (PhD dissertation, Florida State University, 1971).

On the Spanish Civil War, see Robert Jackson, *The Red Falcons* (London, 1970); Peter Elstob, *Condor Legion* (New York: Ballantine, 1973); A. I. Gusev, *Gnevnoe nebo Ispanii* (Moscow: Voenizdat, 1973); José Larios, *Combat Over Spain* (New York: Macmillan, 1966); Jesus Salas Larrazabal, *Air War Over Spain* (London: Ian Allan, 1969); Raymond L. Proctor, *Hitler's Luftwaffe in the Spanish Civil War* (Westport, CT: Greenwood, 1983); Boris N. Smirnov, *Ispanskii veter: zapiski letchika* (Moscow: 1963); M. N. Yakushin and G. M. Prokofiev, *Pod znamenem ispanskoi respubliki* (Moscow: 1965).

On the effects of the purges on military aviation, see Roger Reese, 'The Impact of the Great Purge on the Red Army', *Soviet and Post-Soviet Review*, 19, 1–3 (1992), 71–90; I. I. Kuznetsov, 'The Generals of 1940' (*Voenno-istoricheskii zhurnal*, October 1988); V. Rapoport and Y. Alexeev, *High Treason: Essays on the History of the Red Army, 1918–1938* (Durham, NC: Duke University Press, 1985). On the Winter War with Finland, see Eino Luukanen, *Fighter Over Finland* (London: Macdonald, 1963); N. I. Baryshnikov, 'The Soviet–Finnish War of 1939–1940', *Soviet Studies in History* 29, 3 (1990–91), 43–60; Eloise Engle and Lauri Paananen, *The Winter War: The Russo-Finnish Conflict 1939–40* (New York: Scribners, 1973); A. I. Radzievsky, *Akademiia imeni M. V. Frunze* (Moscow: Voenizdat, 1973).

CHAPTER 3: SOVIET FRONTAL AVIATION DURING
THE GREAT PATRIOTIC WAR, 1941–45

The literature on the Soviet Air Forces in the Great Patriotic War is relatively large in Russian but not so voluminous in English. This essay is based on my previous article, 'The Great Patriotic War, 1941–1945', which appeared in Robin Higham and Jacob Kipp (eds), *Soviet Aviation and Air Power: A Historical View* (Boulder, CO: Westview Press, 1977, pp. 69–136); and my article with Von Hardesty, 'Soviet Air Forces in World War II', in Paul J. Murphy (ed.), *The Soviet Air Forces* (Jefferson, NC: McFarland, 1984), pp. 29–69. The 'Research Notes' section of my article (pp. 133–6) and the footnotes to my piece with Von Hardesty (pp. 322–5) contain a basic starting point for any study of the VVS during the war. I will not repeat these references here but will rather focus on sources that have appeared since the publication of these articles and that I have used to prepare this article on Soviet tactical aviation.

Among the English-language materials, Von Hardesty's *Red Phoenix: The Rise of Soviet Air Power, 1941–1945* (Washington, DC: Smithsonian Institution Press, 1982) is by far the best source available on the wartime trials and triumphs of the VVS. His extensive bibliography (pp. 260–86) is the departure point for any serious examination of the subject. Also see his article 'Roles and Missions: Soviet Tactical Air Power in the Second Period of the Great Patriotic War', in Carl W. Reddel (ed.), *Transformation in Russian and Soviet Military History. Proceedings of the Twelfth Military History Symposium , United States Air Force Academy, 1–3 October 1986* (Washington, DC: US Air Force Academy and Office of Air Force History, 1990, pp. 151–71). Richard P. Hallion in his *Strike from the Sky: The History of Battlefield Air Attack, 1911–1945* (Washington, DC: Smithsonian Institution Press, 1990) covers both the *Luftwaffe* and VVS experience in a general overview in his Chapter 15, 'Battlefield Air Support in the East: The Case of Kursk' (pp. 228–60). Martin van Creveld with Steven L. Canby and Kenneth S. Brower, *Air Power and Maneuver Warfare* (Maxwell AFB, AL: Air University Press, 1994), has chapters on the 1941 German campaign in Russia (pp. 61–101) and the Soviet experience (pp. 109–52). B. F. Cooling's edited collection of essays, *Case Studies in the Development of Close Air Support* (Washington, DC: Office of Air Force History, 1990), contains Kenneth R. Whiting's 'Soviet Air–Ground Coordination, 1941–1945' (pp. 115–51). Also see Kenneth Whiting's *Soviet Air Power* (Boulder, CO: Westview Press, 1985, pp. 11–36). On Allied experience, Mark J. Conversiro, *Fighting with the Soviets: the failure of Operation FRANTIC, 1944–1945* (1997) provides a multinational view. The story of the French Normandy–Niemen squadron is to be found in various issues of the excellently illustrated French airline pilots' magazine *Icare*. Detailed photographic and technical accounts of specific Soviet aircraft, such as Hans-Heiri Stapfer's *Il-2 Stormovik in Action* (Carrolton, TX: Squadron/Signal Publications, 1995), often contain critical information and insights into Soviet aircraft and operations.

Recent publications on the *Luftwaffe*'s experience in Russia include: Richard Muller, *The German Air War in Russia* (Baltimore, MD: The Nautical and Aviation Publishing Company, 1992); Williamson Murray, *Strategy for Defeat: The Luftwaffe, 1933–1945* (Maxwell AFB, AL: Air University Press, 1983); and German Military History Research Office, *Das Deutsche Reich und der Zweite Weltkrieg* (The German Reich and the Second World War), Vol. 4: *Der Angriff auf die Sowjetunion* (The Attack on the Soviet Union) (Stuttgart: Deutsche Verlags-Anstalt, 1983), especially see Horst Boog's 'Die *Luftwaffe*' in Part 2: *Der Krieg gegen die Sowjetunion bis zur Jahreswende 1941/42* (The War Against the Soviet Union until the End of 1941).

As for Russian-language sources, the best single Soviet source on the VVS's operational art during the war remains I. V. Timokhovich's classic

Operativnoye iskusstvo Sovetskikh VVS v Velikoy Otechestvennoy voyne (The Operational Art of the Soviet Air Force in the Great Patriotic War) (Moscow: Voyenizdat, 1976).

During the 1980s a series of excellent detailed accounts appeared on the organization and tactics of the VVS. Of these, the best is Marshal of Aviation N.M. Skomorokhov and Colonel V. N. Chernetskiy, *Taktika v boyevykh primerakh: aviatsionniy polk* (Tactics in Combat Examples: The Aviation Regiment) (Moscow: Voyenizdat, 1984) which provides excellent coverage of the development of bomber, ground-attack, and fighter tactics at the regimental level with numerous actual operational examples thrown in. Y. I. Maslennikov's *Taktika v boyevykh primerakh: Eskadrilya-Ekipazh* (Tactics in Combat Examples: Squadron-Crew) (Moscow: Voyenizdat, 1985) surveys the tactical development of fighter aviation. Marshal of Aviation G. V. Zimin's *Taktika v boyevykh primerakh: Istrebitel'naya aviatsionnaya diviziya* (Tactics in Combat Examples: The Fighter Air Division) (Moscow: Voyenizdat, 1982) is an exhaustive review of the operations and tactical development of the largest VVS fighter organization, the Fighter Air Division.

M.N. Kozhevnikov's *Komandovaniye i shtab VVS Sovetskoy Armii v Velikoy Otechestvennoy voyne 1941–1945gg.* (The Command and Staff of the VVS of the Soviet Army in the Great Patriotic War, 1941–1945) (Moscow: Izdatel'stvo Nauka, 1977) has become the standard work on the subject. For a look at an operational air army, see P. S. Anishchenkov and V. Ye. Shurinov, *Tret'ya vozdushnaya* (The Third Air Army) (Moscow: Voyenizdat, 1983) or K. A. Vershinin, *Chetvertaya vozdushnaya* (The Fourth Air Army) (Moscow: Voyenizdat, 1975) – Vershinin was the 4th Air Army's wartime commander.

By far the best source of periodical historical literature on the VVS during the war appears in the pages of the Ministry of Defense's *Voenno-istoricheskii zhurnal* (*VIZh*) (Journal of Military History). Just a few examples of the many important articles that have appeared in *VIZh* are V. Babich, '*Vlianie razvitiya aviatsionnoy tekhniki i oruzhiya na taktiku frontovoy aviatsii*' (The Influence of Aviation Technology and Armament on the Tactics of Frontal Aviation), 25, 8 (August 1983) and '*Osnovnye napravleniya razvitiya taktiki frontovoy bombardirovochnoy aviatsii*' (The Basic Directions of the Development of the Tactics of Frontal Bombardment Aviation), 19, 5 (May 1978); L. Miryukov, '*Upravlenie istrebitel'yami v vozdushnom boyu*' (Control of Interceptors in Air Combat), 18, 9 (September 1977); V. Myagkov, '*Razvitie taktiki bombardirovochnoy aviatsii*' (The Development of Bombardment Aviation Tactics), 16, 3 (March 1974); M. Novikov, '*Razvitie tekhniki bombardirovochnoy aviatsii v gody voyny*' (The Development of Bomber Aviation Techniques during the War), 19, 4 (April 1978); and Ye. Simakov, '*Boyevoy i chislenniy sostav VVS v tretyem periode voyny*' (Combat Strength of the VVS in the Third Period of the War), 17, 7 (July 1975); O. Frantsev, '*Vstrecha s proslavlennym letchikom trizhdy geroem Sovetskogo Soyuza Marshalom Aviatsii A.I. Pokryshkinym*' (A Meeting with the Famous Flyer, Three-time Hero of the Soviet Union, Marshal of Aviation A. I. Pokryshkin), 25, 3 (March 1983).

One of the most significant books to appear recently has been F. G. Krivosheev *et al.* (eds), *Grif sekretnosti snyat: poteri vooruzhennykh sil SSSR v voynakh, boevykh deistviiakh i voennykh konfliktakh* (The Stamp of Secrecy Removed: The Losses of the Armed Forces of the USSR in Wars, Combat Operations, and Military Conflicts (Moscow: Voyenizdat, 1993) which is a detailed account of all Soviet personnel and equipment losses since the Civil War of 1918. Although still far from the level of detail that we expect in the West, Krivosheev's work does contain some of the most detailed loss figures yet to appear in Russian on the VVS during the war. See especially, Table 95: '*Svodnaya Tablitsa nalichiya, postupleniya i poter' vooruzheniya i tekhniki v period Velikoi Otechestvennoy voyny 1941–1945gg (po godam)*' (Summary Table of Availability, Receipts, and Losses of Armament and Equipment during the Great Patriotic War, 1941–1945 (By Year)), Part IV: '*Samolety*' (Aircraft), pp. 359–60.

CHAPTER 4: AVIATION AND THE TRANSFORMATION
OF COMBINED-ARMS WARFARE, 1941–45

Several general accounts of the Soviet Air Force history cover the war years in varying degrees of emphasis: Alexander Boyd, *The Soviet Air Force Since 1918* (London: Macdonald and Jane's, 1977); and Robert A. Kilmarx, *A History of Soviet Air Power* (New York: Frederick A. Praeger, 1962). Still the most authoritative account for the actual war years is Von Hardesty, *Red Phoenix: The Rise of Soviet Air Power, 1941–1945* (Washington, DC, and London: Smithsonian Institution Press, 1982, 2nd edn 1994). Ray Wagner, (ed.), *The Soviet Air Forces in World War II*, trans. Leland Fetzer (Garden City, NY: Doubleday, 1973), is an English translation of the official Soviet account of the air war in the East.

Many general accounts of the Eastern Front provide selected coverage of air activities. David M. Glantz and Jonathan M. House, *When Titans Clash: How the Red Army Stopped Hitler* (Lawrence, KS: University Press of Kansas, 1995), provides a recent account based on archival material made available over the past decade. John Erickson's *The Road to Stalingrad* (New York: Harper & Row, 1975) and Alan Clark's *Barbarossa: The Russian–German Conflict* (New York: Morrow, 1975) remain authoritative accounts of the war. For an insightful account of the entire air operations in the Second World War, see R. J. Overy, *The Air War 1939–1945* (New York: Stein and Day Publishers, 1980).

The Soviet design bureaux remain a vital, if largely neglected aspect, of aviation literature in the West. L. L. Kerber, *Stalin's Aviation Gulag: A Memoir of Andrei Tupolev and the Purge Era*, edited and translated by Von Hardesty (Washington, DC, and London: Smithsonian Institution Press, 1996), is the first English translation of the Soviet underground classic *Tupolevskaya sharaga*, which appeared in 1971. Kerber provides a candid account of the various Soviet designers and their shifting fortunes before and during the Second World War. A. S. Yakovlev, *Aim of a Lifetime* (Moscow: Progress Publishers, 1972) provides in English Yakovlev's highly personal (and, for his enemies, self-serving) autobiography of Soviet aircraft design. For two standard accounts of Sergei V. Ilyushin, see *Iz istorii Sovetskoi aviatsii samolety OKB imeni S. V. Ilyushina* (The History of Sergei V. Ilyushin's Aircraft Designs) (Moscow: Mashinostroyeniye, 1985 and 1990) and A. N. Ponomarev, *Konstruktor S. V. Ilyushin* (Aircraft Designer S. V. Ilyushin) (Moscow: Voyenizdat, 1988).

The combat role of Shturmovaya aviatsiya (ground-attack aviation) is chronicled in the various histories of the air armies: G. K. Prussakov, *16-ya vozdushnaya, Voyenno-istoricheskiy ocherk o boyevom puti 16-y vozdushnoy armii, 1942–1945* (The 16th Air Army: A Military Historical Sketch of the 16th Air Army in Combat, 1942–1945) (Moscow: Voyenizdat, 1973); S.I. Rudenko and others, *Sovetskiye Voyenno-vozdushnyye sily v Velikoy Otechestvennoy voyne, 1941–1945* (The Soviet Air Force in the Great Patriotic War, 1941–1945) (Moscow: Voyenizdat, 1968); and K. A. Vershinin, *Chetvertaya vozdushnaya* (The Fourth Air Army) (Moscow: Voyenizdat, 1975). Soviet memoir literature is immense. To obtain a comprehensive list of war memoirs, see the bibliography contained in *Red Phoenix*, pp. 260–88. One representative memoir of a Shturmovik pilot is V. B. Yemel'yanenko, *Zapiski letchika-Shturmovika* (A Ground Attack Pilot's Notes) (Moscow: Voyenizdat, 1944). See also, Von Hardesty, 'The Soviet Air Force: Doctrine, Organization, and Technology', *The Conduct of the Air War in the Second World War, An International Comparison*, edited by Horst Boog (New York and Oxford: Berg, 1992).

For an excellent illustrated series on wartime Soviet aircraft, see Carl-Fredrik Geust *et al.*, *Red Stars in the Sky*, 3 vols (Volume Two covers the Il-2 Shturmovik) (Helsinki: Tietoteos, 1979); this same book appeared in a slightly altered format as *Red Stars: The Soviet Air Force in World War Two* (Finland: Forssan Kirjapaino, 1993). A standard English-language reference for Soviet aircraft designs is J. P. Alexander, *Russian Aircraft Since 1940* (London: Putnam, 1975). For the authoritative Russian reference, see V. B. Shavrov, *Istoriya konstruktsii samoletov v SSSR do 1938* (History of Soviet Aircraft Design to 1938) (Moscow: Mashinostroyeniye, 1969, 1975 and 1985); a second volume covers the period 1938–50.

For a more comprehensive treatment of Soviet military aircraft, one that blends aircraft with operational history, there is R.A. Mason and John W. Taylor, *Aircraft, Strategy, and Operations of the Soviet Air Force* (London: Jane's, 1986). Paul J. Murphy edited an excellent anthology also dealing with operations: *The Soviet Air Forces* (Jefferson, NC: McFarland & Company, 1984). John Erickson's *Soviet Combined Arms, Theory and Practice* (College Station: Texas A & M University Press, 1981) provides the larger framework for ground-attack aviation.

CHAPTER 5: RUSSIAN AND SOVIET NAVAL AVIATION 1908–96

Since the publication of *Soviet Aviation and Air Power* in 1977, a number of articles have appeared dealing with the Soviet naval air force. Despite the plethora of monographs on the subject, Jacob W. Kipp remains the most authoritative specialist in the field and his writings serve as a foundation for anyone interested in the subject. Kipp's original essay, 'The Development of

Naval Aviation, 1908–75', which appeared in *Soviet Aviation and Air Power* continues as an influential tool in understanding the history of the Soviet naval air arm. Likewise his other articles, 'Soviet Naval Aviation' in *Soviet Naval Influence: Domestic and Foreign Dimensions* edited by Michael MccGwire and John McDonnell (New York: Praeger, 1977), and 'Soviet Naval Aviation' in *Soviet Armed Forces Review Annual*, edited by David Jones (Gulf Breeze, FL: Academic International Press, 1983), endure as benchmarks in the field.

The works of P. D. Duz', *Istoriia vozdukhoplavaniya i aviatsii v SSSR: period pervoi mirovoi voiny* (Moscow: Naucho-tekhnicheskoe Izdatelstvo, 1960) and *Istoriia vozdukhoplavaniya i aviatsii v SSSR do 1914* (Moscow: Naucho-tekhnicheskoe Izdatelstvo, 1976) supply vital information during the formative years of Russian aviation before and during the First World War. R. D. Layman, *To Ascend from a Floating Base, 1793–1914* (Rutherford, NJ: Farleigh Dickinson University Press, 1979) and his joint article with Boris V. Drashpil, 'Early Russian Shipboard Aviation', *USNI Proceedings* 4 (April 1971), 56–63) chronicles the early development of Russian seaplane tenders during the late Imperial era. A recent reference is Andrei Alexandrov, *Aircraft of the Imperial Russian Navy: 1844–1917*, II, *Naval Aircraft of Russian Origin* (St Petersburg, Russia, 1997). Norman Polmar also made a valuable contribution in 'The Soviet Aircraft Carrier', *USNI Proceedings* 5 (May 1974), 144–61 too.

Soviet sources, especially those that cover the Civil War and NEP (New Economic Policy), are particularly valuable to the scholar. The reader is directed to S. Stoliarskii, 'Iz boevogo proshlogo morskoi aviatsii voenno-morskoi flota SSSR', *Morskoi sbornik*, 4 (April 1938), 67–75; *Istoriia grazh-danskoi voiny v SSSR* (Moscow: Polizdat, 1959), and A. Dorokhov, 'Rozhedenie aviatsii iuzhnykh morei', 8 (August 1981), 56–8, which highlights early Soviet naval aviation at that time. In English, Neil M. Heyman's original article, 'NEP and the Industrialization to 1928', in *Soviet Aviation and Air Power*, remains as significant today as it was in 1977 for information concerning Soviet aviation during Lenin's effort to revitalize the Soviet economy.

Robert Herrick's pioneering study, *Soviet Naval Strategy* (Annapolis: Naval Institute Press, 1968), and the revised edition, *Soviet Naval Theory and Policy* (Newport, RI: Naval War College Press, 1988), provides the best overview of the political and military milieu in which the Soviet Navy matured following the Revolution until the late 1980s. Anyone venturing to understand the Soviet naval experience must become familiar with Herrick's scholarship.

Admiral S. Gorshkov's numerous articles on the navy and naval aviation are vital to comprehend the Soviet naval art. Interested readers are encouraged to examine the following among Gorshkov's works, 'Voenno-morskie floty v voinakh i v mirnoye vremia', *Morskoi sbornik* 6 (June 1972), 11–21, 'Razvitie voenno-morskogo iskusstva', *Voenno-istoricheskii Zhurnal*, 7 (July 1982), 10–18, and *Morskaia mosch' gosudarstva*, 2nd edn (Moscow: Voenizdat, 1979), to appreciate Gorshkov's role in the resurgence of Soviet naval power.

For information concerning Soviet naval aviation before and during the Great Patriotic War, readers are directed to P. N. Ivanov, *Kryl'ia nad morem* (Moscow: Voenizdat, 1981) for a Soviet account of the naval air arm's achievements. When considering the Winter War, the reader should balance Ivanov with Eloise Engle and Lauri Paananen, *The Winter War: The Soviet Attack on Finland 1939–1940* (Harrisburg, PA: Stockpole Books, 1973) in order to counter Soviet excesses. John Erickson's *The Soviet High Command* (New York: St Martin's Press, 1962) continues as the best single-volume account of the Soviet command structure to date.

Likewise, Alvin D. Coox in his two major studies, *The Anatomy of a Small War: The Soviet–Japanese Struggle for Chankufeng/Khasan, 1939* (Westport, CT: Greenwood Press, 1977) and *Nomoham: Japan Against Russia, 1939* (Stanford, CA: Stanford University Press, 1985) provides a concise summary of the border clashes with Japan, in which Soviet air power made a vital contribution in achieving Soviet objectives.

Other sources reviewing Soviet naval combined-arms operations during the Second World War are V. I. Akchasov and N. B. Pavlovich, *Soviet Naval Operations in the Great Patriotic War* (Annapolis: Naval Institute Press, 1981); G. I. Khor'kov, *Sovetskie nadvodnye korabli v Velikoi Otechestvennoi voine* (Moscow: Voenizdat, 1981); and M. N. Lavrent'ev *et al.*, *Aviatsiia VMF v Velikoi Otechestvennoi voine* (Moscow: Voenizdat, 1983), which probes the role of naval air power in those operations. A. F. Kalinichenko, *Geroi neba: dokymental'nye ocherki* (Kaliningrad: Kaliningradskoe knizhnoe izdatel'stvo, 1982) provides individual pilot accounts which are helpful. Petr Khokhlov recounts his participation in the first Soviet air strike on Berlin in his memoir, *Nad tremia moriami* (Moscow: Sovetskaia Rossiia, 1982).

General histories of naval aviation in the Baltic, Black Sea, and Northern Fleets are available in *Morskoi sbornik*. For instance, Iu. Khramov, 'Aviatsiia Krasnoznamennogo baltiiskogo flota v Vostochno-prusskoi operatsiia', *Morskoi sbornik* 8 (August 1986), 20–6, K. D. Denisov, 'Letchiki v boiakh za krym', *Morskoi sbornik* 7 (July 1966), 90–2; Iu. S. Yumashev, 'Tikhookeanskii flot v boiakh za rodinu', *Morskoi sbornik* 8 (August 1965), 6–15, and M. Teret'ev and V. Obukov, 'Aviatsiia Tikhookeanskogo flota v voine s Iaponiei', 8 (August 1985), 36–9, are representative of the sources that are available to document the Soviet naval air war.

With the collapse of the Soviet Union more and more information is becoming accessible to the scholar. For Stalin's objectives following the Second World War, the reader is directed to Vojtech Mastny, *The Cold War and Soviet Insecurity: The Stalin Years* (New York: Oxford, 1996). Prior to 1991, many in the West had speculated about Soviet involvement in the Korean War; however, Sergei N. Goncharov, John W. Lewis, and Xue Litai, *Uncertain Partners: Stalin, Mao, and the Korean War* (Stanford University Press, 1993); William Stueck, *The Korean War: An International History* (Princeton, NJ: Princeton University Press, 1995); Jon Halliday, 'Air Operations in Korea: The Soviet Side of the Story', in *A Revolutionary War: Korea and the Transformation of the Postwar World*, edited by William J. Williams (Chicago, IL: Imprint Publications, 1993); and Yefim Gordon and Vladimir Pigmant, *MiG-15: Design, Development and Korean War Combat*

History (Osceola, WI: Motorbooks, 1993) have all made significant contributions to our current knowledge of Soviet involvement in Korea.

With Gorshkov's quest for a balanced fleet, a debate occurred in *Morskoi sbornik* dealing with the necessity of constructing aircraft-carriers. For a concise summary of the various arguments between Vice Admiral K. A. Stalbo and his critics, readers are directed to Jacob W. Kipp and Christopher C. Lovett's, 'Naval Aviation', in *Soviet Armed Forces Review Annual: Vol. 7, 1982–1983*, edited by David Jones (Gulf Breeze, FL: Academic International Press, 1984). Admiral Gorshkov believed, after reviewing the Falklands War, that the Soviet Navy had followed the appropriate course of carrier development, while monitoring the British victory. Jacob W. Kipp, *Naval Art and the Prism of Contemporaneity: Soviet Naval Officers and the Lessons of the Falklands Conflict* (College Station, TX: The Center for Strategic Technology, 1983) provides the most concise analysis of the Soviet interpretation of the naval war in the South Atlantic to date and the impact it had upon the Soviet Navy.

As was the case in the Spanish Civil War and Korea, the Soviets played an important role in the Vietnam War, which allowed Moscow to reap the rewards following the American withdrawal. For recently published material on the Soviet role in the Vietnam War, readers are directed to Ilya V. Giduk, *The Soviet Union and the Vietnam War* (Chicago: Ivan R. Dee, 1996); and Oleg Sarin and Lev Dvoretsky, *Alien Wars: The Soviet Union's Aggressions Against The World 1919–1989* (Novato, CA: Presidio Press, 1996). The Soviet Pacific Fleet was a principal beneficiary of the American defeat, and P. Lewis Young, 'The Soviet Naval Base at Cam Ranh Bay: An Investigation', *Asian Defence Journal* 9 (September 1987), 44–55; and Gene D. Tracey, 'Soviet Pacific Fleet: Overview for the 1990s', *Asian Defence Journal*, 11 (November 1989), 12–20, furnish an accurate overview of the Soviet position in Vietnam.

The training of pilots, particularly for service at sea, highlighted the problems the Russians confronted in adopting a forward naval strategy. John Barron, *MiG Pilot: The Final Escape of Lt Belenko* (New York: Avon, 1976); and Alexander Zuyev, *Fulcrum Pilot: A Top Gun Pilot's Escape from the Soviet Empire* (New York: Warner Books, 1992) have sections which detail Soviet flight training programs. For further information concerning flight training, see my early work, *Soviet Naval Aviation: Continuity and Change* (College Station, TX: Center for Strategic Technology, 1983); and 'The Training of Soviet Naval Aviators in the 1980s', in *The Soviet Armed Forces Review Annual* edited by George M. Mellinger (Gulf Breeze, FL: Academic International Press, 1995).

To understand the August Coup and the military's role, readers should turn to John B. Dunlop, *The Rise of Russia and the Fall of the Soviet Empire* (Princeton, NJ: Princeton University Press, 1993); Mark Galeotti, 'Decline and Fall, September 1991 Farewell, Soviet Union', *Jane's Intelligence Review* 10 (October 1991), 434–7, Natalie Gross, 'The Military in Coup and Revolution', *Jane's Intelligence Review* 10 (October 1991), 442–9, and David C. Isby, 'The Soviet Armed Forces After the Coup', *Jane's Intelligence Review* 10 (October 1991), 450–4.

By 1992, the Soviet Navy was in serious decline. Authoritative defense and intelligence journals furnish the most reliable information involving the downsizing of the Russian Navy. Richard Woff, 'The Black Sea Fleet', *Jane's Intelligence Review* 11 (November 1992), 492–5, describes the issues at stake between Russia and the Ukraine concerning the Black Sea Fleet, and Taras Kuzio, 'Ukraine since the Elections – From Romanticism to Pragmatism', *Jane's Intelligence Review* 12 (December 1994), 567–71 further highlights the differences between Moscow and Kiev in regard to the division of the former Soviet military in the Ukraine.

Russia's economic plight has forced Moscow both to cut back naval construction and sell off naval vessels. Captain Richard Sharpe RN, *Jane's Fighting Ships 1995–1996*, 98th edn (London: Jane's, 1995) details the current Russian scrapping policy and indicates that carriers listed in reserve will never put to sea again. Stephen J. Blank, 'Russia Arms Exports and Asia', *Asian Defence Journal* 3 (March 1994), 72–9 offers insights into possible Russian naval sales to China and India. Soviet aircraft-carriers had caused considerable consternation in the West during the Cold War. Siegfried Breyer, 'Admiral Kuznetsov: Climax and End of Soviet Carrier Design', *Military Technology* 5 (May 1992), 98–106; and Norman Cigar, 'Soviet Aircraft Carriers: Unfortunate Timing for a Long-Held Dream', *Naval War College Review* 2 (Spring 1992), 20–34, explain the decline of the Russian carrier program; particularly after the loss of the Ukrainian shipyards.

The size of the Soviet naval air arm, in comparison with the Air Force, can be found in *Military Balance* published by the Institute for Strategic Studies since 1962. The figures I cite concerning naval air strength can be located in that publication; those estimates often differ from data found in *Jane's Fighting Ships*. I am indebted to Ray Wagner of the San Diego Aerospace Museum for the recent unpublished statistics concerning Soviet naval aviation strength on the eve of the German invasion.

CHAPTER 6: THE AVIATION INDUSTRY, 1917–97

In the preparation of this chapter, Russian and Soviet source materials were heavily relied upon. The most critical of these were the classic two-volume work of V. B. Shavrov, *Istoriya konstruktsii samoletov v SSSR do 1938g.* (The History of Aircraft Design in the USSR to 1938) (Moscow: Mashinostroyeniye, 3rd edn, 1986) and *Istoriya konstruktsii samolëtov v SSSR 1938–1950gg. (Materialy k istorii samolëtostroyeniya)* (The History of Aircraft Design in the USSR, 1938–50 [Materials for a History of Aircraft Construction]) (Moscow: Mashinostroyeniye, 2nd edn, 1988).

Possibly even more valuable, and much more difficult to obtain, is the two-volume work edited by G. S. Byushgyens *et al.*, *Samolëtostroyeniye v SSSR 1917–45* (Aircraft Construction in the USSR, 1917–45) (Moscow: TsAGI and N. Ye. Zhukovskiy Scientific-Memorial Museum, 1992, 1994). The first volume, *Kniga I* (Book I), carries the story of the overall development of Soviet aviation science, technology, and industry from 1917 through approximately 1939. *Kniga II* (Book II), which takes the story through 1945, appeared in 1994 in an extremely small print run of 1,200 copies. Together these two volumes provided by far the best and most complete coverage of the Soviet aviation industry yet produced in Russia. The books are sponsored and published by TsAGI and the Zhukovskiy Museum, and the authors are recognized technical aviation experts and their individual chapters are based on exhaustive primary research and extensive technical knowledge. Production figures used in Chapters 6 and 7 are largely taken from this source, although even these figures are inconsistent and often contradictory.

An indispensable reference source for anyone working in the subject is G. P. Svishchyov's *Aviatsiya Entsiklopediya* (Aviation Encyclopedia) (Moscow: 1994) which the Scientific Publishing House *Bol'shaya Rossiiskaya Entsiklopediya* (The Great Russian Encyclopedia) and TsAGI collaborated in publishing. Also useful is *Razvitiye aviatsionnoy nauchki i tekhniki v SSSR. Istoriko-tekhnicheskiye ocherki* (The Development of Aviation Science and Technology in the USSR. Historical-Technical Studies) (Moscow: 1980). P. D. Duz', *Istoriya vozdukhoplavaniya i aviatsiya Rossii: Period pervoy mirovoy voiny* (The History of Aeronautics and Aviation in Russia: The Period of the First World War) (Moscow: 1960), covers the years of Russian aviation during the First World War and sets the scene for Soviet developments.

Books by and about the leading aircraft designers were largely limited to Aleksandr S. Yakovlev's volumes and uncritical biographical sketches until the past ten years or so. Yakovlev produced several books of note, especially his autobiographical accounts: *Tsel' zhizni* (The Aim of a Lifetime) (Moscow: 1969) and *Rasskazy aviakonstruktora* (Tales of an Aeronautical Engineer) (Moscow: 1957), which appeared in English, respectively, as *The Aim of a Lifetime: The Story of Alexander Yakovlev, Designer of the YAK Fighter Plane* (Moscow: Progress Publishers, 1972) and *Notes of an Aircraft Designer* (New York: Arno, 1972). His *50 let Sovyetskogo samolëtostroyeniya* (Fifty Years of Soviet Aircraft Construction) (Moscow: 1968) was subsequently translated into English and published for the US National Aeronautics and Space Administration and the National Science Foundation by the Israel Program for Scientific Translations (Jerusalem: 1970). In 1989, Yakovlev's *Sovyetskiye samolëty* (Soviet Aircraft) (Moscow) appeared.

A. N. Ponomarev's *Sovyetskiye aviatsionnyye konstruktory* (Soviet Aircraft Designers) (Moscow: 1977) contains brief, uncritical biographies of the major Soviet aircraft and engine designers. The continuing series of occasional works published under the general title *Iz istorii aviatsii i kosmonavtiki* (From the History of Aviation and Cosmonautics) contains many important articles, including both biographical and autobiographical pieces.

Specific books on Tupolev, Ilyushin, Polikarpov, Kamov, Myasishchev, Kalinin, Sukhoi, and other important aircraft and engine designers and aeronautical engineers have appeared in Russian in the last twenty years. Among them are V. P. Ivanov's *Aviakonstruktor N. N. Polikarpov* (Aircraft Designer N. N. Polikarpov) (St Petersburg: 1995); M. B. Saukke, *Neizvestnyi Tupolev* (The Unknown Tupolev) (Moscow: 1993), by someone who worked closely with Tupolev; D. Gai, *Nebesnoye prityazheniye: zhizn' vydayushchegosiya konstruktora samolëtov Vladimira Mikhailovicha Myasishcheva* (Heavenly Attraction: The Life of the Eminent Aircraft Designer Vladimir

Mikhailovich Myasishchev) (Moscow: 1985); L. M. Kuz'mina, *General'nyi konstrucktor Pavel Sukhoi: stranitsy zhizni* (General Designer Pavel Sukhoi: Pages of His Life) (Moscow: 1983); N. I. Kamov, *Sozdaniye pervogo sovyet-skogo vertoleta* (Creation of the First Soviet Helicopter) (Moscow: 1972); L. M. Kuz'mina, *Konstruktor vertoletov: stranitsy zhizni N. I. Kamova* (Helicopter Designer: Pages from the Life of N. I. Kamov) (Moscow: 1988); A. N. Ponomarev, *Konstruktor S. V. Ilyushin* (Designer S. V. Ilyushin) (Moscow: 1988); Feliks Chuyev, *Stechkin* (Stechkin) (Moscow: 1978); L.L. Kerber, *Tu – chelovek i samolet* (Tu – A Man and His Aircraft) (Moscow: 1973); *Andrei Nikolayevich Tupolev. Zhizn' i deyatel'nost* (Andrei Nikolayevich Tupolev: Life and Work) (Moscow: 1989).

An important recent addition to the works on the designers is an English-language edition of L. L. Kerbe's *Tupolevskaya sharaga* (Tupolev's Prison Workshop) that was first published by Possev-Verlag in West Germany in 1972 under the pseudonym A. Sharagin. In 1996 this work appeared in an English edition, *Stalin's Aviation Gulag: A Memoir of Andrei Tupolev and the Purge Era* (Washington, DC: Smithsonian Institution Press), edited and with an introduction by Dr Von Hardesty of the National Air and Space Museum. Kerber's memoir account covers much more than just Tupolev's time in the NKVD *sharaga* and provides interesting perspectives on Soviet aircraft research, design, development, testing, and production.

Works on the OKBs and their aircraft include: *Samolëty OKB imeni S. V. Ilyushin* (Aircraft of the S. V. Ilyushin OKB) (Moscow: 1990); G. V. Novozhilov (ed.), *Iz istorii sovyetskoi aviatsii: samolëty OKB imeni S. V. Ilyushina* (From the History of Soviet Aviation: Aircraft of the S. V. Ilyushin OKB) (Moscow: 1985); R. A. Belyakov and J. Marmain, *MiG: Fifty Years of Secret Aircraft Design* (Annapolis, MD: Naval Institute Press, 1994); Piotr Butowski with Jay Miller, *OKB MiG: A History of the Design Bureau and its Aircraft* (Leicester: Midland Publishing, 1991); Vladimir Antonov, Yefim Gordon, Nikolai Gordyukov, Vladimir Yakovlev, and Vyacheslav Zenkin, with Lenox Carruth and Jay Miller, *OKB Sukhoi: A History of the Design Bureau and its Aircraft* (Leicester: Midland Publishing, 1996); *60 let OKB A.N. Tupoleva* (60 Years of the A. N. Tupolev OKB) (Moscow: 1982); Bill Gunston, *Tupolev Aircraft Since 1922* (Annapolis, MD: Naval Institute Press, 1995); Paul Duffy and Andrei Kandalov (L. L. Kerber), *Tupolev: The Man and His Aircraft* (Warrendale, PA: SAE International (Society of Automotive Engineers), 1996; and Bill Gunston and Yefim Gordon, *Yakovlev Aircraft Since 1924* (Annapolis, MD: Naval Institute Press, 1997); Richard D. Ward, *Soviet Military Aircraft Design and Procurement* (Fort Worth, TX: General Dynamics Corporation, 1983) and 'The Structured World of the Soviet Designer', *Air Force Magazine*, March 1984; Sergei I. Sikorsky, 'Make it Simple. Make it Producible. Make it Work', *Air Force Magazine*, March 1984; and J.W. Kehoe and K.S. Brower, 'US and Soviet Weapon System Design Practices', *International Defense Review*, June 1982; Clyde Autio, 'Soviet Aircraft Design', in Paul Murphy (ed.), *The Soviet Air Forces* (Jefferson, NC: McFarland, 1984).

Few general works exist on the organization and operations of the Soviet aircraft industry within the Soviet economy in general and the military-industrial complex, either in Russian or in English. The best and most important work on the aviation industry through 1945 is the above-mentioned two-volume set edited by G. S. Byushgyens, *Samolëtostroyeniye v SSSR 1917–45*. A. I. Shakhurin, former head of the Commissariat of Aviation Industry (1940–46), wrote *Kryl'ya pobedy* (The Wings of Victory) (Moscow: 1985) and an article, 'Aviatsionnaya promyshlennost' nakanunye Velikoy Otechestvennoy voyny' (The Aviation Industry on the Eve of the Great Patriotic War), *Voprosy istorii* (Problems of History), 2 (February 1974), 81–99. One of the few available studies in English is the now somewhat dated *The Soviet Aircraft Industry* that was published by the University of North Carolina's Institute for Research in Social Science in 1955. Anton G. Dobler wrote the concise chapter, 'The Soviet Aviation Industry', in Paul Murphy, *The Soviet Air Forces* (Jefferson, NC: McFarland, 1984). David R. Johnson's 'Russia's Military Aviation Industry: Strategy for Survival', *Airpower Journal* (Summer 1997), 45–57, provides an excellent resumé of important recent developments in the military sector of the industry.

Other works that contain information on the economy, industrial development and organizations, the aviation industry, and the Russian and Soviet political and military context are: E. H. Carr's series, *A History of Soviet Russia: The Bolshevik Revolution, 1917–1923* (2 vols) (Baltimore, MD: Penguin Books, 1966); *The Interregnum, 1923–1924* (Baltimore, MD: Penguin Books, 1969); *Socialism in One Country, 1924–1926* (2 vols) (Baltimore, MD: Penguin Books, 1970); *Foundations of a Planned Economy, 1926–1929* (Vol. 2) (New York: Macmillan, 1971); and with R. W. Davies *Foundations of a Planned Economy, 1926–1929* (Vol. 1, Pts 1 and 2) (New York: Macmillan, 1970); Maurice Dobb, *Soviet Economic Development Since 1917* (New York: International Publishers, 1948); Abram Bergson, *The Economics of Soviet Planning* (New Haven, CT: 1964); Alec Nove, *An Economic History of the USSR* (Baltimore, MD: 1972); Eugene Zaleski, *Planning for Economic Growth in the Soviet Union, 1918–1932* (Chapel Hill, NC: University of North Carolina Press, 1967) and *Stalinist Planning for Economic Growth, 1933–1952* (Chapel Hill, NC: University of North Carolina Press, 1980); Naum Jasny, *Soviet Industrialization, 1928–1952* (Chicago, IL: University of Chicago Press, 1961); I. A. Gladkov (ed.), *Sovetskaya ekonomika v period velikoy otechchestvennoy voiny 1941–1945 gg.* (The Soviet Economy during the Great Patriotic War, 1941–1945) (Moscow: 1970); G. S. Kravchenko, *Ekonomika SSSR v gody velikoy otechestvennoy voiny (1941–1945 gg.)* (The Economy of the USSR during the Great Patriotic War (1941–1945) (Moscow: 1970); Christopher Donnelly, *Red Banner: The Soviet Military System in Peace and War* (London: Jane's Information Group, 1988); Harriet Fast and William F. Scott, *The Armed Forces of the USSR* (Boulder, CO: Westview Press, 1984); William F. Scott, 'Moscow's Military-Industrial Complex', *Air Force Magazine*, March 1987; Raymond E. Zickel (ed.), Soviet Union: A Country Study (Washington, DC: USGPO, 1989); Robert Conquest, *The Great Terror: Stalin's Purge of the Thirties* (New York: Macmillan Company, 1968); for prewar and wartime aircraft production, S. M. Shtemenko, *The Soviet General Staff at War, 1941–1945* (Moscow, 1970); for Stalin's role, Robert C. Tucker, *Stalin in Power: The Revolution from Above, 1928–1941* (New York: W. W. Norton & Company, 1992) and Dmitrii Volkoganov, *Stalin: Triumph and Tragedy* (New York: Grove Weidenfeld, 1988); David Holloway, *Stalin and the Bomb: The Soviet Union and Atomic Energy 1939–1956* (New Haven, CT: Yale University Press, 1994); and John Erickson, *The Soviet High Command: A Military-Political History, 1918–1941* (London: St Martin's Press, 1962); for an overview of TsAGI's current state and other of the former 'defense/science cities', see Dr Marina Kalashnikova, 'Russian Naukogrady as the Focal Point of Russia's Drive towards the Future', *The RUSI Journal*, April 1995.

General works on Russian and Soviet aircraft and aviation that contain important details include: Bill Gunston, *The Osprey Encyclopedia of Russian Aircraft: 1875–1995* (London: 1996); Paul J. Murphy, *The Soviet Air Forces* (Jefferson, NC: McFarland & Company, Inc., 1984); Alexander Boyd, *The Soviet Air Force Since 1918* (New York: Stein and Day, 1977); Kenneth R. Whiting, *Soviet Air Power* (Boulder, CO: Westview Press, 1986); Jean Alexander, *Russian Aircraft Since 1940* (London: Putnam, 1975); Lennart Andersson, *Soviet Aircraft and Aviation, 1917–1941* (Annapolis, MD: Naval Institute Press, 1994); Robert A. Kilmarx, *A History of Soviet Air Power* (New York: Praeger, 1962); and Heinz J. Nowarra and G. R. Duval, *Russian Civil and Military Aircraft, 1884–1969* (Fallbrook, CA: Aero Publishers, 1976); Vaclav Nemecek, *The History of Soviet Aircraft from 1918* (London: Willow Books, 1986); Asher Lee, *The Soviet Air Force* (New York: Praeger, 1962); and John Stroud, *Soviet Transport Aircraft since 1945* (New York: Funk & Wagnalls, 1968).

Although much information on the critical interaction between Western aviation technology and the Soviet aviation industry is sprinkled throughout many of these works, four accounts in particular tell this important story. Ulrich Albrecht's *The Soviet Armaments Industry* (Switzerland: Harwood Academic Publishers, 1993) is an exceptionally detailed review of Soviet use of Western technology since 1918 and contains significant coverage of aircraft and engine technological borrowing. Also worthy of review are the three detailed works by Anthony C. Sutton, *Western Technology and Soviet Economic Development, 1917–1930* (Stanford, CA: Hoover Institution Press, 1968); *Western Technology and Soviet Economic Development, 1930–1945* (Stanford, CA: Hoover Institution Press, 1971); and *Western Technology and Soviet Economic Development, 1945–1965* (Stanford, CA: Hoover Institution Press, 1973).

Essential sources for the most current information on the aviation industry are Russian and Western journals and aviation magazines. The magazine *Aviatsiya i kosmonavtika* (Aviation and Cosmonautics), which contained much good information on all aviation matters, was published monthly from 1918 to the mid-1990s, when *Vestnik vozdushnogo flota* (Herald of the Air Fleet) replaced it. The monthly *Vestnik* carries the same wide range of articles as its predecessor, and is produced by the same publishing house.

An especially useful source is the monthly *Kryl'ya Rodiny* (The Country's Wings). Two new bi-monthly journals have appeared in the last several years – *Military Parade* and *Aerospace Journal* – both of which appear in Russian and English editions. Both have current information on Russian military, Air

Force, aircraft, and industry developments. *Military Parade* is the official magazine of major Russian arms dealer, *Rossiiskoye vooruzheniye* (*Rosvooruzheniye*, Russian Arms), and of the Russian military-industrial complex – *Voyenno-promyshlennyi kompleks* (VPK). *Aviatransportnoye obozreniye* (Air Transport Review) is another journal that contains useful information on commercial aviation and the aircraft industry. The Western periodicals *Flight International* and *Air International* provide the most up-to-date information on the aviation industry, governmental policies, and aircraft developments in Russia. The monthly *Air International* provides short monthly updates along with occasional longer articles such as Paul Duffy's 'Silver Service' marking the 25th anniversary of the introduction of the Tu-154 that appeared in the June 1997 issue. *Flight International*, published weekly, carries current news on aviation worldwide and periodically features special coverage of the status of Russian aviation, such as that in its issue of 6–12 August 1997.

The World Wide Web now has a number of sites that provide a wealth of often unstructured information on Russian and Soviet aviation. Among these sites, Evgeni Dvurechenski's home page on Russian and Soviet aviation (http://www.aviation.ru) has a wide range of information and links that supply valuable leads, and his section on OKBs, designers, and their aircraft is particularly to be noted; Aleksei Gretichkhine's *Russian Aviation Page* (http://aeroweb.lucia.it/~agretch/) also has an excellent collection of information, links to other sites, and a very useful bibliography, but, like many of these home pages, emphasizes photos of aircraft and lacks organization; Evgeni Krivenkov's *Red Star Aviation Page* (http://www.dlc.fi/~krive) is another home page, smaller than Dvurechenski's or Gretichkhine's and concentrates on contemporary Russian aircraft and helicopters; Dr Savin's *Russian Aviation Museum* (http://www.physics.arizona.edu/~savin) has an extensive collection of information on Russian and Soviet aircraft; Dennis Newkirk's *Russian Aerospace Guide* (http://www.mcs.net/~rusaerog/) focuses primarily on space-related matters but has some useful links; *Rosvooruzheniye* has placed the contents of some issues of *Military Parade* on http://www.milparade.ru which is also accessible through http://www.vkp.ru; selected issues of *Aviatransportnoye obozreniye* (Air Transport Review) are available at http://www.infoart.ru for those with Russian-language software capabilities and so are issues of the Russian Air Force's *Vestnik vozdushnogo flota* (Herald of the Air Fleet) at http://www.vpk.ru, but at least one issue is available in English (March–April 1996) at http://www.online.ru/sp/afherald. Also available via the Internet at http://bicc.uni-bonn.de/industry is an excellent study by Yevgeny Kuznetsov (ed.), Igor Musienko, and Aleksandr Vorobyev, *Learning to Restructure: Studies in Transformation in the Russian Defense Sector* (Bonn: Bonn International Center for Conversion, BICC, 1996), see especially Part III: 'An Industry Perspective: The Aviation Industry' by Aleksandr Vorobyev. All the major Russian OKBs and production organizations, such as Tashkent, Irkutsk, Antonov, Mil' (Rosvertol), Kamov, MAPO-MiG, etc., have home pages that can be reached for basic information and sometimes details and photographs of their latest aircraft. *AeroWorldNet* (http://www.aeroworldnet.com) also carries items on Russian aviation and aircraft along with its other news. In addition to these sites, there are many more that contain information on Russian and Soviet aircraft and more are certain to be on-line by the time this book is published.

CHAPTER 7: THE DESIGNERS: THEIR DESIGN BUREAUX AND AIRCRAFT

For general information and sources, refer to Chapter 6 notes, above.

CHAPTER 8: THE DEFENSE OF RUSSIAN AEROSPACE

A great deal of information can be located in the Smithsonian Web Browser (SWB), Soviet Series (SU) (dates are those on which the material was entered).

For an excellent study of this period, see R. A. Mason and John W. R. Taylor, *Soviet Air Force* (London: Jane's Publishing Company, 1986, pp. 23–8; and for a detailed historical view of Soviet Air Power up to the mid-1970s see Robin Higham and Jacob W. Kipp, eds, *Soviet Aviation and Air Power: A Historical View* (London: Brassey's, 1977, pp. 195–201). Also see Asher Lee, *Soviet Air and Rocket Forces* (New York: Praeger, 1959), and Robert A. Kilmarx, *A History of Soviet Air Power* (London: Faber & Faber, 1962), for developments up to 1960.

For general studies of Soviet air power, see P. J. Murphy, ed., *The Soviet Air Forces* (Jefferson, NC: McFarland, 1984); W. F. Scott and H. F. Scott, *Armed Forces of the USSR* (New York: Westview, 1983). The Soviet military press contains a wealth of research material: to avoid misunderstanding or misinterpretation, interested researchers should consider the superb facilities available at the Conflict Studies Research Centre, Camberley, England. In addition, see Christopher N. Donnelly, *Red Banner: The Soviet Military System in Peace and War* (London: Jane's Information Group, 1988), and James T. Quinlan, *Soviet Strategic Air Defence: A Long Past and an Uncertain Future* (RAND, 1989).

For a first-class appreciation of Soviet military training, see E. S. Williams, *The Soviet Military: Political Education, Training and Morale* (London: Macmillan Press, 1987). For more detailed analysis of formalism, stereotyping, and stagnation in Soviet training practices, see Dennis J. Marshall-Hasdell, 'Soviet Military Reform: The Training System', *Soviet Studies Research Centre Paper No. C85* (Camberley: SSRC, 1994).

For the impact of peace, note Alexander Kennaway, 'The Effect on People of Restructuring Industry and Reducing the Armed Forces', *Conflict Studies Research Centre Paper No. E77* (CSRC: Camberley, 1995); Dennis J. Marshall-Hasdell, 'Soviet Military Reform and the Afghan Experience – Military Lessons', *Soviet Studies Research Centre Paper No. P12* (Camberley: SSRC, 1993); A. Grokhov, 'The Mooring at Pestryalovo', *Pravda* (1 April

1990, SWB, SU/1605/C-17, 5 February 1993); and Colonel V.I. Timokhin, 'Four Questions to the Head of the Military Academy', *Voyennyi vestnik*, (October 1992).

For detailed analysis of the strategic issues affecting the Baltic States, see Major A. M. Zaccor, *Instabilities in Post-Communist Europe* (Camberley: CSRC, 1994, chapter 10).

See also, Lt-Gen. V. M. Smirnov and Colonel V. F. Grinko, 'Missile Early Warning Systems: Development Trends and Problems', *Voyennaya mysl'*, (June 1992), 15–19; Maj.-Gen. of Aviation N. Kozlov, 'Air Defence: *Quo Vadis*', *Vestnik protivovozdushnoy oborony* (April 1992), 21–2.

The Caucasus question is the subject of an excellent contempory study by Charles Blandy, 'The Caucasus', *Instabilities in Post-Communist Europe* (Camberley: CSRC, 1994). For the military balance in the Caspian region, see M. J. Orr, 'The Regional Military Balance: Conventional and Unconventional Military Forces Around the Caspian', *Conflict Studies Research Centre Paper No. K20* (Camberley: CSRC, February 1995).

For a review of the restructuring debate, see an exhaustive study by Major Brian J. Collins, USAF, 'Russian Air Power: the First Year (Spring 1992–Spring 1993): Blueprint for the Future?', *Conflict Studies Research Centre Research Paper No B54* (Camberley: CSRC, Oct. 1993, pp. 1–34). Also see A. Sumin, 'Russian Air Defence Forces', *Military Technology*, 7/93, pp. 27–8; and Col.-Gen. V. Prudnikov, 'The Integration of the Air Defences of the States of the Commonwealth of Independent States: A Demand of the Times' (*Krasnaya Zvezda*, 3 Feb. 1996). Fundamental problems are discussed in Colonel V. Ya Meleshin and Colonel I. S. Rosnak, 'Air Defence: Problems and Opinions', *Voyennyy vestnik* (April 1993), 21–5; Col.-Gen. V. P. Sinitsyn, 'The Big Task Before Us', *Vestnik protivovozdushnoy oborony* (Jan. 1992), 3–5; and Colonel General P. S. Deynekin, 'Problematic Questions on Building Russia's Air Force', *Aviatsiya i kosmonavtika* (May 1996), 2–11.

CHAPTER 9: AIR COMBAT ON THE PERIPHERY: THE SOVIET
AIR FORCE IN ACTION DURING THE COLD WAR, 1945–89

Until recently, Soviet involvement in the Cold War has been a difficult topic to examine in any depth. Historians are slowly gaining access to Soviet-era archives which is allowing a gradual redressing of the imbalance in Cold War histories. There are, as yet, no comprehensive studies dealing with the activities of the Soviet Air Force during this critical era, however, Tony Mason's *Air Power: A Centennial Appraisal* (London: Brassey's (UK), 1994) has some of the best information on recent Soviet and Russian aviation available. Most of what has been written over the past decades has, out of necessity, concentrated on the numbers and types of aircraft supplied to eastern European and Third World clients. Of these, Bruce Porter's *The USSR in Third World Conflicts: Soviet Arms and Diplomacy in Local Wars, 1945–1980* (Cambridge: Cambridge University Press, 1984) is dated, but still one of the most useful works. An operational history of the SAF can only be pulled together from rather fragmentary evidence.

The Russian government has declassified some of its archival holdings on the early Cold War. Two excellent works on this era are Vojtech Mastny's *The Cold War and Soviet Insecurity: The Stalin Years* (New York: Oxford University Press, 1996); and Vladislav Zubok and Constantine Pleshkov's *Inside the Kremlin's Cold War: From Stalin to Khrushchev* (Cambridge, MA: Harvard University Press, 1996). The Woodrow Wilson Center's *Cold War International History Project Bulletin* is the best source for newly released (and translated) documentary evidence from all sides of the Cold War and is available free of charge. David Holloway's book, *Stalin and the Bomb: The Soviet Union and Atomic Energy, 1939–1956* (New Haven, CT: Yale University Press, 1994), is an award-winning study of the most critical breakthrough for Soviet science and the origins of the Cold War.

The Soviet diplomatic involvement in the Korean War has been treated in several articles by Kathryn Weathersby – 'The Soviet Role in the Early Phase of the Korean War: New Documentary Evidence', *The Journal of American–East Asian Relations*, 2, 4 (Winter 1993), 425–58; 'Korea 1949–1950: to Attack, or not to Attack? Stalin, Kim Il Sung, and the Prelude to War', *Cold War International History Project Bulletin*, 5 (Spring 1995), 1–9; and 'New Russian Documents on the Korean War', *CWIHPB*, Issues 6–7 (Winter 1995/1996), 30–84, are among the most useful. Shu Guang Zhang's *Mao's Military Romanticism: China and the Korean War, 1950–1953* (Lawrence, KS: University Press of Kansas, 1995) has the best published treatment of Soviet air involvement with the PRC in Korea to date. My own unpublished dissertation, *The Other Side of the Yalu: Soviet Pilots in the Korean War* (Tallahassee, FL: History Department, Florida State University, 1996), is based on Soviet military documents and covers the SAF's early engagement in great detail.

Documentary evidence gets thinner, on both sides, the closer one gets to the present, but there has been some excellent work done. Oleg Sarin and Lev Dvoretsky (Sarin worked for years with *Krasnaia Zvezda*) have published two books, *The Afghan Syndrome: The Soviet Union's Vietnam* (Novato, CA: Presidio Press, 1993) and *Alien Wars: The Soviet Union's Aggressions Against the World, 1919 to 1989* (Novato, CA: Presidio Press, 1996), which are more journalistic than historical, but offer some tantalizing evidence of the Soviet role in operations from the war with Poland to Afghanistan. The most thorough treatment to date of the SAF's involvement in Egypt, from the Egyptian perspective, is Lon O. Nordeen and David Nicolle's *Phoenix Over the Nile: A History of Egyptian Air Power, 1932–1994* (Washington, DC: Smithsonian Institution Press, 1996); while Scott R. McMichael provides an excellent analysis of the SAF, and the Soviet military in general, in *The Stumbling Bear: Soviet Military Performance in Afghanistan* (London: Brassey's (UK), 1991). The Brezhnev era is still in need of a thorough treatment, but has received a lot of attention from the US perspective.

CHAPTER 10: THE RISE AND FALL OF AEROFLOT:
CIVIL AVIATION IN THE SOVIET UNION, 1920–91

Vasilii G. Afanas'ev, *Mezhdunarodnye otnosheniia v oblasti grazhdanskoi aviatsii* (Moscow, 1983); Jean Alexander, *Russian Aircraft since 1940* (London, 1975); V. V. Bakanov, N. I. Modrinski, and V. V. Shlopak, 'Development of Large-scale Surveys in the USSR in the Last Fifty Years', *Geodesy, Mapping, and Photogrammetry*, 15 (March 1973); E. V. Altunin, *Kryl'ia Siberii* (Irkutsk, 1981); Lennart Andersson, *Soviet Aircraft and Aviation, 1917–1941* (London, 1994); David M. Bachler, 'Aviation Development in Russia's Far East' (Master's Thesis: University of Hawaii, 1996); A. Boyd, *The Soviet Air Force Since 1918* (London, 1977); A. Buckholtz, *Photogrammetry* (Berlin, 1960); A. Buckholtz, 'Photogrammetry in the Soviet Union', *Photogrammetria*, 18 (1962); Vladimir F. Danilenko, *Kryl'ia Dalnego Vostoka* (Khabarovsk, 1972); R. E. G. Davies, *Aeroflot: An Airline and Its Aircraft. An Illustrated History of the World's Largest Airline* (Rockville, MD, 1992); R. E. G. Davies, *A History of the World's Airlines* (London, 1964); F. V. Drobyshev, 'Soviet Stereophotogrammetric Instruments', *Photogrammetria*, 17 (1960); A. Egorev and V. Kliucharev, *Grazhdanskaia aviatsiia SSSR* (Moscow, 1937); Viktor V. Gavrilenko, *Znakom'tes – Aeroflot* (Moscow, 1968); A. Yu. Gazinazarov, *Grazhdanskaya aviatsiia v ekonomike Uzbekistana* (Tashkent, 1970); A. L. Getman, ed., *Krasnoznamennoe oborone. Kniga o DOSAAF, o vozniknovenii i razvitii obshchestva, ego voenno-patrioticheskoi deiatel'nosti, ego vklade v ukreplenie oboronnogo mogushchestva strany* (1st edn, Moscow, 1971; 2nd edn, Moscow, 1975); Abram Z. Gol'tsman, *Grazhdanskaia aviatsiia* (Moscow, 1932); A. M. Goriashko, *Grazhdanskaia aviatsiia Ukrainy* (Kiev, 1982); *Grazhdanskaia aviatsiia SSSR 1917–1967* (Moscow, 1967); John Grierson, *Through Russia by Air* (London, 1934); A. Gromakov, 'Rastit' zashchitnikov Rodiny', *Voennoistoricheskii zhurnal*, 6 (1984); N. N. Gromov *et al.*, *Ekonomika vozdushnogo transporta* (Moscow, 1971); Bill Gunston, *Tupolev Aircraft since 1922* (London, 1995); Robin Higham and Jacob Kipp (eds), *Soviet Aviation and Air Power: A Historical View* (Boulder, CO, 1977); *Istoriia Dobroleta* (Moscow, 1928); David R. Jones and George M. Mellinger (eds), *Soviet Armed Forces Review Annual*, vols 1–13 (Gulf Breeze, FL, 1976–95); *Istoriia grazhdanskoi aviatsii SSSR* (Moscow, 1983); S. I. Kharlamov (ed.), *Samolety i sport v SSSR* (Moscow, 1978); Robert A. Kilmarx, *A History of Soviet Air Power* (London, 1962); Vsevolod P. Kliucharev, *Grazhdanskii flot SSSR: Statistiko-ekonomiko-ekonomicheskii spravochnik za 1923–1934* (Moscow, 1936); Vsevolod P. Kliucharev, *Grazhdanskaia aviatsiia* (Moscow, 1933); "Komsomolets na samolet!" *Kryl'ia Rodiny*, 10 (1976); I. Kostenko *et al.*, *Boevye vzlety* (Moscow, 1976); Valentin M. Koval'chuk, *Leningrad i Bol'shaia Zemlia. Istoriia Ladozhskoi kommunikatsii blokirovannogo Leningrada v 1941–1943 gg.* (Leningrad, 1975); Jill Lion, *Long Distance Passenger Travel in the Soviet Union* (Cambridge, MA, 1967); Hugh MacDonald, *Aeroflot: Soviet Air Transport since 1923* (London, 1975); V. P. Malko-Skvoz, 'Rol grazhdanskoi aviatsii v oborone Leningrada', *Trudy vysshego aviatsionnogo uchilishcha GVF*, 18 (Leningrad, 1963); Joseph P. Mastro, *USSR Calendar of Events Annual, 1987–1991* (Gulf Breeze, FL,

1988–92); A. P. Molodtsov, *Aeroflot* (Moscow, 1987); *Narodnoe Khoziaistvo SSSR* (Moscow, annual); V. S. Molokov, *Soviet Civil Aviation* (Moscow, 1939); Vaclav Nemecek, *The History of Soviet Aircraft from 1918* (London, 1986); Umberto Nobile, *My Five Years with Soviet Airships* (Akron, OH, 1987); Aleksandr A. Novikov, *V nebe Leningrada. Zapiski komanduiushchego aviatsiei* (Moscow, 1970); H. J. Nowarra and G. R. Duval, *Russian Civil and Military Aircraft: 1884–1969* (London, 1971); V. F. Odintsova, *Ordena Lenina grazhdanskaia aviatsiia* (Moscow, 1967); W. E. Odom, *The Soviet Volunteers: Modernization and Bureaucracy in a Public Mass Organization* (Princeton, NJ, 1973); Ralph Ostrich, 'Aeroflot', in Paul J. Murphy (ed.), *The Soviet Air Forces* (Jefferson, NC, 1984); Georgii N. Pakilev, *Sovetskaia voenno-transportnaia aviatsiia* (Moscow, 1974); Boris M. Parakhonskii, *Tekhniko-ekonomicheskie problemy vozdushnogo transporta* (Moscow, 1961); Boris M. Parakhonskii, *Transport SSSR: Itogi za piat'desiat let i perspektivy razvitiia* (Moscow, 1967); B. N. Pastukhov, *Slavnyi put' Leninskogo Komsomola. Istoriia VLKSM, 1918–1978* (Moscow, 1978); Harriet E. Porch, *Russian Commercial Aviation – Friend or Foe?* (Denver, 1958); V. Puzeikin and A. Proskurin, 'Vozdushnyi most'', *Voenno-istoricheskii zhurnal*, 2 (1973), 88–92; Lawrence Robertson *et al.*, *USSR Facts and Figures Annual, 1977–1995* (Gulf Breeze, FL, 1978/1996); M. M. Roussinov and A. Ch. Chalkhvordov, 'Les Objectifs Photogrammétriques Soviétiques', *Bulletin de la Société Française de Photogrammetrie*, 39 (1970); A. K. Rozovskii, *Krylate agitatory* ((Moscow–Leningrad, 1935); Evgenii V. Sofronov, *V vozdukhe – samolety Aeroflota* (Moscow, 1967); L. Safronov, 'Iz opyta aerofotorazvedki v gody Velikoi Otechestvennoi voiny', *Voenno-istoricheskii zhurnal*, 5 (1979); P. Serebriannikov *et al.*, *Nashi kryl'ia. Molodezhi o Sovetskoi aviatsii. Sbornik* (Moscow, 1959); A. I. Shakhurian, *Krylia pobeda* (Moscow, 1987); V. B. Sharov, *Istoriia konstruktsii samoletov v SSSR do 1938* (3rd edn; Moscow, 1986); Aleksei L. Shepelev, *V nebe i na zemle* (Moscow, 1974); *40 let GVF. Sbornik Leningrade* (Moscow, 1968); John Stroud, *European Transport Aircraft since 1910* (London, 1966); John Stroud, *Soviet Transport Aircraft since 1945* (London, 1968); Leslie Symons, 'Soviet Air Transport: Geographic, Technical, and Organizational Problems', in John Ambler, Denis J. B. Shaw and Leslie Symons (eds), *Soviet and East European Transport Problems* (London, 1985); Leslie Symons, 'Soviet Civil Aviation – Objectives and Aircraft', in Z. W. Fallenbuchl, *Technology and Development in the Soviet Union* (New York, 1975), 1; Leslie Symons, and C. W. White, *Russian Transport: An Historical and Geographical Survey* (London, 1975); J. W. R. Taylor, *Jane's All the World's Aircraft* (London, annual); Michael Taylor (ed.), *Brassey's World Aircraft and Systems Directory 1996/97* ((London, 1996); *Transport i sviaz' SSSR* (Moscow, 1972); United States Air Force, *Soviet Aerospace Handbook* (Washington, DC, n.d.); F. Vazhin and R. Wagner (eds), *Den Aeroflota*, translated as *A Day with Aeroflot* (Moscow, 1973); A. D. Vinokurov, *Aviatsionnyi sport* (Moscow, 1955); R. Wagner (ed.), *The Soviet Air Force in World War II: The Official History*, trans. L. Fetzer (New York, 1973); Andrew Wilson and Nina Bachkatov, *Russia and the Commonwealth: A to Z* (New York, 1992); Aleksandr S. Yakovlev, *Sovetskie samolety. Kratkii ocherk* (Moscow, 1982).

JOURNALS

Aviation Week and Space Technology (weekly); *Flight International* (weekly); *Grazhdanskaia aviatsiia* (monthly); *Interavia Aerospace Review* (monthly); *Kryl'ia Rodiny* (monthly); *The Aeroplane* (monthly); *Vozdushnyi transport* (three/week).

Index